Game Theory Basics

Game theory is the science of interaction. This textbook, derived from courses taught by the author and developed over several years, is a comprehensive, straightforward introduction to the mathematics of non-cooperative games. It teaches what every game theorist should know: the important ideas and results on strategies, game trees, utility theory, imperfect information, and Nash equilibrium. The proofs of such results – in particular, existence of an equilibrium via fixed points and an elegant direct proof of the minimax theorem for zero-sum games – are presented in a self-contained, accessible way. Complementary to that are chapters on combinatorial games such as Go, and – as introductions to algorithmic game theory – traffic games and the geometry of two-player games. This detailed and lively text requires minimal mathematical background and includes many examples, exercises, and pictures. It is suitable for self-study or introductory courses in mathematics, computer science, or economics departments.

Bernhard von Stengel, educated in Germany and the US, is a mathematical game theorist at the London School of Economics and Political Science, and an authority on computational and geometric methods for solving games. He chaired the 2016 World Congress of the Game Theory Society, and is Co-Editor of the *International Journal of Game Theory*.

"This looks like a fine introduction to game theory, *inter alia* emphasizing methods for computing equilibria, and mathematical aspects in general. Especially worthy of note is the chapter devoted to correlated equilibria, a topic of central importance not normally covered in introductory texts."

Robert Aumann, *The Hebrew University of Jerusalem,*
Nobel Memorial Prize in Economic Sciences 2005

"This book is a delightful adventure into the mathematics of game theory. Without any heavy apparatus, it lets us into the secrets of a whole range of exciting results that are usually thought too advanced for the common herd. It is not only undergraduate students who will benefit from reading this book. Professional game theorists will find it very useful too."

Ken Binmore, *University College London*

"This is a rather reader-friendly, engaging, and polished superior creation. It illustrates, explains, motivates every definition, theorem, proof. Interesting and unique choice of topics, such as a delightful introductory chapter on combinatorial games. Highly recommended."

Aviezri Fraenkel, *Weizmann Institute of Science, Israel*

"A masterful presentation of mathematical game theory in all its beauty and elegance, from basic notions to advanced techniques. It fills the gaps left by the many textbooks that cover concepts and applications, but devote only the bare minimum to the mathematical tools and insights, without which game theory would not have become the success it is today."

Sergiu Hart, *The Hebrew University of Jerusalem*

"Game theory is the child of mathematicians, as this textbook demonstrates through self-contained, elegant proofs of all seminal theorems. The lively and rigorous exposition of carefully selected models, such as bargaining, combinatorial, and congestion games (the latter two rarely the stuff of textbooks), explains its success far beyond mathematics. To reach deep results on both sides of the theory, Bernhard von Stengel's marvellous learning tool uses uncompromising, yet accessible mathematics, and chooses examples to maximal effect."

Hervé Moulin, *University of Glasgow*

"This book is a gem. The presentation is clear and well structured, often with nice geometric illustrations. It moves step by step from basics to powerful concepts, methods, and results. It is ideal for students of mathematics, computer science, and economics who are curious about what game theory is and how it can be used."

Jörgen Weibull, *Stockholm School of Economics*

"An exceptionally lucid introduction to the fundamentals of game theory, enlivened by examples that are sure to captivate students."

Peyton Young, *University of Oxford*

"This is a rigorous, yet accessible introduction to mathematical non-cooperative game theory. In addition to the coverage of the basic concepts and results, it includes special and advanced topics and applications usually not contained in game theory textbooks, such as combinatorial games, congestion games, and inspection games. The special emphasis on algorithmic and computational techniques makes this textbook, just like its author, a valuable bridge between game theory and computer sciences."

Shmuel Zamir, *The Hebrew University of Jerusalem*

Game Theory Basics

Bernhard von Stengel
London School of Economics and Political Science

CAMBRIDGE
UNIVERSITY PRESS

CAMBRIDGE
UNIVERSITY PRESS

University Printing House, Cambridge CB2 8BS, United Kingdom

One Liberty Plaza, 20th Floor, New York, NY 10006, USA

477 Williamstown Road, Port Melbourne, VIC 3207, Australia

314–321, 3rd Floor, Plot 3, Splendor Forum, Jasola District Centre, New Delhi – 110025, India

103 Penang Road, #05–06/07, Visioncrest Commercial, Singapore 238467

Cambridge University Press is part of the University of Cambridge.

It furthers the University's mission by disseminating knowledge in the pursuit of
education, learning, and research at the highest international levels of excellence.

www.cambridge.org
Information on this title: www.cambridge.org/9781108843300
DOI: 10.1017/9781108910118

First published 2022

Printed in the United Kingdom by TJ Books Limited, Padstow, Cornwall

A catalogue record for this publication is available from the British Library.

Library of Congress Cataloging-in-Publication Data
Names: Stengel, Bernhard von, author.
Title: Game theory basics / Bernhard von Stengel, London School of
    Economics and Political Science, Department of Mathematics.
Description: New York : Cambridge University Press, 2021. | Includes
    bibliographical references and index.
Identifiers: LCCN 2021024598 (print) | LCCN 2021024599 (ebook) | ISBN
    9781108843300 (hardback) | ISBN 9781108843300 (ebook)
Subjects: LCSH: Game theory. | BISAC: MATHEMATICS / Discrete Mathematics
Classification: LCC QA269 .S699 2021 (print) | LCC QA269 (ebook) | DDC
    519.3–dc23
LC record available at https://lccn.loc.gov/2021024598
LC ebook record available at https://lccn.loc.gov/2021024599

ISBN 978-1-108-84330-0 Hardback
ISBN 978-1-108-82423-1 Paperback

Contents

Preface

This book is an introduction to the mathematics of non-cooperative game theory. Each concept is explained in detail, starting from a main example, with a slow-paced proof of each theorem. The book has been designed and tested for self-study, and as an undergraduate course text with core chapters and optional chapters for different audiences. It has been developed over 15 years for a one-semester course on game theory at the London School of Economics and the distance learning program of the University of London, attended each year by about 200 third-year students in mathematics, economics, management, and other degrees. After studying this book, a student who started from first-year mathematics (the basics of linear algebra, analysis, and probability) will have a solid understanding of the most important concepts and theorems of non-cooperative game theory.

The intended audience are primarily students in mathematics, computer science, and mathematical economics. For mathematicians, we provide complete self-contained proofs (in an economics course, these may be used as reliable background and reference material). For computer scientists, we introduce important ideas in algorithmic game theory, such as traffic equilibria, the "parity argument" for the existence and computation of Nash equilibria, and correlated equilibria. For economists, we use some important economic models as examples, such as Cournot's quantity competition, which is the oldest formal definition of a game. Commitment games are applied to Stackelberg leadership, and in Chapter 11 to the iterated-offers bargaining model. However, given the many good introductions to game theory for economists (for example, Gibbons, 1992, or Osborne, 2004), economic applications are not central to this book.

Aims and Contents

The first aim of this book is to let the student become fluent in game theory. The student will learn the modeling tools (such as game trees and the strategic form) and methods for analyzing games (finding their equilibria). This is provided in the core chapters on non-cooperative games: Chapter 3 on games in strategic form introduces important games such as the Prisoner's Dilemma or Matching Pennies, and the concept of Nash equilibrium (in the book nearly always just called "equilibrium" for brevity). Chapter 4 treats game trees with perfect information. Chapter 6 explains mixed strategies and mixed equilibrium. Game trees with imperfect information are covered in Chapter 10.

A second aim is to provide the conceptual and mathematical foundations of the theory. Chapter 5 explains how the concept of expected utility represents a consistent preference for risky outcomes (and not "risk neutrality" for monetary payoffs, a common confusion). Chapter 7 proves Brouwer's fixed-point theorem, used by Nash (1951) to show the existence of an equilibrium point. Chapter 8 gives a self-contained two-page proof of von Neumann's minimax theorem for zero-sum games. These chapters form independent "modules" that can be used for reference and that can be omitted in a taught course that has less emphasis on mathematical proofs. If only one proof from these chapters is presented, it should be the short proof of the minimax theorem.

A third aim is to introduce ideas that *every game theorist should know* but which are not normally taught, and which make this book special. They are mathematical highlights, typically less known to economists:

- An accessible introduction to combinatorial games in Chapter 1, pioneered by Berlekamp, Conway, and Guy (2001–2004), with a focus on impartial games and the central game of Nim.

- Congestion games, with the famous Braess paradox where increasing network capacity can worsen congestion, and where an equilibrium is found with the help of a potential function (Chapter 2).

- An elegant constructive proof of Sperner's lemma, used to prove Brouwer's fixed-point theorem, and the Freudenthal simplicial subdivision that works naturally in any dimension (Sections 7.4 and 7.7 in Chapter 7).

- The geometry of two-player games (which is my own research specialty) and the algorithm by Lemke and Howson (1964) for finding a Nash equilibrium, which implies that generic games have an odd number of equilibria (Chapter 9).

- An introduction to correlated equilibria in Chapter 12, with an elegant existence proof due to Hart and Schmeidler (1989) that does not require a fixed-point theorem but only von Neumann's minimax theorem.

These chapters are optional. I teach the course in variations but always include Chapter 1 on combinatorial games; they are much enjoyed by students, who would not encounter them in an introductory economics course on game theory. These chapters give also a short introduction to *algorithmic game theory* for computer science students.

The general emphasis of the book is to teach *methods, not philosophy*. Game theorists tend to question and to justify the approaches they take, for example the concept of Nash equilibrium, and the assumed common knowledge of all players about the rules of the game. These questions are of course very important. In fact, these are probably the very issues that require a careful validation in a practical game-theoretic analysis. However, this problem is not remedied by a lengthy discussion of why one should play Nash equilibrium.

I think that a student of game theory should first learn the central models of the strategic and extensive form, and be fluent in analyzing them. That toolbox will then be useful when comparing different game-theoretic models and the solutions that they imply.

Mathematical and Scholarly Level

General mathematical prerequisites are the basic notions of linear algebra (vectors and matrices), probability theory (independence, conditional probability), and analysis (continuity, closed sets). Rather than putting them in a rarely read appendix, important concepts are recalled where needed in text boxes labeled as **Background material**.

For each chapter, the necessary mathematical background and the required previous chapters are listed in a first section on prerequisites and learning objectives.

The mathematics of game theory is not technically difficult, but very conceptual, and requires therefore a certain mathematical maturity. For example, combinatorial games have a recursive structure, for which a generalization of the mathematical induction known for natural numbers is appropriate, called "top-down induction" and explained in Section 1.3.

Game-theoretic concepts have precise and often unfamiliar mathematical definitions. In this book, each main concept is typically explained by means of a detailed introductory example. We use the format of definitions, theorems and proofs in order to be precise and to keep some technical details hidden in proofs. The proofs are detailed and complete and can be studied line by line. The main idea of each proof is conveyed by introductory examples, and geometric arguments are supported by pictures wherever possible.

Great attention is given to details that help avoid unnecessary confusions. For example, a random event with two real values x and y as outcomes will be described with a probability p assigned to the *second* outcome y, so that the interval $[0, 1]$ for p corresponds naturally to the interval $[x, y]$ of the expectation $(1 - p)x + py$ of that event.

This book emphasizes *accessibility, not generality*. I worked hard (and enjoyed) on presenting the most direct and overhead-free proof of each result. In studying these proofs, students may encounter for the first time important ideas from topology, convex analysis, and linear programming, and thus may become interested in the more general mathematical theory of these subjects.

Bibliographic references, historical discussions, and further extensions are deferred to a final section in each chapter, to avoid distractions in the main text. References to further reading may include popular science books that I found relevant and interesting. This book is primarily directed at undergraduate students and not at researchers, but I have tried to be historically accurate. For important works in French or German I have also cited English translations where I could find them.

References to statements inside this book are in upper case, like "Theorem 2.2", and to other works in lower case, like "theorem 13.6 of Roughgarden (2016)".

I use both genders to refer to players in a game, in particular in two-player games to distinguish them more easily. Usage of the pronoun "she" has become natural, and (like "he") it may stand collectively for "he or she".

Each chapter concludes with a set of exercises. They test methods (such as how to find mixed equilibria), or the understanding of mathematical concepts used in proofs (in particular in Chapters 5 and 7). Complete solutions to the exercises are available from the publisher for registered lecturers.

Acknowledgments

Some of the material for this book has been adapted and developed from a course guide originally produced for the distance learning International Programme offered by the University of London (http://london.ac.uk/). Rahul Savani and George Zouros gave valuable comments on that first course guide. Many students, some of them anonymously, have pointed out errors and needs for clarification in the subsequent versions of lecture notes. Rahul Savani proof-read the final version with great care. Vanessa Wells drew the nice animal pictures on the cover and in Figures 8.1 and 9.6. I thank Robert Aumann, Paul Dütting, Aviezri Fraenkel, Sergiu Hart, Urban Larsson, Abraham Neyman, Richard Nowakowski, David Tranah, Peyton Young, and Shmuel Zamir for further suggestions.

Martin Antonov is the current lead programmer of the *Game Theory Explorer* software (http://www.gametheoryexplorer.org), a joint project with Rahul Savani and (as part of the *Gambit* software) Theodore Turocy, for solving games in strategic and extensive form. From Game Theory Explorer, one can export games into the LaTeX formats used in this book for drawing payoff tables and game trees (available on request).

1

Nim and Combinatorial Games

Combinatorial game theory is about perfect-information two-player games, such as Checkers, Go, Chess, or Nim, which are analyzed using their rules. It tries to answer who will win in a game position (assuming optimal play on both sides), and to quantify who is ahead and by how much. The topic has a rich mathematical theory that relates to discrete mathematics, algebra, and (not touched here) computational complexity, and highly original ideas specific to these games.

Combinatorial games are not part of "classical" game theory as used in economics. However, they nicely demonstrate that game theory is about rigorous, and often unfamiliar, mathematical *concepts* rather than complex techniques.

This chapter is only an introduction to combinatorial games. It presents the theory of *impartial games* where in any game position both players have the same allowed moves. We show the powerful and surprisingly simple result (Theorem 1.14), independently found by Sprague (1935) and Grundy (1939), that every impartial game is equivalent to a "Nim heap" of suitable size.

In Section 1.8 we give a short glimpse into the more general theory of partizan games, where the allowed moves may depend on the player (e.g., one player can move the white pieces on the game board and the other player the black pieces).

For a deeper treatment, the final Section 1.9 of this chapter lists some excellent textbooks on combinatorial games. They treat impartial games as a special case of general combinatorial games. In contrast, we first treat the simpler impartial games in full.

1.1 Prerequisites and Learning Outcomes

The combinatorial games that we consider are finite, and dealing with them uses a type of mathematical induction that we explain further in Section 1.3. It is helpful to have seen induction about the natural numbers before. Although not essential, is is useful to know the algebraic concept of an abelian group, in particular addition modulo 2. We assume familiarity with the *binary* system where numbers are written in base 2, using only the two digits 0 and 1, rather than in the familiar base 10.

After studying this chapter, you should be able to:

- play Nim optimally;
- explain the different concepts of options, game *sums*, equivalent games, Nim values, and the mex rule;
- apply these concepts to play other impartial games like those described in the exercises;
- understand how some (not all) game positions in partizan games can be expressed as *numbers* that state how many moves the Left player is safely ahead, and why these numbers may be fractions with powers of two in the denominator.

1.2   Nim

Combinatorial games are two-player win-lose games of *perfect information*, that is, every player is perfectly informed about the state of play (unlike, for example, the card games Bridge or Poker that have hidden information). The games do *not* have chance moves like rolling dice or shuffling cards. When playing the game, the two players always alternate in making a move. Every play of the game ends with a win for one player and a loss for the other player (some games like Chess allow for a draw as an outcome, but not the games we consider here).

The game has a (typically finite) number of *positions*, with well-defined rules that define the allowed moves to reach the next position. The rules are such that play will always come to an end because some player is unable to move. This is called the *ending condition*. We assume the *normal play* convention that a player unable to move loses. The alternative to normal play is *misère play*, where a player who is unable to move wins (so the previous player who has made the last move loses).

We study *impartial* games where the available moves in a game position do not depend on whose turn it is to move. If that is not the case, as in Chess where one player can only move the white pieces and the other player the black pieces, the game is called *partizan*.

For impartial games, the game *Nim* plays a central role. A game position in Nim is given by some *heaps* of *tokens*, and a move is to remove some (at least one, possibly all) tokens from *one* of the heaps. The last player able to move wins the game, according to the normal play convention.

We analyze the Nim position 1, 2, 3, which means three heaps with one, two, and three tokens, respectively. One possible move is to remove two tokens from the heap of size three, like here:

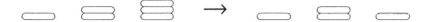

which we write as a move from $1,2,3$ to $1,2,1$. Because the move can be made in any one heap, the order of the heap sizes does not matter, so the position $1,2,1$ could also be written as $1,1,2$. The *options* of a game position are the positions that can be reached by a single legal move (according to the game rules) from the player to move. We draw them with moves shown as downward lines, like here,

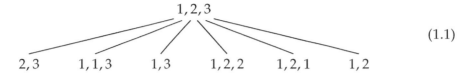

(1.1)

where the first option $2,3$ is obtained by removing from $1,2,3$ the entire heap of size 1, the second option $1,1,3$ by removing one token from the heap of size 2, and so on. The *game tree* is obtained by continuing to draw all possible moves in this way until play ends (game trees are studied in much more detail in Chapter 4). We may conflate options with obvious equal meaning, such as the positions $1,1,2$ and $1,2,1$ that can be reached from $1,2,2$. However, we do not draw moves to the same position from two different predecessors, such as $1,1,2$ that can be reached from $1,1,3$ and $1,2,2$. Instead, such a position like $1,1,2$ will be repeated in the game tree, so that every position has a unique history of moves.

In an impartial game, the available moves in a game position are by definition independent of the player to move. A game position belongs therefore to exactly one of two possible *outcome classes*, namely it is either a *winning* or a *losing position*. "Winning" or "losing" applies to the player whose turn it is to move, assuming optimal play. A winning position means that the player can force a win with a suitable first "winning" move (and subsequent winning moves at all later positions). A losing position means that *every* move from the current position leads to a winning position of the other player, who can then force a win, so that the current player will lose.

When drawn in full with all possible subsequent moves, the Nim position $1,2,3$ has already a rather large game tree. However, we can tell from its options in (1.1) that $1,2,3$ is a losing position, using the following simple observations of how to play optimally in a Nim position with *at most two heaps*:

- If there are no Nim heaps left, this defines a losing position, by the normal play convention.

- A single Nim heap is a winning position; the winning move is to remove the entire heap.

- Two Nim heaps are a winning position if and only if the heaps have different sizes; the winning move is to equalize their sizes by suitably reducing the larger heap, which creates a losing position of two equal heaps. If the two heaps are equal, then a Nim move means reducing one of the heaps, so that the two heaps become unequal, which is a winning position; hence, two equal heaps define a losing position, as claimed.

Every option of $1, 2, 3$ shown in (1.1) is a winning position, because it allows moving to a two-heap losing position, as follows:

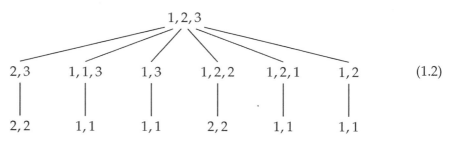

$$(1.2)$$

This picture shows sequences of two moves from position $1, 2, 3$, but not all of them, so this is not the top part of the full game tree. Only from the first position $1, 2, 3$, which is a losing position, all moves are shown. For the second move, only a single, winning move is shown, which leads again to a losing position. This suffices to prove that $1, 2, 3$ is a losing position in Nim, by the following observation.

Lemma 1.1. *In an impartial game, a game position is losing if and only if all its options are winning positions. A game position is winning if and only if at least one of its options is a losing position; moving there is a winning move.*

This lemma applies also to the game position that has no options at all, so trivially ("vacuously") all these options are winning (because there aren't any) and hence the position is losing. The lemma can also be taken as a definition of losing and winning positions, which is not circular but *recursive* because the options of a game position are *simpler* in the sense of being closer to the end of any play in the game. In Section 1.3 we will make this precise, and will give a formal proof of Lemma 1.1.

The concepts of winning and losing position assume optimal play. Binmore (2007, p. 49) writes: "A dull art movie called *Last Year in Marienbad* consists largely of the characters playing Nim very badly." In that movie (from 1961, directed by Alain Resnais), they play several times misère Nim, always from the starting position $1, 3, 5, 7$. There is a scene where three successive positions are

$$\begin{array}{ccc} \text{player I} & & \text{player II} \\ 2, 3, 6 \quad \longrightarrow & 2, 3, 5 \quad \longrightarrow & 2, 3, 1 \end{array}$$

and then player I, whose turn it is, eventually loses. We can conclude from this that player I indeed played badly. Namely, either $2, 3, 1$ is a winning position, in which case player I failed to force a win. Or $2, 3, 1$ is a losing position (which is in fact the case), in which case player I failed to move there from position $2, 3, 6$.

On the other hand, a game is more entertaining if it is not known how to play it optimally. In Chess, for example, which allows for a draw, one can show similarly to Lemma 1.1 that either White can force a win, or Black can force a win, or both players can force a draw. However, Chess is not yet "solved" in the sense that we know which outcome results from optimal play, which is why it is still an interesting game because players do not play perfectly.

Nim, however, has been solved by Bouton (1901). We demonstrate his winning strategy for the already formidable Nim position

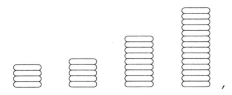

which is $4, 5, 9, 14$ and a winning position. The trick is to write the heap sizes in the *binary* system. The following tables show that one winning move is to reduce the Nim heap of size 5 to size 3, which leads to the losing position $4, 3, 9, 14$:

$$
\begin{array}{rcllll}
 & & 8 & 4 & 2 & 1 \\
\hline
4 & = & 0 & 1 & 0 & 0 \\
5 & = & 0 & 1 & 0 & 1 \\
9 & = & 1 & 0 & 0 & 1 \\
14 & = & 1 & 1 & 1 & 0 \\
\hline
 & & 0 & 1 & 1 & 0
\end{array}
\qquad \rightarrow \qquad
\begin{array}{rcllll}
 & & 8 & 4 & 2 & 1 \\
\hline
4 & = & 0 & 1 & 0 & 0 \\
\boxed{3} & = & 0 & \boxed{0} & \boxed{1} & 1 \\
9 & = & 1 & 0 & 0 & 1 \\
14 & = & 1 & 1 & 1 & 0 \\
\hline
 & & 0 & 0 & 0 & 0
\end{array}
\qquad (1.3)
$$

In (1.3), the top row represents the powers of 2 used to represent each heap size with binary digits, such as $5 = 0 \cdot 8 + 1 \cdot 4 + 0 \cdot 2 + 1 \cdot 1$. The bottom row is called the *Nim sum* of the heap sizes and is obtained by "addition modulo 2 without carry" for each column; that is, the Nim sum has a 0 if the number of 1's in that column is even, and a 1 if that number is odd.

We claim that a Nim position is losing if and only if that Nim sum is zero for all columns, like on the right in (1.3); we call such a position a *zero position*. In order to show this, we have to show, in line with Lemma 1.1, that

(a) every move from a zero position leads to a position which is not zero and therefore winning, and

(b) from every winning position (with a Nim sum that is not zero) there is a move to a zero position.

It easy to see (a): A move changes exactly one Nim heap, which corresponds to one row in the table of binary numbers, and changes at least one binary digit of the binary representation of the heap size. Therefore, the resulting Nim sum is also changed in the corresponding digits and becomes nonzero.

Condition (b) amounts to finding at least one winning move for a position that is not zero, like on the left in (1.3). Such a winning move is found by these steps:

- Choose the leftmost "1" column in the Nim sum (which exists because not all columns are 0), which represents the highest power of two used in the binary representation of the Nim sum, here $4 = 2^2$.

- By definition, there is an odd number of Nim heaps which use that power of two in their binary representation. In (1.3), these are the heaps of sizes 4, 5, and 14. Choose one of the corresponding rows; we choose the heap of size 5.

- In the chosen row, "flip" the binary digits for *each* 1 in the Nim-sum, here for the columns 4 and 2; the resulting changes are shown with boxes on the right in (1.3). The largest of the changed powers of two in that row has digit 1, which is changed to 0. Even if all subsequent digits were changed from 0 to 1, the resulting binary number gives a new, *smaller* integer (possibly zero, which will mean removing the entire heap). The winning move is to reduce the Nim heap to that smaller size, here from 5 to 3.

As a result of these steps, the Nim sum is changed from 1 to 0 exactly in the columns where there was a 1 before, with the digits 0 untouched, so the new Nim sum has a 0 in each column. The resulting Nim position is therefore a zero position, as required in (b).

This winning strategy in Nim can be done with pen and paper, but is probably difficult to perform in one's head in actual play. For small heap sizes, a helpful mental image is to group the tokens in each Nim heap (now shown as dots) into distinct powers of two:

$$
\begin{array}{c}
4 \qquad \vcenter{\hbox{$\because$}} \\[2mm]
5 \qquad \vcenter{\hbox{$\because$}} \qquad \bullet \\[2mm]
9 \quad \text{:::::} \qquad \bullet \\[2mm]
14 \quad \text{:::::} \;\; \vcenter{\hbox{$\because$}} \;\; \bullet
\end{array}
\tag{1.4}
$$

This is an equivalent representation to the left table in (1.3), but without binary digits. It shows that the powers of two that occur an odd number of times are 4 and 2. A winning move has to change one heap so that afterwards they occur an even number of times. This can be done by changing the four dots in the heap of size 5 to two dots, as in $\vcenter{\hbox{$\because$}} \to \vcenter{\hbox{$\colon$}}$. This is the winning move in (1.3) that removes two tokens from the heap of size 5; the same winning move can be made in the heap of size 4. The third possible winning move is to remove six tokens

(⠒⠒ and ⠒) from the heap of size 14. Caution: The resulting powers of two in a heap size must be *distinct*. The change ⠒⠒ → ⠒ is not allowed in the heap of size 14, because the intended result ⠒⠒⠒⠒ ⠒ ⠒ is not a binary representation because 2 appears twice; rather, the result of removing two tokens from the heap of size 14 is ⠒⠒⠒⠒ ⠒⠒, and then the overall created Nim position is not zero. Once the heap for the winning move is chosen, the number of tokens that have to be removed is unique.

1.3 Top-down Induction

When talking about combinatorial games, we will often use for brevity the word *game* for "game position". Every game G has finitely many *options* $G_1, \ldots, G_m$ that are reached from G by one of the allowed moves in G, as in this picture:

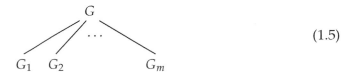

$$(1.5)$$

If $m = 0$ then G has no options. We denote the game with no options by 0, which by the normal play convention is a losing game. Otherwise the options of G are themselves games, defined by their respective options according to the rules of the game. In that way, any game is completely defined by its options. In short, the starting position defines the game completely.

We introduce a certain type of mathematical induction for games, which is applied to a *partial order* (see the background material text box on the next page).

Consider a set S of games, defined, for example, by a starting game and all the games that can reached from it via any sequence of moves of the players. For two games G and H in S, call H *simpler* than G if there is a sequence of moves that leads from G to H. We allow $G = H$ where this sequence is empty. The relation of being "simpler than" defines a partial order which for the moment we denote by $\leq$. Note that $\leq$ is antisymmetric because it is not possible to reach G from G by a nonempty sequence of moves because this would violate the ending condition. The ending condition for games implies the following property:

$$\text{Every nonempty subset of } S \text{ has a minimal element.} \qquad (1.8)$$

If there was a nonempty subset T of S without a minimal element, then we could produce an infinite play as follows: Start with some G in T. Because G is not minimal, there is some H in T with $H < G$, so there is some sequence of moves from G to H. Similarly, H is not minimal, so another game in T is reached from H. Continuing in this manner creates an infinite sequence of moves, which contradicts the ending condition.

Background material: Partial orders

Definition 1.2 (Partial order, total order). A binary relation $\leq$ on a set S is called a *partial order* if the following hold for all x, y, z in S:

$$x \leq y \text{ and } y \leq z \quad \Rightarrow \quad x \leq z \qquad (\leq \text{ is } \textit{transitive}),$$

$$x \leq x \qquad\qquad\qquad\qquad\qquad\quad (\leq \text{ is } \textit{reflexive}), \text{ and}$$

$$x \leq y \text{ and } y \leq x \quad \Rightarrow \quad x = y \qquad (\leq \text{ is } \textit{antisymmetric}).$$

If, in addition, for all x, y in S

$$x \leq y \quad \text{or} \quad y \leq x \qquad\qquad\quad (\leq \text{ is } \textit{total})$$

then $\leq$ is called a *total order*. $\qquad\qquad\qquad\qquad\qquad\qquad\qquad\qquad\qquad$ □

For a given partial order $\leq$, we often say "x is less than or equal to y" if $x \leq y$. We then define $x < y$ ("x is less than y") as follows:

$$x < y \qquad \Leftrightarrow \qquad x \leq y \quad \text{and} \quad x \neq y. \qquad\qquad (1.6)$$

This relation $<$ is also called the *strict order* that corresponds to $\leq$. Exercise 1.1 asks you to show how a partial order $\leq$ can be defined from a relation $<$ on S with suitable properties (transitivity and "irreflexivity") that define a strict order. Then $\leq$ is obtained from $<$ according to

$$x \leq y \qquad \Leftrightarrow \qquad x < y \quad \text{or} \quad x = y. \qquad\qquad (1.7)$$

An element x of S is called *minimal* if there is no y in S with $y < x$.

Any set S of rational or real numbers has the familiar relation $\leq$, which is a total order. The most important partial order that is usually not total is the *set inclusion* $\subseteq$ on a set S of sets. Then elements x, y of S as used in Definition 1.2 are usually written with capital letters like A, B. The partial order is written with the symbol $\subseteq$ where $A \subseteq B$ holds if A is a subset of B (which allows for $A = B$). The inclusion order is not total because two sets may be *incomparable*, that is, neither $A \subseteq B$ nor $B \subseteq A$ hold, for example if $A = \{1, 2\}$ and $B = \{2, 3, 4\}$. If S is the set of *all* subsets of, say, the finite set $\{1, \ldots, n\}$ for some positive integer n, then S, partially ordered by $\subseteq$, has the empty set $\emptyset$ as a unique minimal element. If S is the set of *nonempty* subsets of $\{1, \ldots, n\}$, then its minimal elements are the singletons $\{1\}, \{2\}, \ldots, \{n\}$.

For any partial order with the property (1.8), the following theorem applies.

Theorem 1.3 (Top-down induction). *Consider a set S with a partial order $\leq$ that fulfills (1.8). Let $P(x)$ be a statement that may be true or false for an element x of S, and assume that $P(x)$ holds whenever $P(y)$ holds for all y of S with $y < x$. Then $P(x)$ is true for all x*

in S. In symbols, where "∀x" means "for all elements x of S":

$$\big(\forall x : (\forall y < x : P(y)) \;\Rightarrow\; P(x)\big) \quad \Rightarrow \quad \big(\forall x : P(x)\big). \tag{1.9}$$

Proof. Suppose, as assumed, that condition $P(x)$ holds whenever it holds for all elements y of S with $y < x$. Consider the set $T = \{\, z \in S \mid P(z) \text{ is false}\,\}$. Then $P(x)$ is true for all x in S if T is the empty set, so assume T is not empty. Let x be a minimal element of T, which exists by (1.8). Then all y in S with $y < x$ do not belong to T, that is, $P(y)$ is true. But, by assumption (1.9), this implies that $P(x)$ is also true, which contradicts $x \in T$. So T is empty, as claimed. $\qquad\square$

Theorem 1.3 is called a method of *induction* because it assumes that a property $P(x)$ can be inferred from the fact that $P(y)$ is true for all y that are smaller than x. Then, assuming (1.8), the property holds for all x in S. An important application of this induction method is for proving the prime number decomposition theorem $P(x)$, which states that if x is an integer and $x \geq 2$, then x can be written as a product of prime numbers. In this induction proof, one first shows that x has a prime factor p, where either $x = p$ (in which case we have shown $P(x)$ and are done) or $x = p \cdot y$ for some integer y with $2 \leq y < x$. In the latter case, $P(y)$ holds by the "inductive hypothesis", that is, $y = p_1 p_2 \cdots p_k$ for some prime numbers $p_1, p_2, \ldots, p_k$. Hence, $x = p \, p_1 p_2 \cdots p_k$, which also shows $P(x)$. By Theorem 1.3, $P(x)$ holds throughout. The induction can be applied because any nonempty set of positive integers has a minimal element.

The method is called "top-down" induction because we start from the "top" (here the element x) and use the property $P(y)$ in some way for elements y that are smaller than x, where we typically encounter y without knowing *how much smaller* y is compared to x, as in the prime number decomposition theorem. In contrast, ordinary induction for natural numbers n proceeds typically by proving $P(1)$ and then the implication $P(n) \Rightarrow P(n + 1)$, that is, one step at a time, which then implies that $P(n)$ holds for all natural numbers n. Top-down induction is more powerful; it also applies to partial orders and not just to the totally ordered natural numbers (where it is often called "strong induction").

Where did the base case go, which is normally proved separately in an induction proof? The base case states that $P(x)$ holds for any minimal element x (if $\leq$ is total, then there is only one minimal element). The answer is that the base case is covered by the assumption in (1.9): The left-hand side of the implication (1.9) says that

$$(\forall y < x : P(y)) \;\Rightarrow\; P(x) \tag{1.10}$$

holds for all x, in particular for all minimal x. But then $(\forall y < x : P(y))$ is vacuously true because there is no y with $y < x$. Therefore, for minimal x, condition (1.10) means that $P(x)$ is true, which *is* the base case.

As a proof technique, we can often use Theorem 1.3 generally without any case distinctions about whether the considered elements are minimal or not. It also allows using a property about games *recursively* by referring to the same property applied to *simpler* games.

Using top-down induction, we can now prove Lemma 1.1, which states that any game G is either winning or losing. Assume that this holds for all games that are simpler than G, in particular the options of G. If all of these options are winning, then G is a losing position, because the other player can force a win no matter which move the player makes in G. If not all options of G are winning, then one of them, say H, is losing, and by moving to H the player forces a win, hence G is a winning position. This argument is not in any way deeper than what we said before. Theorem 1.3 merely asserts that it is rigorous (and that it uses the ending condition).

1.4   Game Sums and Equivalence of Games

Combinatorial games often "decompose" into parts in which players can move independently, and the players then have to decide in which part to make their move. This is captured by the important concept of a *sum* of games.

Definition 1.4. Suppose that G and H are game positions with options (positions reached by one move) $G_1, \ldots, G_k$ and $H_1, \ldots, H_m$, respectively. Then the options of the *game sum* $G + H$ are

$$G_1 + H, \ldots, G_k + H, \qquad G + H_1, \ldots, G + H_m. \tag{1.11}$$

$\square$

The first list of options $G_1 + H, \ldots, G_k + H$ in (1.11) simply means that the player makes his move in G, the second list $G + H_1, \ldots, G + H_m$ that he makes his move in H; the other part of the game sum remains untouched. As an example, a Nim position is simply the game sum of its individual Nim heaps, because the player moves in exactly one of the heaps.

Definition 1.4 is a recursive definition, because the game sum is defined in terms of its options, which are themselves game sums (but they are simpler games).

The sum of games turns out to define an *abelian group* on the (appropriately defined) set of games. It is a commutative and associative operation: for any games G, H, J,

$$G + H = H + G \qquad \text{and} \qquad (G + H) + J = G + (H + J). \tag{1.12}$$

The first condition (commutativity) holds because the order of the options of a game, used in (1.11), does not matter. The second condition (associativity) holds because both $(G + H) + J$ and $G + (H + J)$ mean in effect that the player decides to move in G, in H, or in J, leaving the other two parts of the game sum unchanged. We can therefore assume the equalities (1.12). More generally, in a sum of several games $G_1, \ldots, G_n$ the player moves in exactly one of these games, which does not depend on how these games are arranged, so that we can write this sum unambiguously without parentheses as $G_1 + \cdots + G_n$.

The losing game 0 which has no options is a *zero* (neutral element) for game sums: It fulfills $G + 0 = G$ for any game G, because the game 0 is "invisible" when added to G.

In order to obtain a group, every game G needs to have a "negative" game $-G$ so that $G + (-G) = 0$. However, this equality cannot hold as stated as soon as the game G has options, because then the game $G + (-G)$ also has options but 0 has none. Instead, we need a more general condition

$$G + (-G) \equiv 0, \tag{1.13}$$

where $G \equiv H$ means that the two games G and H are *equivalent*, according to the following definition.

Definition 1.5. Two games G, H are called *equivalent*, written $G \equiv H$, if and only if for any other game J, the game sum $G + J$ is losing if and only if $H + J$ is losing. $\square$

Lemma 1.6. *The relation $\equiv$ in Definition 1.5 is indeed an equivalence relation between games, that is, it is reflexive, symmetric and transitive.*

Proof. Reflexivity means $G \equiv G$ for any game G, which holds trivially. Symmetry means that if $G \equiv H$ then $H \equiv G$, which holds also by definition. Suppose that $G \equiv H$ and $H \equiv K$ for games G, H, K. Transitivity means that then $G \equiv K$. Consider any other game J. If $G + J$ is losing then $H + J$ is losing because $G \equiv H$, and then $K + J$ is losing because $H \equiv K$. Conversely, if $K + J$ is losing then $H + J$ is losing because $H \equiv K$, and then $G + J$ is losing because $G \equiv H$. Hence, $G + J$ is losing if and only if $K + J$ is losing, that is, $G \equiv K$. This shows that $\equiv$ is transitive. $\square$

By taking $J = 0$ in Definition 1.5, if the games G and H are equivalent then either both are losing or both are winning. Equivalence of G and H means that the games G and H always have the same outcome (winning or losing) and that *this property is preserved in any game sum $G + J$ and $H + J$ for any other game J*. This is much stronger than just belonging to the same outcome class. For example, if G and H are Nim heaps of different sizes, then both games are winning but they are not equivalent. We state this as a lemma.

Lemma 1.7. *Two Nim heaps are equivalent if and only if they have equal size.*

Proof. Let G and H be two Nim heaps of different sizes, and consider $J = G$ in Definition 1.5. Then $G + G$ is a losing game, but $H + G$ is winning. This shows that G and H are not equivalent. If G and H are Nim heaps of equal size then they are the same game and therefore equivalent.   □

However, any two *losing* games are equivalent, because they are all equivalent to the game 0 that has no options.

Lemma 1.8. *G is a losing game if and only if $G \equiv 0$.*

Proof. If $G \equiv 0$ then G is losing because 0 is losing, using $J = 0$ in Definition 1.5. So it remains to prove the "only if" part.

Let G be a losing game. We want to show $G \equiv 0$, that is, for any other game J, the game $G + J$ is losing if and only if $0 + J$ is losing. Because $0 + J = J$, this amounts to the two implications

$$G + J \text{ is losing} \quad \Rightarrow \quad J \text{ is losing} \tag{1.14}$$

and (we add the assumption that G is losing for symmetry)

$$G \text{ and } J \text{ are losing} \quad \Rightarrow \quad G + J \text{ is losing}. \tag{1.15}$$

Condition (1.15) is rather obvious: If G and J are both losing games, there is no way to win the game sum $G + J$ because the second player always has a counter-move. We prove (1.15) formally by top-down induction. Let G and J be losing. We want to show that $G + J$ is losing, which means that every option of $G + J$ is a winning game. Any such option is of the form $G' + J$ where is G' is an option of G, or of the form $G + J'$ where J' is an option of J. Consider the first type of option $G' + J$. Because G is a losing game, its option G' is a winning position, so G' has some option G'' that is a losing position. Because G'' is a losing game that is simpler than G, we can assume that (1.15) holds, that is, G'' and J are losing $\Rightarrow G'' + J$ is losing. So the option $G' + J$ is winning because it has the losing option $G'' + J$, and this was an arbitrary option of $G + J$ with a move to $G' + J$ in the first game G of the sum $G + J$. The same applies for any option of $G + J$ of the form $G + J'$ where J' is an option of J. So all options of $G + J$ are winning games, and $G + J$ is therefore a losing game, as claimed. This concludes the proof of (1.15).

We now show (1.14). Suppose that $G + J$ is losing but that J is winning. Then there is a winning move to some option J' of J where J' is a losing game. But then by (1.15), which we just proved, this implies that $G + J'$ is a losing game, which is an option of $G + J$, and would represent a winning move from $G + J$, contradicting that $G + J$ is losing. This shows that (1.14) holds, which completes the proof.   □

The next lemma shows that adding a game preserves equivalence between two games G and H.

Lemma 1.9. *For all games G, H, K,*

$$G \equiv H \quad \Rightarrow \quad G + K \equiv H + K. \tag{1.16}$$

Proof. Let $G \equiv H$. We show that $G + K \equiv H + K$, that is, $(G + K) + J$ is losing $\Leftrightarrow$ $(H + K) + J$ is losing for any other game J. By (1.12), this means $G + (K + J)$ is losing $\Leftrightarrow H + (K + J)$ is losing, which holds because G and H are equivalent. $\qquad\square$

Using equivalence, *any losing game* Z can take the role of a zero (neutral) element for game sums, that is,

$$Z \text{ is losing} \quad \Rightarrow \quad G + Z \equiv G \qquad\qquad (1.17)$$

for any game G. The reason is that any move in the component Z of the game sum $G + Z$ has an immediate counter-move back to a losing position, which will eventually become the game 0 without options. Therefore, adding Z does not affect if G is winning or losing, and this is also preserved when adding any other game J. In particular, $Z \equiv 0$ for any losing game Z, by taking $G = 0$ in (1.17). We state (1.17) as a lemma, which is proved easily with the help of the previous lemmas.

Lemma 1.10. *Let Z be a losing game. Then $G + Z \equiv G$ for any game G.*

Proof. By Lemma 1.8, $Z \equiv 0$, and by (1.16) $G + Z = Z + G \equiv 0 + G = G$. $\qquad\square$

The next lemma implies that any game G has a negative game $-G$ so that (1.13) holds. For impartial games, this negative game is G itself, that is, $G + G$ is always a losing position. This is the *copycat principle* for impartial games: the second player always has a move left, by copying the move of the first player. If G' is an option of G and the first player moves in $G + G$ to $G' + G$ or $G + G'$, then the second player responds by moving to $G' + G'$. This continues until the first player has run out of moves and loses.

Lemma 1.11 (The copycat principle). $G + G \equiv 0$ *for any impartial game G.*

Proof. Let G be an impartial game. By Lemma 1.8, we have to show that $G + G$ is a losing game. We prove this by top-down induction, by showing that every option of $G + G$ is a winning game. Any such option of $G + G$ is of the form $G' + G$ or $G + G'$ for an option G' of G. Then the next player has the winning move to the game $G' + G'$ which is losing, by the inductive assumption because $G' + G'$ is simpler than $G + G$. So $G + G$ is indeed a losing game. $\qquad\square$

Lemma 1.10 states that any losing game in a game sum is irrelevant for the outcome class of that sum. For example, by Lemma 1.11 a losing game of this sort may be the sum of two equal games in a larger game sum. This can simplify game sums substantially. For example, in the Nim position $1, 2, 3, 5, 5, 5$, the game sum of two Nim heaps of size 5 is a losing game, and the position $1, 2, 3$ is also losing, as we saw earlier, so the overall position is equivalent to a single Nim heap of size 5.

The following condition is very useful to prove that two games are equivalent.

Lemma 1.12. *Two impartial games G, H are equivalent if and only if $G + H \equiv 0$.*

Proof. If $G \equiv H$, then by (1.16) and Lemma 1.11, $G + H \equiv H + H \equiv 0$. Conversely, $G + H \equiv 0$ implies $G \equiv G + H + H \equiv 0 + H = H$. $\square$

The following lemma states that two games G and H are equivalent if all their options are equivalent.

Lemma 1.13. *Consider two games G and H so that for every option of G there is an equivalent option of H and vice versa. Then $G \equiv H$.*

Proof. We show that $G + H$ is losing, by showing that every option of $G + H$ is a winning game. Consider such an option $G' + H$ where G' is an option of G. By assumption, there is an option H' of H with $G' \equiv H'$, that is, $G' + H' \equiv 0$ by Lemma 1.12, so $G' + H'$ is losing. Moving there defines a winning move in $G' + H$. By the same argument, any option $G + H'$ of $G + H$ is winning. Hence, every option of $G + H$ is winning, so $G + H$ is a losing game. Therefore, $G + H \equiv 0$, and $G \equiv H$ by Lemma 1.12. $\square$

Note that Lemma 1.13 states only a sufficient condition for the equivalence of G and H. Games can be equivalent without that property. For example, $G + G$ is equivalent to 0, but $G + G$ has in general many options and 0 has none.

We conclude this section with a comment on the abelian group of impartial games. The group operation is the sum + of games. In order for the group to work, we have to identify equivalent games, which are "compatible" with game sums by Lemma 1.9. Any losing game (which is equivalent to the game 0 with no options) is a zero element of the group operation. Every game is its own negative. Essentially (although we will not prove it that way), in our context the only groups with this property are given by the set of binary vectors with component-wise addition modulo 2. This may serve as an intuition why the optimal strategy of Nim as in (1.3) uses the binary representation based on powers of two (and not powers of three, say). Moreover, Nim plays a central game for all impartial games. We will prove this directly, without any reference to abelian groups, in the next section.

A second comment concerns misère games, where the last player to move loses. One losing position in misère Nim is a single heap with one token in it. However, adding such a heap Z, say, to a game G completely reverses the outcome class of G because the token can just be removed, so we certainly do not have $G + Z \equiv G$ as in Lemma 1.10, no matter how the equivalence $\equiv$ is defined. Also, $1, 1$ is a winning position but $2, 2$ is a losing position in misère Nim, so there is no easy copycat principle for misère games. Although misère Nim is well understood (see Exercise 1.3), general misère games are not. In contrast, game sums have a nice structure with the normal play convention.

1.5   Nim, Poker Nim, and the Mex Rule

In this section, we prove a sweeping statement (Theorem 1.14): *Every impartial game is equivalent to some Nim heap.* Note that this means a single Nim heap, even if the game itself is rather complicated, for example a general Nim position, which is a game sum of several Nim heaps. This can proved without any reference to playing Nim optimally. As we will see at the end of this section with the help of Figure 1.2, the so-called *mex rule* that underlies Theorem 1.14 can be used to *discover* the role of powers of two for playing sums of Nim heaps.

We use the following *notation for Nim heaps.* If G is a single Nim heap with n tokens, $n \geq 0$, then we denote this game by $*n$. This game is completely specified by its n options, and they are defined recursively as follows:

$$\text{options of } *n: \quad *0, *1, *2, \ldots, *(n-1). \tag{1.18}$$

Note that $*0$ is the empty heap with no tokens, that is, $*0 = 0$; we will normally continue to just write 0.

We can use (1.18) as the definition of $*n$. For example, the game $*4$ is defined by its options $*0, *1, *2, *3$. It is very important to include $*0$ in that list of options, because it means that $*4$ has a winning move. Condition (1.18) is a recursive definition of the game $*n$, because its options are also defined by reference to such games $*k$, for numbers k smaller than n. This game fulfills the ending condition because the heap gets successively smaller in any sequence of moves.

A general Nim position is a game sum of several Nim heaps. Earlier we had written such a position by just listing the sizes of the Nim heaps, such as $1, 2, 3$ in (1.1). The fancy way to write this is now $*1 + *2 + *3$, a sum of games.

The game of *Poker Nim* is a variation of Nim. Suppose that each player is given, at the beginning of the game, some extra "reserve" tokens. Like Nim, the game is played with heaps of tokens. In a move, a player can choose, as in ordinary Nim, a heap and remove some tokens, which he can add to his reserve tokens. A second, new kind of move is to *add* some of the player's reserve tokens to some heap (or even to create an entire new heap with these tokens). These two kinds of moves are the only ones allowed.

Suppose that there are three heaps, of sizes $1, 2, 5$, and that the game has been going on for some time, so that both players have accumulated substantial reserves of tokens. It is player I's turn, who moves to $1, 2, 3$ because that is a good move in ordinary Nim. But then player II adds 50 tokens to the heap of size 2, creating the position $1, 52, 3$, which seems complicated.

What should player I do? After a moment's thought, he just removes the 50 tokens that player II has just added to the heap, reverting to the previous position. Player II may keep adding tokens, but will eventually run out of them, no matter how many she acquires in between, and then player I can proceed as in ordinary Nim.

Hence, a player who can win in a position in ordinary Nim can still win in Poker Nim. He replies to the opponent's heap-reducing moves just as he would in ordinary Nim, and reverses the effect of any heap-increasing move by using a heap-reducing move to restore the heap to the same size again. Strictly speaking, the ending condition is violated in Poker Nim because in theory the game could go on forever. However, a player in a winning position wants to end the game with his victory, and never has to put back any tokens; then the losing player will eventually run out of reserve tokens that she can add to a heap, so that the game terminates.

Consider now an impartial game where the options of player I are games that are equivalent to the Nim heaps $*0, *1, *2, *5, *9$. This can be regarded as a rather peculiar Nim heap of size 3 which can be reduced to any of the sizes $0, 1, 2$, but which can also be increased to size 5 or 9. The Poker Nim argument shows that this extra freedom is of no use to turn a losing game into a winning game.

The *mex rule* says that if the options of a game G are equivalent to Nim heaps with sizes from a set S (like $S = \{0, 1, 2, 5, 9\}$ above), then G is equivalent to a Nim heap of size m, where m is the *smallest nonnegative integer not contained in S*. This number m is written mex(S), where *mex* stands for *minimum excluded number*. That is,

$$m = \text{mex}(S) = \min\{k \geq 0 \mid k \notin S\}.$$

For example, mex($\{0, 1, 2, 3, 5, 6\}$) = 4, mex($\{1, 2, 3, 4, 5\}$) = 0, and mex($\emptyset$) = 0.

If $G \equiv *m$ for an impartial game G, then m is called the *Nim value* of G. The Nim value is unique because Nim heaps of different sizes are not equivalent by Lemma 1.7. Because the concept has been discovered independently by Sprague (1935) and Grundy (1939), other common names for the Nim value are also Sprague–Grundy value, sg-value, or Grundy value.

Theorem 1.14 (The mex rule). *Consider an impartial game G. Then G has Nim value m (that is, $G \equiv *m$), where m is uniquely determined as follows: For each option H of G, let H have Nim value s_H, and let $S = \{s_H \mid H$ is an option of $G\}$. Then $m = \text{mex}(S)$, that is, $G \equiv *(\text{mex}(S))$.*

Proof. The statement is true if $S = \emptyset$ because then $G = 0 = *0$. Assume, per top-down induction, that each option H of G has Nim value s_H, where s_H is itself the mex of the Nim values of the options of H, and that S is the set of these Nim values s_H as described. Let $m = \text{mex}(S)$. We show $G + *m \equiv 0$, which proves that $G \equiv *m$ by Lemma 1.12. The game $G + *m$ is losing if each of its options is winning.

First, consider such an option of the form $G + *h$ for some $h < m$, that is, the player moves in the Nim heap $*m$ of the game sum $G + *m$. Then $h \in S$ because

$m = \text{mex}(S)$, and by construction of S there is an option H of G so that $h = s_H$, that is, $H \equiv *h$. Hence, the other player has the counter-move from $G + *h$ to the losing position $H + *h$, which shows that $G + *h$ is indeed winning.

Second, consider an option of $G + *m$ of the form $H + *m$ where H is an option of G, where $H \equiv *s_H$ and $s_H \in S$, and either $s_H < m$ or $s_H > m$ (the case $s_H = m$ is excluded because $m \notin S$). If $s_H < m$, then the other player counters by moving from $H + *m$ to the losing position $H + *s_H$. If $s_H > m$, we use the inductive hypothesis that s_H is the mex of the Nim values of the options of H, so one of these Nim values is m, with $H' \equiv *m$ for an option H' of H. Then the other player counters by moving from $H + *m$ to the losing position $H' + *m$ (this is the Poker Nim argument). This shows that $H + *m$ is also a winning position. Hence, $G + *m$ is a losing game and $G + *m \equiv 0$, as required.   □

A special case of the preceding theorem is that $\text{mex}(S) = 0$, which means that all options of G are equivalent to positive Nim heaps, so they are all winning positions, or that G has no options at all. Then G is a losing game, and indeed $G \equiv *0$.

> $\Rightarrow$ The digraph game in Exercise 1.9 is excellent for understanding the mex rule.

Note that the mex rule has nothing to do with game sums (apart from the way Theorem 1.14 is proved, which is irrelevant). The mex rule evaluates the options of a game G, and they are not added when looking at them. If the options are themselves game sums (for example for the game Kayles in (1.22) below or the game in Exercise 1.7), then this is a separate matter that is used to find the Nim value of each option, and is specific to the rules of that game. The mex rule is general.

By Theorem 1.14, any impartial game can be played like Nim, provided the Nim values of the positions of the game are known. The Nim values of the positions can be evaluated recursively by the mex rule. This is illustrated in the *Rook-move* game in Figure 1.1. Place a rook on a Chess board that extends arbitrarily to the right and bottom but ends at the left and at the top. In one move, the rook is moved either horizontally to the left or vertically upwards, for any number of squares (at least one) as long as it stays on the board. The first player who can no longer move loses, when the rook is on the top left square of the board. We number the rows and columns of the board by $0, 1, 2, \ldots$ starting from the top left.

Figure 1.2 gives the Nim values for the positions of the rook on the Chess board. The top left square is equivalent to 0 because the rook can no longer move. The square below only allows reaching the square with 0 on it, so it is equivalent to $*1$ because $\text{mex}(\{0\}) = 1$. The square below gets $*2$ because its options are equivalent to 0 and $*1$. From any square in the leftmost column in Figure 1.2, the rook can only move upwards, so any such square in column 0 and row i corresponds obviously to a Nim heap $*i$. Similarly, the topmost row 0 has entry $*j$ in column j.

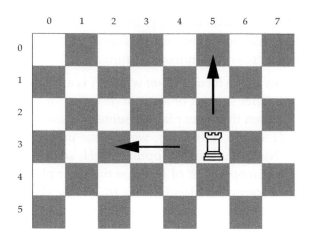

Figure **1.1** Rook-move game, where the player may move the rook on the Chess board in the direction of the arrows.

In general, a position on the board is evaluated knowing all Nim values for the squares to the left and the top of it, which are the options of that position. As an example, consider the square in row 3 and column 2. To the left of that square, the entries $*3$ and $*2$ are found, and to the top the entries $*2, *3, 0$. So the square itself is equivalent to $*1$ because $\mathrm{mex}(\{0, 2, 3\}) = 1$. The square in row 3 and column 5, where the rook is placed in Figure 1.1, gets entry $*6$.

	0	1	2	3	4	5	6	7	8	9	10
0	0	$*1$	$*2$	$*3$	$*4$	$*5$	$*6$	$*7$	$*8$	$*9$	$*10$
1	$*1$	0	$*3$	$*2$	$*5$	$*4$	$*7$	$*6$	$*9$	$*8$	$*11$
2	$*2$	$*3$	0	$*1$	$*6$	$*7$	$*4$	$*5$	$*10$	$*11$	$*8$
3	$*3$	$*2$	$*1$	0	$*7$	$*6$	$*5$	$*4$	$*11$	$*10$	$*9$
4	$*4$	$*5$	$*6$	$*7$	0	$*1$	$*2$	$*3$	$*12$	$*13$	$*14$
5	$*5$	$*4$	$*7$	$*6$	$*1$	0	$*3$	$*2$	$*13$	$*12$	$*15$
6	$*6$	$*7$	$*4$	$*5$	$*2$	$*3$	0	$*1$	$*14$	$*15$	$*12$
7	$*7$	$*6$	$*5$	$*4$	$*3$	$*2$	$*1$	0	$*15$	$*14$	$*13$
8	$*8$	$*9$	$*10$	$*11$	$*12$	$*13$	$*14$	$*15$	0	$*1$	$*2$
9	$*9$	$*8$	$*11$	$*10$	$*13$	$*12$	$*15$	$*14$	$*1$	0	$*3$
10	$*10$	$*11$	$*8$	$*9$	$*14$	$*15$	$*12$	$*13$	$*2$	$*3$	0

Figure 1.2   Equivalent Nim heaps $*n$ for positions of the Rook-move game.

Figure 1.2 shows that the only losing squares on the Chess board are on the diagonal, which are equivalent to the losing game 0. It is also easy to see directly that the diagonal squares are exactly the losing positions of the Rook-move game:

These include the top left square where the rook can no longer move. Whenever the rook is on the diagonal the next move has to move the rook away from the diagonal, which is a winning position because at the next move the rook can always be moved back to the diagonal.

The astute reader will have noticed that the Rook-move game is just Nim with two heaps. A rook positioned in row i and column j can either reduce i by moving up, or j by moving left. So this position is the sum of Nim heaps $*i + *j$. It is a losing position if and only if $i = j$, where the rook is on the diagonal.

In addition to distinguishing winning and losing positions, the Nim values become very useful as soon as we place *multiple rooks* on the board, assuming that they are allowed to share any square, which corresponds to a sum of Rook-move games. For example, it is not obvious that two rooks, one placed in row 2 and column 0, and another in row 4 and column 6, define a losing position because both squares have Nim value 2.

Theorem 1.14 states that *any* impartial game is equivalent to a *single* Nim heap of a certain size. In particular, the sum of two Nim heaps is equivalent to a single Nim heap. Assuming we don't know how to play Nim using the binary system as explained with the example (1.3), so far we only know that the game sum of two Nim heaps is winning if they have different sizes, and losing if they are equal. Figure 1.2 represents the computation of the Nim values of a pair of Nim heaps $*i + *j$.

There is a certain pattern to these Nim values: Clearly, they are symmetric when reflected at the diagonal because rows and columns can be interchanged. In a single row (or column), all the values are different because for any set S of integers, $\text{mex}(S)$ does not belong to S, so the value in a square is certainly different from the values to the left of that square. Now consider some power of two, say 2^k, and look at the first 2^k rows and columns, which are each numbered $0, 1, \ldots, 2^k - 1$. This quadratic top left part of the table (for example, the top left 4×4 squares if $k = 2$) only has the entries $0, 1, \ldots, 2^k - 1$ in each row and column in some permutation. The role of powers of two for computing the Nim value of a sum of two Nim heaps is made precise in the next section, and will explain the use of the binary representation as in (1.3).

1.6 Sums of Nim Heaps

In this section, we derive how to compute the Nim value for a general Nim position, which is a sum of different Nim heaps. This will be the Nim sum that we have defined using the binary representation, now cast in the language of game sums and equivalent games, and without assuming the binary representation.

For example, we know that $*1 + *2 + *3 \equiv 0$, so by Lemma 1.12, $*1 + *2$ is equivalent to $*3$. In general, however, the sizes of the Nim heaps cannot simply be added to obtain the equivalent Nim heap, because $*2 + *3$ is also equivalent to $*1$, and $*1 + *3$ is equivalent to $*2$.

If $*k \equiv *n + *m$, then we call k the *Nim sum* of n and m, written $k = n \oplus m$. The following theorem states that the Nim sum of distinct powers of two is their arithmetic sum. For example, $1 = 2^0$ and $2 = 2^1$, so $1 \oplus 2 = 1 + 2 = 3$.

Theorem 1.15. *Let* $n \geq 1$, *and* $n = 2^a + 2^b + 2^c + \cdots$, *where* $a > b > c > \cdots \geq 0$. *Then*

$$*n \equiv *(2^a) + *(2^b) + *(2^c) + \cdots . \tag{1.19}$$

We first discuss the implications of this theorem, and then prove it. The expression $n = 2^a + 2^b + 2^c + \cdots$ is an arithmetic sum of distinct powers of two. Any n is uniquely given as such a sum. It amounts to the binary representation of n, which, if $n < 2^{a+1}$, gives n as the sum of all powers $2^a, 2^{a-1}, 2^{a-2}, \ldots, 2^0$ where each power of two is multiplied with 0 or 1, the binary digit for the respective position. For example,

$$9 = 8 + 1 = 1 \cdot 2^3 + 0 \cdot 2^2 + 0 \cdot 2^1 + 1 \cdot 2^0,$$

so that 9 in decimal is written as 1001 in binary. Theorem 1.15 uses only the distinct powers of two $2^a, 2^b, 2^c, \ldots$ that correspond to the digits 1 in the binary representation of n.

The right-hand side of (1.19) is a *game sum*. Equation (1.19) states that the single Nim heap $*n$ is equivalent to a game sum of Nim heaps whose sizes are distinct powers of two. If we visualize the tokens in a Nim heap as dots like in (1.4), then an example of this equation is

$$*14 \quad = \quad \vdots\vdots\vdots\ \vdots\vdots\ \vdots \quad \equiv \quad \vdots\vdots\vdots\vdots \quad + \quad \vdots\vdots \quad + \quad \vdots$$

for $n = 14$. In addition to $*n$, consider a second Nim heap $*m$ and represent it as its equivalent game sum of several Nim heaps, all of which have a size that is a power of two. Then $*n + *m$ is a game sum of many such heaps, where equal heaps cancel out in pairs because a game sum of two identical games is losing and can be omitted. The remaining heap sizes are all distinct powers of two, which can be added to give the size of the single Nim heap $*k$ that is equivalent to $*n + *m$. As an example, let $n = 9 = 8 + 1$ and $m = 14 = 8 + 4 + 2$. Then $*n + *m \equiv *8 + *1 + *8 + *4 + *2 \equiv *4 + *2 + *1 \equiv *7$, which we can also write as $9 \oplus 14 = 7$. In particular, $*9 + *14 + *7$ is a losing game, which would be very laborious to show without the theorem.

One consequence of Theorem 1.15 is that the Nim sum of two integers never exceeds their arithmetic sum. Moreover, if both integers are less than some power of two, then so is their Nim sum.

Lemma 1.16. *Suppose that Theorem 1.15 holds for any n with $n < 2^d$. Let $0 \leq p, q < 2^d$. Then $*p + *q \equiv *r$ where $0 \leq r < 2^d$, that is, $r = p \oplus q < 2^d$.*

Proof. Both p and q are sums of distinct powers of two, all smaller than 2^d. By Theorem 1.15, r is also a sum of such powers of two, where those powers that appear in both p and q, say 2^k, cancel out in the game sum because $*(2^k) + *(2^k) \equiv 0$. Hence, the resulting sum r is also less than 2^d. □

The following proof may be best understood by considering it along with an example, say $n = 7$.

Proof of Theorem 1.15. We proceed by induction. Consider some n, and assume that the theorem holds for all smaller n. Let $n = 2^a + q$ where $q = 2^b + 2^c + \cdots$. If $q = 0$, the claim holds trivially (n is just a single power of two), so let $q > 0$. We have $q < 2^a$. By inductive assumption, $*q \equiv *(2^b) + *(2^c) + \cdots$, so all we have to prove is that $*n \equiv *(2^a) + *q$ in order to show (1.19). We show this using Lemma 1.13, that is, by showing that each option of $*n$ is equivalent to some option of the game $*(2^a) + *q$ and vice versa. The options of $*n$ are $0, *1, *2, \ldots, *(n-1)$.

The options of $*(2^a) + *q$ are of two kinds, depending on whether the player moves in the Nim heap $*(2^a)$ or $*q$. The first kind of options are given by

$$
\begin{aligned}
0 + *q &\equiv *r_0 \\
*1 + *q &\equiv *r_1 \\
&\vdots \\
*(2^a - 1) + *q &\equiv *r_{2^a - 1}
\end{aligned}
\tag{1.20}
$$

where the equivalence of $*i + *q$ with some Nim heap $*r_i$, for $0 \leq i < 2^a$, holds by the inductive assumption even if we did not know the mex rule. Moreover, by Lemma 1.16 (with $d = a$, because for $n < 2^a$ Theorem 1.15 holds by the inductive assumption), both i and q are less than 2^a, and therefore also $r_i < 2^a$. On the right-hand side in (1.20), there are 2^a many Nim heaps $*r_i$ for $0 \leq i < 2^a$. We claim they are all different, so that these options form exactly the set $\{0, *1, *2, \ldots, *(2^a - 1)\}$. Namely, by adding the game $*q$ to the heap $*r_i$, (1.16) implies $*r_i + *q \equiv *i + *q + *q \equiv *i$, so that $*r_i \equiv *r_j$ implies $*r_i + *q \equiv *r_j + *q$, that is, $*i \equiv *j$ and hence $i = j$ by Lemma 1.7, for $0 \leq i, j < 2^a$.

The second kind of options of $*(2^a) + *q$ are of the form

$$
\begin{aligned}
*(2^a) + 0 &\equiv *(2^a + 0) \\
*(2^a) + *1 &\equiv *(2^a + 1) \\
&\vdots \\
*(2^a) + *(q - 1) &\equiv *(2^a + q - 1),
\end{aligned}
$$

where the heap sizes on the right-hand sides are given again by the inductive assumption. These heaps form the set $\{*(2^a), *(2^a + 1), \ldots, *(n - 1)\}$. Together with the first kind of options, they are exactly the options of $*n$. This shows that the options of $*n$ and of the game sum $*(2^a) + *q$ are indeed equivalent, which completes the proof.   □

In the proof of Theorem 1.15, if $n = 7$ then $n = 2^a + q$ with $a = 2$ and $q = 3$. Then the Nim heaps $*r_0, *r_1, *r_2, *r_3$ in (1.20) are the entries in the first 2^a rows $0, 1, 2, 3$ in column $q = 3$ of the Rook-move table in Figure 1.2.

This step in the proof of Theorem 1.15, which is based on Lemma 1.16, also explains to some extent why powers of two appear in the computation of Nim sums. The options of moving from $*(2^a) + *q$ in (1.20) neatly produce exactly the numbers $0, 1, \ldots, 2^a - 1$, which would not work when replacing 2^a with something else.

The Nim sum of any set of numbers can be obtained by writing each number as a sum of *distinct* powers of two and then canceling repetitions in pairs. For the example in (1.3) and (1.4), we have

$$4 \oplus 5 \oplus 9 \oplus 14 \;=\; 4 \oplus (4 + 1) \oplus (8 + 1) \oplus (8 + 4 + 2) \;=\; 4 + 2 = 6. \qquad (1.21)$$

Hence, the Nim position $4, 5, 9, 14$ is equivalent to a single Nim heap of size 6. This equivalence holds because adding this Nim heap to the game sum of the four Nim heaps produces a losing game:

$$(*4 + *5 + *9 + *14) + *6 \equiv 0 \,.$$

The same can be seen from the left table in (1.3). That is, Nim sums amount to writing the heap sizes in binary and adding the binary digits separately in each column (power of 2) modulo 2. The column entry of the Nim sum is 0 if the respective power of 2 appears an even number of times (all pairs cancel) and 1 if it appears an odd number of times.

1.7   Finding Nim Values

In this section, we analyze some impartial games using the mex rule in Theorem 1.14.

A game similar to the Rook-move game is the *Queen-move* game shown in Figure 1.3 where the rook is replaced by a Chess queen, which may move horizontally, vertically, and diagonally (left or up). The squares on the main diagonal are therefore no longer losing positions. This game can also be played with two heaps of tokens where in one move, the player may either remove tokens from one heap as in Nim, or reduce *both* heaps by the *same* number of tokens (so this is no longer a sum of two Nim heaps!). In order to illustrate that we are not just interested in the winning and losing squares, we add to this game a Nim heap of size 4.

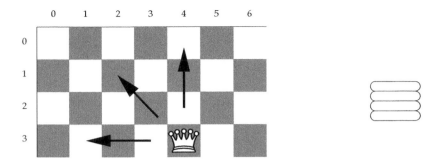

Figure 1.3 Game sum of a *Queen-move* game and a Nim heap of size four. The player may *either* move the queen in the direction of the arrows *or* take some of the four tokens from the heap.

Figure 1.4 shows the equivalent Nim heaps for the positions of the Queen-move game, determined by the mex rule. The square in row 3 and column 4 occupied by the queen in Figure 1.3 has entry $*2$. So a winning move is to remove two tokens from the Nim heap to turn it into the heap $*2$, creating the losing position $*2 + *2$. Because 2 is the mex of the Nim values of the options of the queen, these may include (as in Poker Nim) higher Nim values. Indeed, the queen can reach two positions equivalent to $*4$, in row 3 column 1, and row 0 column 4. If the queen moves there this creates the game sum $*4 + *4$ which is losing, so these are two further winning moves in Figure 1.3.

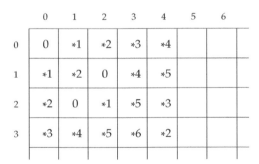

	0	1	2	3	4	5	6
0	0	$*1$	$*2$	$*3$	$*4$		
1	$*1$	$*2$	0	$*4$	$*5$		
2	$*2$	0	$*1$	$*5$	$*3$		
3	$*3$	$*4$	$*5$	$*6$	$*2$		

Figure 1.4 Equivalent Nim heaps for positions of the Queen-move game.

The impartial game *Kayles* is played as follows: Given a row of n bowling pins (numbered $1, 2, \ldots, n$), a move knocks out one or two *consecutive* pins, as in this example of 5 pins where pins 2 and 3 are knocked out:

$$\text{🎳🎳🎳🎳🎳} \rightarrow \text{🎳 _ _ 🎳🎳} \tag{1.22}$$

If this game is called K_n, then knocking out a single pin p creates the game sum $K_{p-1} + K_{n-p}$, and knocking out pins p and $p+1$ (in the picture, $p = 2$ and $n = 5$) creates the game sum $K_{p-1} + K_{n-p-1}$ (here $K_1 + K_2$). The options of K_n are these game sums for all possible p, where p only needs to be considered up to the middle pin due to the symmetry of the row of pins. Note that in this game, options happen to be *sums* of games. For the mex rule, only their Nim value is important. The following picture shows all options of Kayles with 1, 2, or 3 pins:

which shows $K_1 \equiv {*}\mathrm{mex}(0) = {*}1$, $K_2 \equiv {*}\mathrm{mex}(1,0) = {*}2$, and $K_3 \equiv {*}\mathrm{mex}(2,1,1 \oplus 1) = {*}3$ because $1 \oplus 1 = 0$. The options of Kayles with 4 pins are

and therefore $K_4 \equiv {*}\mathrm{mex}(3,2,1 \oplus 2,1 \oplus 1) = {*}\mathrm{mex}(3,2,3,0) = {*}1$ which leads to an unexpected winning move for 6 pins:

To check this, it is illustrative to show directly that $K_1 + K_4$ is a losing position, by showing that all its options are winning by finding for each one a counter-move to a losing position.

Further impartial games are described in the exercises.

1.8 A Glimpse of Partizan Games

A combinatorial game is called *partizan* if the available moves in a game position may be different for the two players. These games have a rich theory of which we sketch the first concepts here, in particular the four outcome classes, and *numbers* as strengths of positions in certain games, pioneered by Conway (2001).

The two players are called *Left* and *Right*. Consider the game *Domineering*, given by a board of squares (the starting position is typically rectangular but does not have to be), where in a move Left may occupy two vertically adjacent free squares with a *vertical* domino, and Right two horizontally adjacent squares with a *horizontal* domino. (At least in lower case, the letter "l" is more vertical than the letter "r" to remember this.) Hence, starting from a 3-row, 2-column board, we

have, up to symmetry,

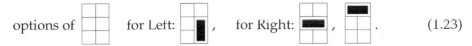

$$\text{options of} \quad \boxed{} \quad \text{for Left:} \quad \boxed{} \ , \quad \text{for Right:} \quad \boxed{} \ , \quad \boxed{} \ . \tag{1.23}$$

This game is very different from the impartial game *Cram* (see Exercise 1.4) where each player may place a domino in either orientation. As (1.23) shows, in Domineering the options of Left and Right are usually different. As before, we assume the normal play convention that a player who can no longer move loses.

We always assume optimal play. In (1.23), this means Right would choose the first option and place the domino in the middle, after which Left can no longer move, whereas the second option that leaves a 2×2 board would provide Left with two further moves.

An impartial game can only be losing or winning as stated in Lemma 1.1. For partizan games, there are four possible *outcome classes*, which are denoted by the following calligraphic letters:

$\mathcal{L}$: Left wins no matter who moves first.

$\mathcal{R}$: Right wins no matter who moves first.

$\mathcal{P}$: The first player to move loses, so the *previous* player wins.

$\mathcal{N}$: The first player to move wins. (Sometimes called the "next player", although this is as ambiguous as "next Friday", because it is the current player who wins.)

Every game belongs to exactly one of these outcome classes. The $\mathcal{P}$-positions are what we have called *losing positions*, and $\mathcal{N}$-positions are what we have called *winning positions*. For partizan games we have to consider the further outcome classes $\mathcal{L}$ and $\mathcal{R}$. In Domineering, a single vertical strip of at least two squares, in its smallest form $\boxminus$, belongs to $\mathcal{L}$, whereas a horizontal strip such as $\boxed{\ \ \ \ }$ belongs to $\mathcal{R}$. The starting 3×2 board in (1.23) belongs to $\mathcal{N}$.

The first move of Right in (1.23) creates two disjoint 1×2 games which represent, as before, a *game sum*, here $\boxed{\ \ } + \boxed{\ \ }$. This game clearly belongs to $\mathcal{R}$, as does the game $\boxed{\ \ }$. In order to distinguish games more finely than just according to the four outcome classes, games G and H are called *equivalent*, written $G \equiv H$, if and only if $G + J$ and $H + J$ always belong to the same outcome class, for any other game J; this extends Definition 1.5. Every losing game (any game in $\mathcal{P}$) is equivalent to the zero game 0 that has no options, as in Lemma 1.8, by the same argument. As before, Lemma 1.10 holds, that is, adding a losing game does not change the outcome class of a game.

For any partizan game G, there is a *negative* game $-G$ so that $G + (-G) \equiv 0$ as stated in (1.13). This negative game is obtained by exchanging the options of Left and Right, and doing so recursively for their options. For Domineering, the negative of any game position G (as shown by the squares that have not yet been occupied by dominos) is obtained by turning the board by 90 degrees. For example,

$$\text{(domino diagram)} + \text{(domino diagram)} \equiv 0 \tag{1.24}$$

because any move by a player in one component can be *copied* by the other player by the corresponding move in the other component. In that way, the second player always has a move left and has won when the position reached is $0 + 0$.

Formally, a game G is defined by two sets of games G^L and G^R which define the options of Left and Right, respectively. Then G is written as $\{G^L \mid G^R\}$. In concrete examples, one lists directly the elements of the sets G^L and G^R rather than using set notation. For example, if $G^L = G^R = \emptyset$, then we write this game as $\{ \mid \}$ rather than $\{\emptyset \mid \emptyset\}$. This is the simplest possible game that has no options for either player, which we have denoted by 0:

$$0 = \{ \mid \} . \tag{1.25}$$

The options in a game are defined recursively. We can write (1.23) as

$$\text{(diagram)} = \{ \text{(diagram)} \mid \text{(diagram)} + \text{(diagram)} , \text{(diagram)} \} \tag{1.26}$$

and continue to define

$$\text{(diagram)} = \{ 0 , \text{(diagram)} \mid \text{(diagram)} \} , \tag{1.27}$$

where the first option of Left is to put the domino in the top left which leaves Right with no further move, the second (inferior) option to put the domino at the bottom; Right has only one choice. Furthermore,

$$\text{(diagram)} = \{ 0 \mid \} , \qquad \text{(diagram)} = \{ \mid 0 \} , \tag{1.28}$$

which are the positions where Left respectively Right have exactly one move left and the other player none. The first of Right's options in (1.26) can also be defined by the recursive definition of game sums as

$$\text{(diagram)} + \text{(diagram)} = \{ \mid \text{(diagram)} \} ,$$

and the second of Right's options in (1.26) is defined by

$$\text{(diagram)} = \{ \text{(diagram)} \mid \text{(diagram)} \} . \tag{1.29}$$

The definitions (1.25) to (1.29) are recursive by referring to simpler games, with the shown board configurations as suitable placeholders. It only matters which player eventually loses in subsequent optimal play.

In (1.27), Left has the *dominated* option ☐☐ that Left should never move to because it unnecessarily gives Right an extra move. Dominated options can be omitted, which simplifies the recursive description of a game in terms of its options. Note: there are *no* dominated options in the board configuration ☐ in the *impartial* game Cram. In Cram, this configuration is equivalent to a Nim heap $*2$ of size two with its two options $*1$ and 0, neither of which dominates the other. For example, in the game sum $*2 + *1$, the move to $*1 + *1$ is a winning move which requires the option $*1$ of $*2$. Impartial games are exactly the games $\{G^L \mid G^R\}$ where $G^L = G^R$, and the option sets themselves contain only impartial games. The simplest impartial game is 0 by (1.25), the next simplest is $\{0 \mid 0\}$ which is the Nim heap $*1$ of size one, which is also denoted by $*$ (just a star).

An important new concept for partizan games is that of a *number*, which applies to some (but not all) partizan games. If a game is identified with a number, then this number states *how many moves player Left is safely ahead* in the game. The simplest numbers are integers, but they can also be certain types of fractions. For example, the Domineering position ☐ can be identified with the integer 1. Furthermore, $1 + 1 = 2$, that is, the game sum of numbers corresponds to the arithmetic sum of these numbers. Similarly, the Domineering position ☐☐ is identified with -1, in agreement with $1 + (-1) \equiv 0$. For a positive integer n, if Left is n moves ahead and makes one of these moves, then Left is still $n - 1$ moves ahead afterwards. By denoting the game itself by n, this gives rise to the recursive definition

$$n = \{n - 1 \mid \ \}, \tag{1.30}$$

which has its base case $1 = \{0 \mid \ \}$ as in (1.28). Games that are negative numbers are defined as negative games $-n$ of positive numbers n. In the game $-n$, player Right is ahead by n moves.

Caution: *No impartial game is a number* (except for the game 0). Nim heaps as defined in (1.18) are very different, and obey different rules such as $*n + *n \equiv 0$. (Nim heaps are sometimes called *nimbers* to distinguish them from numbers.)

In order to define a number more generally, we need a partial order $\geq$ on games (which is *not* the "simpler than" order), defined as follows:

$$G \geq H \quad \Leftrightarrow \quad \begin{aligned} (\forall J : H + J \in \mathcal{L} \cup \mathcal{N} &\Rightarrow G + J \in \mathcal{L} \cup \mathcal{N}, \\ \forall J : H + J \in \mathcal{L} \cup \mathcal{P} &\Rightarrow G + J \in \mathcal{L} \cup \mathcal{P}) \end{aligned} \tag{1.31}$$

where a game belongs to $\mathcal{L} \cup \mathcal{N}$ if Left wins by moving first, and to $\mathcal{L} \cup \mathcal{P}$ if Left wins by moving second. With this understanding, (1.31) can be written as

$$G \geq H \quad \Leftrightarrow \quad (\forall J \, : \, \text{Left wins in } H + J \quad \Rightarrow \quad \text{Left wins in } G + J)\,.$$

That is, $G \geq H$ if replacing H by G never hurts Left, no matter what the context (always assuming optimal play). Then $\geq$ is a partial order on games because it is obviously reflexive and transitive, and antisymmetric up to equivalence of games:

Lemma 1.17. $G \geq H$ and $H \geq G \Leftrightarrow G \equiv H$.

Proof. Recall that $G \equiv H$ means that $G + J$ and $H + J$ belong to the same outcome class $\mathcal{L}, \mathcal{R}, \mathcal{P}, \mathcal{N}$ for any game J, so "$\Leftarrow$" is obvious. To show "$\Rightarrow$", assume $G \geq H$ and $H \geq G$, and let J be any game. If $H + J \in \mathcal{L}$ then $G + J \in (\mathcal{L} \cup \mathcal{N}) \cap (\mathcal{L} \cup \mathcal{P}) = \mathcal{L}$ by (1.31), because the outcome classes are disjoint. By symmetry, if $G + J \in \mathcal{L}$ then $H + J \in \mathcal{L}$. Next, consider the outcome class $\mathcal{P}$: If $H + J \in \mathcal{P}$ then $G + J \in \mathcal{L} \cup \mathcal{P}$ by (1.31). If we had $G + J \in \mathcal{L}$ then $H + J \in \mathcal{L}$ as just shown, but $H + J \in \mathcal{P}$ which is disjoint from $\mathcal{L}$, so we have $G + J \in \mathcal{P}$. By symmetry, if $G + J \in \mathcal{P}$ then $H + J \in \mathcal{P}$. Similar arguments apply for the outcome classes $\mathcal{R}$ and $\mathcal{N}$. □

Define $G < H$ as $H \geq G$ and not $G \equiv H$. This allows the following recursive definition of a partizan game that is a number.

Definition 1.18. A game $G = \{\mathcal{G}^L \mid \mathcal{G}^R\}$ is a *number* if all left options $G^L \in \mathcal{G}^L$ of G and all right options $G^R \in \mathcal{G}^R$ of G are numbers and fulfill $G^L < G < G^R$. □

The simplest number is $0 = \{ \mid \}$ because it has no left or right options, so the condition $G^L < G < G^R$ for $G = 0$ holds vacuously. Similarly, the game defined in (1.30) for a positive integer n is a number because its left option fulfills $n - 1 < n$, and it has no right options.

The intuition of the inequalities $G^L < G < G^R$ in Definition 1.18 is that by moving in a number, a player gives up a safe advantage and thereby weakens his position, so moving to a left option G^L weakens the position of Left ($G^L < G$), and moving to a right option G^R weakens the position of Right ($G < G^R$).

The advantage that is given up by moving in a number can be small, because games can also be numbers that are *fractions*. For example, the Domineering position in (1.27) is given by $\{0, -1 \mid 1\}$ (or, because -1 is dominated, $\{0 \mid 1\}$), where $0 = \{ \mid \}$ and $1 = \{0 \mid \}$ according to (1.30), and is therefore a number by by Definition 1.18. That number is the fraction $\frac{1}{2}$, that is, Left is *half a move* ahead in this game, because $\frac{1}{2} + \frac{1}{2} + (-1) \equiv 0$, that is,

$$\text{} \quad + \quad \text{} \quad + \quad \text{} \quad \equiv \quad 0\,. \tag{1.32}$$

This can be inferred from the (essentially unique) sequence of optimal moves

Left $\longrightarrow$ $+$ $+$ Right $\longrightarrow$ $+$ $+$

which shows that Left as starting player loses, and similarly that Right as starting player loses.

In terms of recursively defined games, starting from the definition of $0 = \{\ |\ \}$, (1.32) states

$$\{0 \mid 1\} + \{0 \mid 1\} + (-1) \equiv 0. \tag{1.33}$$

This holds for the same reason as for the Domineering moves in (1.32), just a bit more abstract. Recall that $-1 = \{\ | 0\}$ (which is one safe move for Right). In the game sum $\{0 \mid 1\} + \{0 \mid 1\} + \{\ | 0\}$ in (1.33), Left can only move in the component $\{0 \mid 1\}$, so that the next game is $0 + \{0 \mid 1\} + \{\ | 0\} = \{0 \mid 1\} + \{\ | 0\}$. Here Right has a choice, moving either in the first component to $1 + \{\ | 0\}$, which is $1 + (-1) = 0$ and a loss for Left who moves next, or in the second component (using his safe move) to $\{0 \mid 1\} + 0 = \{0 \mid 1\}$, which is a win for Left. Clearly, Right should choose the former. Hence, Left as starting player loses in $\{0 \mid 1\} + \{0 \mid 1\} + \{\ | 0\}$. By a very similar reasoning, Right as starting player loses, which proves (1.33).

As an extension of this argument, there is a game that can be identified with the fraction $\frac{1}{2^p}$ for every positive integer p, defined recursively as follows:

$$\frac{1}{2^p} = \left\{ 0 \,\middle|\, \frac{1}{2^{p-1}} \right\} \qquad \text{and hence} \qquad \frac{-1}{2^p} = \left\{ \frac{-1}{2^{p-1}} \,\middle|\, 0 \right\}. \tag{1.34}$$

In order to show

$$\frac{1}{2^p} + \frac{1}{2^p} + \frac{-1}{2^{p-1}} \equiv 0, \tag{1.35}$$

one can see that in this game sum it is better for both players to move in the component $\frac{1}{2^p}$ rather than in $\frac{-1}{2^{p-1}}$ where the player gives up more, similar to the reasoning for (1.33). Hence, no matter who starts, after two moves the position (1.35) becomes the losing position $\frac{1}{2^{p-1}} + \frac{-1}{2^{p-1}}$, which shows (1.35).

The following is the more general, recursive definition of a game that is a fraction with a power of two in its denominator.

Definition 1.19. Let p be a positive integer and m be an odd integer. Define the game $\frac{m}{2^p}$ by

$$\frac{m}{2^p} = \left\{ \frac{m-1}{2^p} \,\middle|\, \frac{m+1}{2^p} \right\}. \tag{1.36}$$

$\square$

On the right-hand side of (1.36), $m - 1$ and $m + 1$ are even integers, and we assume that the resulting fractions are maximally canceled as fractions normally are, so they have a smaller denominator, or are integers as in (1.30). Hence, (1.36) is a recursive definition. An example of (1.36) is $\frac{17}{8} = \{\frac{16}{8} \mid \frac{18}{8}\} = \{2 \mid \frac{9}{4}\}$.

Equation (1.33) justifies the definition $\frac{1}{2} = \{0 \mid 1\}$. In a similar way, one has to show that the game $\frac{m}{2^p}$ in (1.36) is a number and that

$$\frac{m}{2^p} + \frac{m}{2^p} + \frac{-m}{2^{p-1}} \equiv 0. \tag{1.37}$$

This is done by using the recursive definition of $G = \frac{m}{2^p}$ in (1.36) as $G = \{G^L \mid G^R\}$ with $G^L = \frac{m-1}{2^{p-1}}$ and $G^R = \frac{m+1}{2^{p-1}}$. According to Definition 1.18, we also have to show $G^L < G$ and $G < G^R$. For details we refer to chapters 4 and 5 of Albert, Nowakowski, and Wolfe (2007). Exercise 1.14 asks to compute some numbers that represent positions of the game Black-White Hackenbush, as special cases of (1.37).

Using Definition 1.19, every rational number with a denominator 2^p for a positive integer p can be represented as a game. In that game, a player who makes a move gives up a safe advantage of $\frac{1}{2^p}$ of a move. In a game sum of numbers, the player will move in the component with largest dominator where he gives up least. A game sum may have components that are not numbers, such as Nim heaps. The *number avoidance theorem* states that a player should move first in those components, and *last* in components that are numbers.

Not all partizan games are numbers. Indeed, there are games which are not numbers because moving there confers an extra advantage. For example, the 2×2 Domineering board in (1.29) is equal to $\{1 \mid -1\}$. This is called a *hot* game because moving there gives the player an extra advantage in the form of additional moves. The 3×2 Domineering board, which by (1.26) we found to be equal to $\{\frac{1}{2} \mid -2\}$, is also a hot game. A number, in contrast, is a *cold* game.

If there are several hot components in a game sum, analyzing their *temperature* is of interest to decide where to move first. This has been applied to some situations of the board game Go with disconnected components. In Chess, different moves are rarely about disconnected parts of the game. However, Chess has the related concept of *tempo* of development which can outweigh a material advantage.

This concludes our peek into the theory of partizan games. Further comments on this topic, with literature references, are given in the next section.

1.9   Further Reading

An undergraduate textbook on combinatorial games is Albert, Nowakowski, and Wolfe (2007), from which we have adopted the name "top-down induction". This book starts from many examples of games, considers in depth the game *Dots and Boxes*, and treats the theory of *short games* where (as we have assumed throughout) every position has only finitely many options and no game position is ever revisited (which in "loopy" games is allowed and could lead to a draw). An authoritative graduate textbook is Siegel (2013), which gives excellent overviews before treating each topic in depth, including newer research developments. It stresses the concept of *equivalence* of games that we have treated in Section 1.4.

The classic text on combinatorial game theory is *Winning Ways* by Berlekamp, Conway, and Guy (2001–2004). Its wit, vivid colour drawings, and the wealth of games it considers make this a very attractive original source. However, the mathematics is hard, and it helps to know first what the topic is about; for example, equivalence of games is rather implicit and written as equality.

All these books consider directly partizan games. We have chosen to focus on the simpler impartial games, apart from our short introduction to partizan games in Section 1.8. For a more detailed study of games as numbers see Albert, Nowakowski, and Wolfe (2007, Section 5.1), where Definition 5.12 is our Definition 1.18.

The winning strategy for the game Nim based on the binary system was first described by Bouton (1901). The Queen move game is due to Wythoff (1907), described as an extension of Nim. The observation that every impartial game is equivalent to a Nim heap is independently due to Sprague (1935) and Grundy (1939). This is therefore also called the *Sprague-Grundy* theory of impartial games, and Nim values are also called "Sprague-Grundy values" or just "Grundy values".

The approach in Section 1.5 to the mex rule is inspired by Berlekamp, Conway, and Guy (2001–2004), chapter 4, which describes Poker Nim, the Rook and Queen move games, Northcott's game (Exercise 1.8), and chapter 5 with Kayles and Lasker's Nim (called Split-Nim in Exercise 1.12). Chomp (Exercises 1.5 and 1.10) was described by Gale (1974); for a generalization see Gale and Neyman (1982). The digraph game in Exercise 1.9 is another great example for understanding the mex rule, due to Fraenkel (1996).

The systematic study of sums of partizan games led John Conway, one of the authors of *Winning Ways*, to the insight that it is possible to define in an elegant way real numbers (in fact, a generalization called *surreal numbers*) as the strengths of positions in certain games. A number is a game G as in Definition 1.18, but where the sets G^L and G^R of left options G^L and right options G^R are allowed to be infinite. The condition $G^L < G < G^R$ generalizes "Dedekind cuts", a standard construction of the real numbers. The original ideas are described in Conway (2001), known as *ONAG* ("On Numbers and Games"), which is another classic. For an introduction see Conway (1977).

An informative and entertaining mathematical study of parlour games, with detailed historical notes, is Bewersdorff (2005). Combinatorial games are treated in part II of that book.

1.10   Exercises for Chapter 1

Exercise 1.1. For this exercise, be careful to use only the definitions and not your intuitions that are familiar to you about the symbols $\leq$ and $<$. Consider a set S with a binary relation $<$ that is transitive and fulfills for all x in S

$$\text{not}\quad x < x \qquad (< \text{ is } \textit{irreflexive}). \qquad (1.38)$$

Define the relation $\leq$ on S by (1.7), that is, $x \leq y \Leftrightarrow x < y$ or $x = y$.

(a) Show that $\leq$ is a partial order, and that $<$ is obtained from $\leq$ via (1.6), that is, $x < y \Leftrightarrow x \leq y$ and $x \neq y$.

(b) Show that if $<$ is obtained from a partial order $\leq$ via (1.6), then $<$ is transitive and irreflexive, and (1.7) holds, that is, $x \leq y \Leftrightarrow x < y$ or $x = y$.

Exercise 1.2. Consider the game Nim with heaps of tokens. The players alternately remove some tokens from one of the heaps. The player to remove the last token wins. Try to prove part (a) without reference to the binary representation for playing Nim optimally.

(a) For all positions with three heaps, where one of the heaps has only *one* token, describe exactly the *losing* positions. Justify your answer, for example with a proof by induction, or by theorems on Nim. [*Hint*: *Start with the easy cases to find the pattern.*]

(b) Consider a Nim game with three heaps whose sizes are consecutive integers, for example $2, 3, 4$. Show that the *only* case where this is a losing position is $1, 2, 3$ and that otherwise this is always a winning position. [*Hint*: *Use* (a).]

(c) Determine all initial winning moves for Nim with three heaps of sizes 8, 11, and 13, using the binary representation of the heap sizes.

Exercise 1.3. Misère Nim is played just like Nim but where the last player to move loses. A losing position is therefore a single Nim heap with a single token in it (which the player then has to take). Another losing position is given by three heaps with a single token each.

(a) Determine the winning and losing positions of misère Nim with one or two heaps, and the winning moves from a winning position.

(b) Are three heaps with sizes $1, 2, 3$ winning or losing in misère Nim?

(c) Try to describe the losing positions in misère Nim for any number of heaps. [*Hint*: *They are closely related to the losing positions in Nim with normal play.*]

Exercise 1.4. The impartial game *Cram* is played on a board of $m \times n$ squares, where players alternately place a domino on the board which covers two adjacent squares that are free (not yet occupied by a domino), vertically or horizontally. The first player who cannot place a domino any more loses. Example play for a 2×3 board:

(a) Who will win in 3×3 Cram? [*Hint: Use the symmetry of the game to investigate possible moves, and remember that it suffices to find one winning strategy.*]

(b) Who will win in $m \times n$ Cram when both m and n are even? (That player has a simple and precisely definable winning strategy which you have to find.)

(c) Who will win in $m \times n$ Cram when m is odd and n is even? (Easy with (b).)

Note (not a question): Because of the known answers from (b) and (c), this game is more interesting for "real play" on an $m \times n$ board where both m and n are odd. Play it with your friends on a 5×5 board, for example. The situation often decomposes into independent parts, like contiguous fields of 2, 3, 4, 5, 6 squares, that have a known winner, which may help you analyse the situation.

Exercise 1.5. Consider the following game *Chomp*: A rectangular array of $m \times n$ dots is given, in m rows and n columns, like 3×4 in the next picture on the left. A dot in row i (from the top) and column j (from left) is named (i, j). A move consists in picking a dot (i, j) and removing it and *all other dots to the right and below it*, which means removing all dots (i', j') with $i' \geq i$ and $j' \geq j$, as shown for $(i, j) = (2, 3)$ in the middle picture, resulting in the picture on the right:

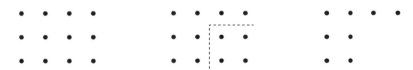

Player I is the first player to move, players alternate, and the last player who removes a dot *loses*.

One can think of these dots as (real) cookies: a move is to eat a cookie and all those to the right and below it, but the top left cookie is poisoned.

In the following questions, do not rush into any quick and false application of Nim values. Part (c) has an elegant but tricky answer.

(a) Assuming optimal play, determine the winning player and a winning move for Chomp of size 2×2, size 2×3, size $2 \times n$, and size $m \times m$, where $m \geq 3$. Justify your answers.

(b) As it is described here, Chomp is a misère game where the last player to make a move loses. Suppose we want to play the *same* game but so that the normal play convention applies, where the last player to move wins. How should the array of dots be changed to achieve this? (Normal play with the array as given would be a boring game, with the obvious winning move of taking the top left dot $(1,1)$.)

(c) Show that when Chomp is played for a game of any size $m \times n$, player I can always win (unless $m = n = 1$). [*Hint: You only have to show that a winning move exists, but you do not have to describe that winning move.*]

Exercise 1.6.

(a) Complete the entries of equivalent Nim heaps for the Queen-move game in columns 5 and 6, rows 0 to 3, in the table in Figure 1.4 (assuming the queen can be anywhere on the board).

(b) Describe *all* winning moves in the game sum of the Queen-move game and the Nim heap in Figure 1.3.

Exercise 1.7. Consider the game Cram from Exercise 1.4, played on a $1 \times n$ board for $n \geq 2$. Let D_n be the Nim value of that game, so that the starting position of the $1 \times n$ board is equivalent to a Nim heap of size D_n. For example, $D_2 = 1$ because the 1×2 board is equivalent to $*1$.

(a) How is D_n computed from smaller values D_k for $k < n$? [*Hint: Use sums of games and the mex rule. The notation for Nim sums is $\oplus$, where $a \oplus b = c$ if and only if $*a + *b \equiv *c$.*]

(b) Give the values of D_n up to $n = 10$ (or more, if you are ambitious – higher values come at $n = 16$, and at some point they even repeat, but before you detect that you will probably have run out of patience). For which values of n in $\{1, \ldots, 10\}$ is Cram on a $1 \times n$ board a losing game?

Exercise 1.8. Consider the following game on a rectangular board where a white and a black counter are placed in each row, like in this example:

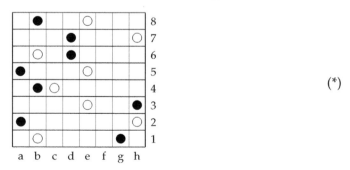

(*)

Player I is White and starts, and player II is Black. Players take turns. In a move, a player moves a counter of his color to any other square within its row, but may not jump over the other counter. For example, in (*) above, in row 8 White may move from e8 to any of the squares c8, d8, f8, g8, or h8. The player who can no longer move loses.

(a) Who will win in the following position?

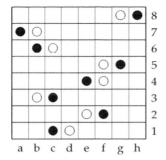

(b) Show that White can win in position (*) above. Give at least two winning moves from that position. Justify your answers. [*Hint: Compare this with another game that is not impartial and that also violates the ending condition, but that nevertheless is close to Nim and ends in finite time when played well.*]

Exercise 1.9. Consider the following network (in technical terms, a directed graph or "digraph"). Each circle, here marked with one of the letters A to P, represents a *node* of the network. Some of these nodes (here A, F, G, H, and K) have counters on them. There can be any number of counters on a node, like currently the two counters on H. In a move, one of the counters is moved to a neighbouring node in the direction of the arrow as indicated, for example from F to I (but not from F to C, nor directly from F to D, say). Players alternate, and the first player no longer able to move loses.

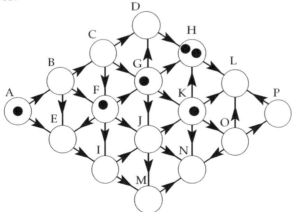

(a) Explain why this game fulfills the ending condition.

(b) Who is winning in the above position? If it is the first player to move, describe *all* possible winning moves. Justify your answer.

(c) How does the answer to (b) change when the arrow from J to K is reversed so that it points from K to J instead?

Exercise 1.10. Consider the game Chomp from Exercise 1.5 of size 2×4, in a game sum with a Nim heap of size 4.

What are the winning moves of the starting player I, if any?   [*Hint*: *Represent Chomp as a game in the normal play convention (see Exercise 1.5(b), by changing the dot pattern), so that the losing player is not the player who takes the "poisoned cookie", but the player who can no longer move. This will simplify finding the various Nim values.*]

Exercise 1.11. In $m \times n$ *Cram* (see Exercise 1.4), a rectangular board of $m \times n$ squares is given. The two players alternately place a domino either horizontally or vertically on two unoccupied adjacent squares, which then become occupied. The last player to be able to place a domino wins.

(a) Find the Nim value (size of the equivalent Nim heap) of 2×3 Cram.

(b) Find all winning moves, if any, for the *game sum* of a 2×3 Cram game and a 1×4 Cram game.

Exercise 1.12. Consider the following variant of Nim called *Split-Nim*, which is played with heaps of tokens as in Nim. Like in Nim, a player can remove some tokens from one of the heaps, or else *split* a heap into two (not necessarily equal) new heaps. For example, a heap of size 4 can be reduced to size 3, 2, 1, or 0 as in ordinary Nim, or be split into two heaps of sizes 1 and 3, respectively, or into two heaps of sizes 2 and 2. As usual, the last player to be able to move wins.

(a) Find the Nim values (size of equivalent Nim heaps) of the single Split-Nim heaps of size 1, 2, 3, and 4, respectively.

(b) Find all winning moves, if any, when Split-Nim is played with three heaps of sizes 1, 2, 3.

(c) Find all winning moves, if any, when Split-Nim is played with three heaps of sizes 1, 2, 4.

(d) Determine if the following statement is true or false: "Two heaps in Split-Nim are equivalent if and only if they have equal size." Justify your answer.

Exercise 1.13. The impartial game *Gray Hackenbush* is played on a figure consisting of nodes and edges that are connected to these nodes or to the ground (the ground is the dashed line in the pictures below). A move is to remove an edge, and with it all the edges that are then no longer connected to the ground. For example, in the leftmost figure below, one can remove any edge in one of the three stalks. Removing the second edge from the stalk that consists of three edges takes the topmost edge with it, leaving only a single edge. All edges are colored gray, which means they can be removed by either player, so this is an impartial game. As usual, players alternate and the last player able to move wins.

(a) Compute the Nim values for the following three Gray Hackenbush figures (using what you know about Nim, and the mex rule):

(b) Compute the Nim values for the following four Gray Hackenbush figures:

(c) Based on the answer in (b), give a rule to compute the Nim value of a "tree" that is constructed by putting several "branches" with known Nim values on top of a "stem" of size n, for example $n = 2$ in the picture below, where three branches of heights 3, 2, and 4 are put on top of the "stem". Prove the correctness of that rule (you may find it advantageous to glue the branches together at the bottom, as in (b)). Use this rule to find the Nim value of the rightmost picture.

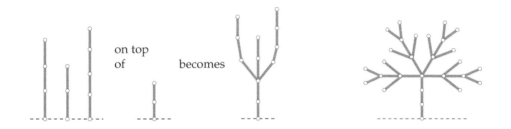

Exercise 1.14. This exercise is about *partizan games* as treated in Section 1.8.

(a) Complete the proof of (1.32) by showing that Right as starting player loses.

The partizan game *(Black-White) Hackenbush* is played on a figure consisting of nodes and black or white edges that are connected to these nodes or to the ground, as in the pictures below. In a move the Left player removes a black edge and the Right player a white edge; if no such edge exists then the player loses. After such a move, all nodes and edges no longer connected to the ground disappear as well. For example, in a Hackenbush figure with a white edge on top of a black edge, Left can remove the bottom (black) edge and creates the zero game, whereas Right can remove the top (white) edge and leaves a single black edge.

(b) Recall that for a combinatorial game G its *negative* is a game $-G$ so that $G + (-G) \equiv 0$. Describe the negative of a Hackenbush figure, and give an example. [*Hint: Compare* (1.24).]

We consider Hackenbush *stalks*, which are just paths of edges of either color. It can be shown that any such stalk is a *number*, with some examples considered here.

(c) Draw the game that represents the number 3 (where Left is 3 moves ahead).

(d) Find the numbers represented by the following Hackenbush stalks. Justify your answers. It helps to study and understand (1.36), and to consider only undominated moves of the players; also, compare the strength of the Left (black) player for these games to get an idea about the answer:

where you can use as shorthand the following notation for these games:

	W	W
	W	W
W	B	W
B	B	B

(e) Construct a Hackenbush game (preferably a single stalk) that represents the number $\frac{7}{4}$.

Exercise 1.15. (Albert, Nowakowski, and Wolfe, 2007, Prep Problems 4.1 and 4.2.)

(a) Show that if G and H are partizan games and Left wins when moving second in G and in H, then Left wins when moving second in $G + H$. [*Hint: Express what it means for Left to win when moving second in terms of outcome classes, as in* (1.31).]

(b) Give an example of games G and H so that Left wins when moving first in G and in H, but cannot win in $G + H$, even if Left can choose to move first or second in the game sum $G + H$. [*Hint: Can both games be numbers? Or impartial games?*]

2

Congestion Games

Congestion means that a shared resource, such as a road, becomes more costly when more people use it. In a congestion game, multiple players decide on which resource to use, with the aim to minimize their cost. This interaction defines a game because the cost depends on what the other players do.

We present congestion networks as a model of traffic where individual users choose routes from an origin to a destination. Each edge of the network is congested by creating a cost to each user that weakly increases with the number of users of that edge. The central result (Theorem 2.2) states that every congestion game has an *equilibrium* where no user can individually reduce her cost by changing her chosen route.

Like Chapter 1, this is another introductory chapter to an active and diverse area of game theory, where we can quickly show some important and nontrivial results. It also gives an introduction to the game-theoretic concepts of strategies and equilibrium before we develop the general theory.

Equilibrium is the result of *selfish routing* by the users who individually minimize their costs. This is typically not the socially optimal outcome, which could have a lower average cost. A simple example is the "Pigou network" (see Section 2.2), named after the English economist Arthur Pigou, who introduced the concept of an *externality* (such as congestion) to economics. Even more surprisingly, the famous *Braess paradox*, threated in Section 2.3, shows that adding capacity to a road network can *worsen* equilibrium congestion.

In Section 2.4 we give the general definition of congestion networks. In Section 2.5 we prove that every congestion game has an equilibrium. This is proved with a cleverly chosen *potential function*, which is a single function that simultaneously reflects the change in cost of every user when she changes her strategy. Its minimum over all strategy choices therefore defines an equilibrium.

In Section 2.6, we explain the wider context of the presented model. It is the *discrete* model of *atomic* (non-splittable) flow with finitely many individual users who choose their traffic routes. As illustrated by the considered examples, the resulting equilibria are often not unique when the "last" user can optimally choose between more than one edge. The limit of many users is the continuous model of *splittable* flow where users can be infinitesmally small fractions of a "mass" of users who want to travel from an origin to a destination. In this model the equilibrium is often unique, but its existence proof is more technical and not given here.

We also mention the *Price of Anarchy* that compares the worst equilibrium cost with the socially optimal cost. For further details we give references in Section 2.7.

2.1   Prerequisites and Learning Outcomes

There are no particular prequisites for this chapter. We use the terminology of directed graphs with nodes and their connecting edges (with possibly several "parallel" edges that connect the same two nodes); they are formally defined in Section 2.4. Standard calculus is used to find the minimum of a function of one variable. In Section 2.5 the flow through a network, given by the collection of flows f_e through each edge e, is denoted by the letter f, which is here a variable.

After studying this chapter, you should be able to:

- understand the concept of *congestion networks* with their cost functions, and how edges in the network have a flow that depends on users' chosen paths;

- find an equilibrium flow in a network, and the socially optimal flow, and compare them;

- understand how a *potential function* helps finding an equilibrium.

2.2   Introduction: The Pigou Network

Getting around at rush hour takes much longer than at less busy times. Commuters take buses where people have to queue to board. There are more buses that fill the bus lanes, and cars are in the way and clog the streets because only a limited number can pass each green light. Commuters can take the bus, walk, or ride a bike, and cars can choose different routes. *Congestion*, which slows everything down, depends on how many people use certain forms of transport, and which route they choose. We assume that people try to choose an optimal route, but what is optimal depends on what others do. This is an interactive situation, which we call a *game*. We describe a mathematical model that makes the rules of this game precise.

Figure 2.1 shows an example of a *congestion network*. This particular network has two nodes, an origin o and a destination d, and two edges that both connect o to d. Suppose there are several users of the network who all want to travel from o to d and can choose either edge. Each edge has a *cost* $c(x)$ associated with it,

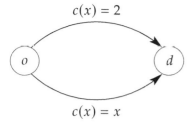

$c(x) = 2$

$c(x) = x$

Figure 2.1 A congestion network with two nodes o and d that define the origin and destination of all users, with two parallel edges between them which have different costs $c(x)$ that depend on the usage x of the respective edge.

which is a function that describes how costly it is for each user to use that edge when it has a *flow* or *load* of x users. The top edge has constant cost $c(x) = 2$, and the bottom edge has cost $c(x) = x$. The cost could for example be travel time, or incorporate additional monetary costs. The cost is the same for all users of the edge. The cost for each user of the top edge, no matter how many people use it, is always 2, whereas the bottom edge has cost 1 for one user, 2 for two users, and so on (and zero cost for no users, but then there is no one to profit from taking that zero-cost route).

Suppose the network in Figure 2.1 is used by two users. They can either both use the top edge, or choose a different edge each, or both use the bottom edge. If both use the top edge, both pay cost 2, but if one of them switches to the bottom edge, that user will travel more cheaply and only pay cost 1 on that edge. We say that this situation is *not in equilibrium* because *at least one user* can improve her cost by changing her action. Note that we only consider the *unilateral deviation* of a *single* user in this scenario. If both would simultaneously switch to the bottom edge, then this edge would be congested with $x = 2$ and again incur cost 2, which is no improvement to the situation that both use the top edge.

If the two users use different edges, this is an equilibrium. Namely, the user of the bottom edge currently has cost 1 but would pay cost 2 by switching to the top edge, and clearly she has no incentive to do so. And the user of the top edge, when switching from the top edge, would increase the load on the bottom edge from 1 to 2 with the resulting new cost 2 which is also no improvement for her. This proves the equilibrium property.

Finally, the situation where both users use the bottom edge is also an equilibrium. Both have cost 2, but that is not worse than the cost on the top edge. So this network, with two users, has two equilibria.

The underlying behavior of the users in this game is called *selfish routing*. That is, every user tries to find the route that is least costly for her. Equilibrium is obtained when every user is in a situation where she cannot find a better route on her own.

Is selfish routing always *socially optimal*, as measured by the average cost per user? In this example, it is better if one user chooses the top edge and the other the bottom edge, with costs 2 and 1, respectively (or 1.5 on average), which is clearly socially optimal. The other equilibrium where both use the bottom edge has the higher cost 2 (paid by both users).

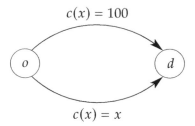

Figure 2.2 The same *Pigou* network as in Figure 2.1, but with a constant cost $c(x) = 100$ on the top edge. We assume 100 users want to travel from o to d.

However, selfish routing may lead to a situation where the average user is worse off than in the socially optimal flow through the network. Figure 2.2 is identical to Figure 2.1 except that the top edge has a constant cost of 100. It is called the *Pigou* network. Assume that there are 100 users who want to travel from o to d. It is easy to see that there are only two equilibria in this congestion game, namely one where 99 users use the bottom edge and only 1 the top edge, or another where all 100 users use the bottom edge. In all other cases, at most 98 users use the bottom edge, so that any one of the remaining users on the top edge would pay at most 99 by switching to the bottom edge, which improves her cost. In other words, the bottom edge will be "saturated" with users until its usage costs reaches or nearly reaches the cost of the top edge.

The average cost for each user in these equilibria is found as follows. In the equilibrium where all users use the bottom edge the average cost is 100. In the equilibrium where one user uses the top edge and 99 the bottom edge the average cost is 99.01, computed as $\frac{99}{100}99 + \frac{1}{100}100 = \frac{99}{100}99 + \frac{1}{100}(99+1) = \frac{99+1}{100}99 + \frac{1}{100}$.

In contrast, if the flow from o to d is split equally between the two edges, then 50 users will use the top edge and each pay cost 100, and 50 users will use the bottom edge and each pay cost 50. So the average cost is $\frac{50}{100}100 + \frac{50}{100}50 = \frac{1}{2}(100+50) = 75$. We show that this 50–50 split has the optimal average cost. Namely, suppose that y users use the bottom edge and correspondingly $100 - y$ use the top edge. Then their average cost is $\frac{100-y}{100}100 + \frac{y}{100}y$, which equals $100 - y + \frac{1}{100}y^2$. Assume first that y can take any real value. Then the derivative of this function of y is $-1 + \frac{2}{100}y$, which is zero if and only if $y = 50$, which is an integer and gives the unique minimum of the function. Equivalently, $100 - y + \frac{1}{100}y^2 = 100 + \frac{1}{100}(y-50)^2 - \frac{2500}{100}$, which has its smallest value 75 for $y = 50$.

However, note that the 50–50 split in this example is definitely not an equilibrium, because any user of the top edge would switch to the bottom edge. This would continue until that edge no longer has a lower cost. The socially optimal 50–50 split would have to be enforced in some way (for example by limiting the usage of the bottom edge), or would require new incentives that "change the game", for example a congestion charge.

2.3   The Braess Paradox

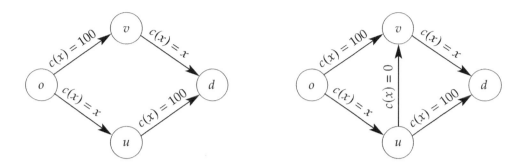

Figure 2.3   The Braess paradox. Each edge has a cost function $c(x)$ that depends on the flow x on that edge (the flow may be different for different edges). The right congestion network has one extra edge uv with zero cost compared to the left congestion network. For 100 users, the equilibrium flow in the left network is optimal, whereas in the right network it is worse, *due to* the extra capacity.

We describe a famous congestion game that shows an unexpected effect of selfish routing. Figure 2.3 shows two congestion networks. In both cases, we assume that there are 100 users who all want to travel from the origin o to the destination d. In the left network in Figure 2.3, there are only two routes, across the top and across the bottom. Suppose that y users take the bottom route via the intermediate node u, and $100 - y$ users take the top route via node v. Both routes use two edges, whose costs are added. With this usage distribution on the edges, the bottom route incurs cost $y + 100$ and the top route incurs cost $100 + (100 - y)$. In equilibrium, both costs should be equal or nearly equal in order to not incentivize any user to switch routes. That is, if $y = 50$, then both routes cost 150, and any switch would only lead to a higher cost on the other route. If, for example, $y = 51$, then the bottom route would cost 151 and the top route would cost 149. Then a user on the bottom route could switch to the top route which has a new, lower cost of 150. So the only equilibrium is an equal split between the two routes. It is easy to see that this is also the *optimal* flow through the network.

The right picture in Figure 2.3 has an extra edge uv (from u to v). This edge provides a "shortcut" with a new route from o via u and v to d along the edges ou, uv, vd. Suppose that z users take this new route, that $y - z$ users for $y \geq z$ take the bottom route from o via u to d (so that y users use the edge ou, and $y - z$ users use the edge ud), and that $100 - y$ users take the top route from o to v to d. The resulting usage of the edge ov is then $100 - y$ and of the edge vd is $100 - y + z$. The left diagram in Figure 2.3 corresponds to the case where the edge uv is blocked and therefore $z = 0$. Now consider the previous equilibrium flow where $y = 50$ and $z = 0$ and suppose that the shortcut uv is opened. In this situation, the cost of getting from o to v via the shortcut is currently 50, much lower than 100 along the edge ov, and similarly getting from u to d via v is also cheaper than via the edge ud with cost 100. This is not in equilibrium. In fact, the only equilibrium flow is given with at most one user on edge ov and at most one user on the edge ud, and everybody else using the shortcut route, in the worst case with $y = z = 100$ where everyone takes the shortcut and nobody uses the edges ov and ud with constant costs 100. Then the overall cost of traveling from o to d is now $100 + 0 + 100 = 200$, whereas it was 150 beforehand.

On closer inspection, the new edge uv with zero cost creates two Pigou networks as shown on the right, if one identifies the two nodes u and v. The first Pigou network goes from o to destination v via two routes, namely along edge ov with constant cost 100, and along the edges ou and uv with cost $c(x) = x$, all of which is incurred

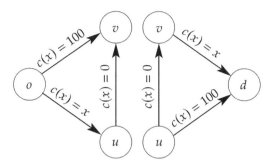

on the first edge ou. The second Pigou network starts at node u with destination d, also via two routes, one along the edge ud with cost 100, and the other along edges uv and vd, of which only the second edge is costly. Like in the Pigou network, neither of the edges ov and ud with constant cost is used in selfish routing (except possibly by one user).

The Pigou network shows that a selfish equilibrium flow may be more costly than a centrally planned socially optimal flow. The added zero-cost edge uv in Figure 2.3 goes beyond this observation, because it shows that *adding extra capacity to the network can make congestion worse* in equilibrium because the new shortcut destroys the balance between the previous two routes. This is called *Braess's paradox*, after Braess (1968).

Similarly, closing a road may *reduce* traffic congestion. This has also been observed in practice. Most evidence on this is anecdotal, but at least some urban planners seem to be aware of this possible effect. In Seoul (Korea), a large expressway over the central Cheonggyecheon riverbed was dismantled in 2002 and replaced by a long recreational park (Hong, Song, and Wu, 2007, p. 235). Traffic improved, although certainly due to many factors in this complex project.

2.4   Definition of Congestion Games

In this section, we give a general definition of congestion games and of the concept of an equilibrium. A *congestion network* has the following components:

- A finite set of *nodes*.

- A finite collection E of *edges*. Each edge e is an ordered pair, written as uv, *from* some node u *to* some node v, which is graphically drawn as an arrow from u to v. *Parallel* edges (that is, with the same pair uv) are allowed (hence the edges form a "collection" E rather than a set, which would not allow for such repetitions), as in Figure 2.1.

- Each edge e in E has a *cost function* c_e that gives a value $c_e(x)$ when there are x users on edge e, which describes the same cost to each user for using e. Each cost function is *weakly increasing*, that is, $x \leq y$ implies $c_e(x) \leq c_e(y)$.

- A number N of *users* of the network. Each user $i = 1, \ldots, N$ has an origin o_i and destination d_i, which are two nodes in the network, which may or may not be the same for all users (if they are the same, they are usually called o and d as in the above examples).

The underlying structure of nodes and edges is called a directed graph or *digraph* (where edges are sometimes called "arcs"). In such a digraph, a *path P from u to v* is a sequence of distinct nodes $u_0, u_1, \ldots, u_m$ for $m \geq 0$ where $u_k u_{k+1}$ is an edge for $0 \leq k < m$, and $u = u_0$ and $v = u_m$. For any such edge $e = u_k u_{k+1}$ for $0 \leq k < m$ we write $e \in P$. Note that a node may appear at most once in a path. Every user i chooses a path (which we have earlier also called a "route") from her origin o_i to her destination d_i.

- A *strategy* of user i is a path P_i from o_i to d_i.

- Given a strategy P_i for each user i, the *load* on or *flow* through an edge e is defined as $f_e = |\{i \mid e \in P_i\}|$, which is the number of chosen paths that contain e, that is, the number of users on e. The *cost* to user i for her strategy P_i, given that the other users have chosen their strategies, is then

$$\sum_{e \in P_i} c_e(f_e). \tag{2.1}$$

The *congestion game* obtained from this network is defined by the strategies of the users (or "players"), and by the resulting cost to each user when all strategies are known. Each user tries, on her own, to minimize her cost. Every edge e has a certain flow f_e, which is number of users on e, and creates the cost $c_e(f_e)$ which is the same for every user of that edge. Users may have different total costs when they use different strategies (that is, different paths; these paths necessarily differ if they do not have the same origin and destination).

Now consider what happens when user i changes her strategy P_i to a different strategy Q_i. Some edges of Q_i are the same as before, which define the set $Q_i \cap P_i$. Other edges in Q_i, those not in P_i, are new, which define the set $Q_i \setminus P_i$. Clearly, every edge in Q_i belongs to exactly one of these two sets. On any edge in $Q_i \cap P_i$, the flow and hence the cost remains the same as before, but every edge e in $Q_i \setminus P_i$ has user i as one additional user and therefore an increased flow of $f_e + 1$. We say P_i is a *best response* of user i to the remaining strategies P_j for $j \neq i$ if for every alternative strategy Q_i of user i

$$\sum_{e \in P_i} c_e(f_e) \;\leq\; \sum_{e \in Q_i \cap P_i} c_e(f_e) + \sum_{e \in Q_i \setminus P_i} c_e(f_e + 1). \tag{2.2}$$

The left-hand side of this inequality is the cost to user i when she uses strategy P_i, and the right-hand side is the possibly changed cost when she *deviates* to strategy Q_i. Then P_i is a best response to the other strategies if no deviation gives rise to a lower cost. (When user i changes from P_i to Q_i, some edges of P_i may no longer be used in Q_i and therefore have a reduced flow; this may help other users of those edges, but is irrelevant for user i.) Note that the concept of best response concerns only deviations of a *single* user.

Definition 2.1. The strategies $P_1, \ldots, P_N$ of all N users define an *equilibrium* if each strategy is a best response to the other strategies, that is, if (2.2) holds for all $i = 1, \ldots, N$. □

We have seen examples of equilibria in congestion games in the previous sections. Next, we prove that such an equilibrium always exists.

2.5   Existence of Equilibrium in a Congestion Game

The following is the central theorem of this chapter. It is proved with the help of a *potential function* Φ. The potential function is constructed in such a way that it defines for each edge the *increase in cost* created by each additional user on the edge, as explained further after the proof.

Theorem 2.2. *Every congestion game (as obtained from a congestion network) has at least one equilibrium.*

Proof. Suppose the N strategies of the users are $P_1, \ldots, P_N$, which defines a flow f_e on each edge $e \in E$, namely the number of users i with $e \in P_i$. We call this the flow f induced by these strategies. We now define the following function $\Phi(f)$ of this flow by

$$\Phi(f) = \sum_{e \in E} \big(c_e(1) + c_e(2) + \cdots + c_e(f_e) \big). \tag{2.3}$$

Suppose that user i changes her path P_i to Q_i. We call the resulting new flow f^{Q_i}. We will prove that

$$\Phi(f^{Q_i}) - \Phi(f) \;=\; \sum_{e \in Q_i} c_e(f_e^{Q_i}) - \sum_{e \in P_i} c_e(f_e). \tag{2.4}$$

The right-hand side is the difference in costs to user i between her strategy Q_i and her strategy P_i, according to (2.1). Her cost for the flow f^{Q_i} has been written on the right-hand side of (2.2) in terms of the original flow f (when she uses P_i) as

$$\sum_{e \in Q_i} c_e(f_e^{Q_i}) = \sum_{e \in Q_i \cap P_i} c_e(f_e) + \sum_{e \in Q_i \setminus P_i} c_e(f_e + 1), \qquad (2.5)$$

and her cost for the original flow f can be expressed similarly, namely as

$$\sum_{e \in P_i} c_e(f_e) = \sum_{e \in P_i \cap Q_i} c_e(f_e) + \sum_{e \in P_i \setminus Q_i} c_e(f_e). \qquad (2.6)$$

Hence, the right-hand side of (2.4) is

$$\sum_{e \in Q_i} c_e(f_e^{Q_i}) - \sum_{e \in P_i} c_e(f_e) = \sum_{e \in Q_i \setminus P_i} c_e(f_e + 1) - \sum_{e \in P_i \setminus Q_i} c_e(f_e). \qquad (2.7)$$

We claim that because of (2.3) (which is why Φ is defined that way), the right-hand side of (2.7) is equal to the left-hand side $\Phi(f^{Q_i}) - \Phi(f)$ of (2.4). Namely, by changing her path from P_i to Q_i, user i increases the flow on any new edge e in $Q_i \setminus P_i$ from f_e to $f_e + 1$, and thus adds the term $c_e(f_e + 1)$ to the sum in (2.3). Similarly, for any edge e in $P_i \setminus Q_i$ which is in P_i but no longer in Q_i, the flow $f^{Q_i}(e)$ is reduced from f_e to $f_e - 1$, so that the term $c_e(f_e)$ has to be subtracted from the sum in (2.3). This shows (2.4).

Equation (2.4) states that any *change in cost to any user i* when she changes her strategy, and thus the flow, is the same as the *change in Φ*. Consider now a flow that achieves the global minimum of $\Phi(f)$ for all possible flows f that result from users' strategies $P_1, \ldots, P_N$. There are only finitely many such strategy combinations, so this minimum exists. For such a minimum, no change in flow will reduce $\Phi(f)$ any further, and hence no individual player can reduce her cost. Therefore, the players' strategies for such a minimum of Φ define an equilibrium. $\qquad \square$

The function Φ introduced in the proof of Theorem 2.2 is of special interest. Its definition (2.3) can be thought of being obtained in the following way. Consider an edge e, and assume that there are no users of the network yet. Then $c(1)$ is the cost when one user travels on edge e. Next, $c(2)$ is the cost when a second user travels on edge e, and so on until $c(f_e)$ for the full flow f_e on edge e when all users are present (recall that f_e is the number of users of the edge). However, the sum $c_e(1) + c_e(2) + \cdots + c_e(f_e)$ in (2.3) is *not* the combined cost to all users of edge e, because the load f_e creates the cost $c(f_e)$ for *every* user of the edge, so their combined cost is $f_e \cdot c(f_e)$, which is larger than the sum $c_e(1) + c_e(2) + \cdots + c_e(f_e)$. Rather, this sum accumulates the additional cost created by each new user of the edge.

Because of this definition, the function Φ has the crucial property (2.4) that any change of strategy by a single user incurs a cost change for that user that equals the change of Φ. A single function Φ achieves this for all users simultaneously. A function Φ with this property (2.4) is called a *potential function*, and a game that has such a potential function is called a *potential game*. By the preceding proof, any potential game has an equilibrium, which can be found via the minimum of the potential function.

The property (2.4) can also be used in the reverse direction to *find* an equilibrium. Namely, a smaller value of the potential function Φ can be found by identifying one user who can reduce her cost by changing her strategy. So we change her strategy accordingly and reduce the value of the potential function. This can be seen as an iterative process where every user (one at a time) adapts her strategy until no user can find further improvements, which then defines an equilibrium. This clearly occurs for *any* equilibrium. Such an equilibrium may only represent a "local" minimum of the potential function in the sense that it does not allow further reduction by changing a single user's strategy at a time. In the example in Figure 2.1, there are two equilibria that have different values of the potential function.

2.6   Atomic and Splittable Flow, Price of Anarchy

The presented model of congestion games with finitely many users on a network is only one of many ways of modeling congestion by selfish routing.

First, the representation of costs in (2.1) is a simplification in the sense that it assumes that user i creates a "flow" of 1 along its chosen path, and thus contributes to the congestion on every edge on that path. This model does not take any temporal aspects into account, like the fact that the user can be only on one edge at a time but not everywhere on the path at once. Nevertheless, this is not completely unrealistic because everywhere along her chosen route, the user will create congestion at some point in time together with the other users. The flow model is very appropriate for internet routing, for example, with a continuous flow of data packets that consume bandwidth along the taken route.

The specific model that we consider here is called *atomic* (or non-splittable) flow where single users decide on their paths from their origin to their destination.

In contrast, *splittable flow* means that a flow unit is just some "mass" of users of which any parts can be routed along different paths. For example, in Figure 2.1 the flow of 2 from o to d could be split into a flow of 0.01 on the top edge and a flow of 1.99 on the bottom edge. This would still not be an equilibrium because any one user from the 0.01 fraction of users on the top edge could improve her cost by moving to the bottom edge (each user is negligibly small). Splittable flows are in some sense simpler because there will be no minor variation of equilibria, such as the distinct equilibria with a flow of either $y = 100$ or $y = 99$ on the bottom edge in Figure 2.2; the only equilibrium flow would be $y = 100$. The equilibrium condition is then simply that there is no alternative path with *currently* smaller cost, without taking into account that there is an increased cost by switching to another path, as in the term $c_e(f_e + 1)$ in (2.2) where the increase of f_e by 1 results from that switch. In the splittable flow model, users have negligible size, so there is no such increase. It can be shown that the *cost* of an equilibrium flow in a splittable flow network is *unique* (however, there may be several equilibria with that cost, for example with an undetermined flow across two parallel edges with the same constant cost). For splittable flow, the proof of Theorem 2.2 can be amended with a potential function Φ defined with an integral $\int_0^{f_e} c_e(x)dx$ that replaces the sum $c_e(1) + c_e(2) + \cdots + c_e(f_e)$ in (2.3); see Roughgarden (2016, section 13.2.2).

We also mention the *Price of Anarchy* in this context. This is defined, for a specific congestion network, as the ratio between the worst average cost to a user in an equilibrium and the optimal average cost in the social optimum. The "anarchy" refers to the selfishness in the routing game where each user chooses her optimal route on her own. In the Pigou network, the Price of Anarchy is $\frac{100}{75} = \frac{4}{3}$. For splittable flow, it can be shown that this is the *worst possible* Price of Anarchy when the cost function on each edge is *affine*, that is, of the form $c(x) = ax + b$ for suitable nonnegative real numbers a and b (which depend on the edge). (For a proof of this result see Roughgarden and Tardos, 2002, theorem 4.5, or Roughgarden, 2016, chapter 11.) In other words, for affine cost functions, selfish routing is never more than 1/3 more costly than what could be achieved with central planning. For non-splittable flow (which has been the topic of this chapter), the Price of Anarchy for affine cost functions can be shown to be at most $\frac{5}{2}$ (Roughgarden, 2016, theorem 12.3). Exercise 2.3 describes a network where this bad ratio is actually achieved, so $\frac{5}{2}$ as the Price of Anarchy for this model is tight.

2.7 Further Reading

We have used material from chapter 13 of the book by Roughgarden (2016), in particular theorem 13.6 and its notation for our proof of Theorem 2.2. Exercise 2.3 is figure 12.3 of that book, rotated so as to display its symmetry. The Pigou network has been described qualitatively by Pigou (1920, p. 194).

The Braess paradox is due to Braess (1968). Exercise 2.2 discusses the actual network used by Braess. Hong, Song, and Wu (2007) do not mention the Braess paradox in the Cheonggyecheon urban regeneration project, but show some nice "before" and "after" pictures on page 235, and add some Fengshui to this book.

The proof of Theorem 2.2 for atomic selfish routing with the help of a potential function Φ was originally described by Rosenthal (1973). More general games where an equilibrium is found with the help of a potential function are the *potential games* studied by Monderer and Shapley (1996).

The model of splittable flow is due to Wardrop (1952), and its equilibrium is often called a *Wardrop equilibrium*.

2.8   Exercises for Chapter 2

Exercise 2.1. Consider the following congestion network with origin and destination nodes o and d, and an intermediate node v. There are ten individual users who want to travel from o to d.

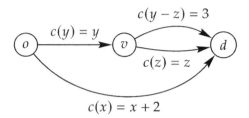

As shown, the cost on the bottom edge is $c(x) = x + 2$ when it has x users, where $x + y = 10$. The edge from o to v has cost function $c(y) = y$. The top edge from v to d has constant cost 3, and the lower edge from v to d has cost function $c(z) = z$ if z of the y users use that edge, where $0 \le z \le y$.

(a) Find all equilibria of this congestion game (note that x, y and z are integers), and their resulting *average cost per user*. [*Hint: Consider first what z can be in an equilibrium.*]

(b) For the same network, find the equilibrium when x, y, z can be fractional for a total flow of 10 from o to d (so this flow is now splittable). Compare its average cost with the costs of the equilibria in (a).

Exercise 2.2. Consider the following two congestion networks with the indicated cost functions for a flow x on the respective edge; for example, $50 + x$ is short for $c(x) = 50 + x$. Note that different edges may have different flows x that depend on the chosen routes by the users.

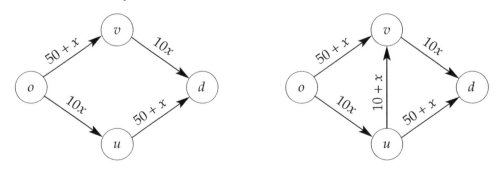

Suppose there are six users who want to travel from o to d.

(a) Find the equilibrium flow in the network on the left.

(b) In the network on the right, start from the equilibrium flow in (a) and improve the route of one user at a time until you find an equilibrium flow. Compare the equilibrium costs in both networks.

Exercise 2.3. Consider the following congestion network with three nodes. As shown, an edge has the cost function $c(x) = x$ if the flow (number of users) on that edge is x, except for the edges vu and wu where $c(x) = 0$. The table on the right shows four users $i = 1, 2, 3, 4$ with different origins o_i and destinations d_i. In this network, each user has two possible routes from her origin to her destination.

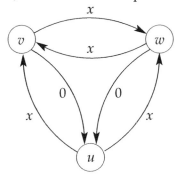

user i	o_i	d_i
1	u	v
2	u	w
3	v	w
4	w	v

Find two equilibria of this congestion game, one with low cost and one with high cost. Explain why the equilibrium property holds in each case. Compare the costs of the two equilibria. What is the socially optimal (possibly non-equilibrium) flow?

3

Games in Strategic Form

In this chapter we start with the systematic development of non-cooperative game theory. Its most basic model is the game in *strategic form*, the topic of this chapter. The available actions of each player, called *strategies*, are assumed as given. The players choose their strategies simultaneously and independently, and receive individual *payoffs* that represent their preferences for *strategy profiles* (combinations of strategies).

For two players, a game in strategic form is a table where one player chooses a row and the other player a column, with two payoffs in each cell of the table. We present a number of standard games such as the Prisoner's Dilemma, the Quality game, Chicken, the Battle of the Sexes, and Matching Pennies.

The central concept of *equilibrium* is a profile of strategies that are mutual best responses. A game may have one, several, or no equilibria. As shown in Chapter 6, allowing for *mixed* (randomized) strategies will ensure that every finite game has an equilibrium, as shown by Nash (1951). An equilibrium without randomization as considered in the present chapter is also known as a *pure Nash equilibrium*.

A strategy *dominates* another strategy if the player strictly prefers it for any fixed strategies of the other players. Dominated strategies are never played in equilibrium and can therefore be eliminated from the game. If iterated elimination of dominated strategies results in a unique strategy profile, the game is called *dominance solvable*. We illustrate this with the "Cournot duopoly" of quantity competition. (In Section 4.7 in Chapter 4, we will change this game to a *commitment game*, known as *Stackelberg leadership*.)

A strategy *weakly dominates* another strategy if the player weakly prefers it for any fixed strategies of the other players, and in at least one case strictly prefers it. Eliminating a weakly dominated strategy does not introduce new equilibria, but may lose equilibria, which reduces the understanding of the game. Unless one is interested in finding just one equilibrium of the game, one should therefore not eliminate weakly dominated strategies, nor iterate that process.

The final Section 3.8 shows that symmetric N-player games with two strategies per player always have an equilibrium.

3.1　Prerequisites and Learning Outcomes

This is a first core chapter of the book. The previous Chapter 2 has introduced the concepts of strategies and equilibrium for the special congestion games and is therefore useful but not a prerequisite. We deal mostly with finite sets. You should be thoroughly familiar with Cartesian products $S_1 \times \cdots \times S_N$ of sets $S_1, \ldots, S_N$. The Cournot game on intervals is analyzed with basic calculus.

After studying this chapter, you should

- know the components of a game in strategic form: strategies, strategy profiles, and payoffs;

- be able to write down and correctly interpret tables for two-player games;

- understand the concept of an equilibrium, use this term correctly (and its plural "equilibria"), and know how it relates to strategies, partial strategy profiles, and best responses;

- be familiar with common 2×2 games such as the Prisoner's Dilemma or the Stag Hunt, and understand how they differ;

- know the difference between dominance and weak dominance and why dominated strategies can be eliminated in a complete equilibrium analysis but weakly dominated strategies cannot;

- know the Cournot quantity game, also when strategy sets are real intervals;

- understand symmetry in games with two players and more than two players.

3.2　Games in Strategic Form

A game in strategic form is the fundamental model of non-cooperative game theory. The game has N players, $N \geq 1$, and each player $i = 1, \ldots, N$ has a nonempty set S_i of *strategies*. If each player i chooses a strategy s_i from S_i, the resulting N-tuple $s = (s_1, \ldots, s_n)$ is called a *strategy profile*. The game is specified by assigning to each strategy profile s a real-valued *payoff* $u_i(s)$ to each player i.

The payoffs represent each player's preference. For two strategy profiles s and $\hat{s}$, player i strictly *prefers* s to $\hat{s}$ if $u_i(s) > u_i(\hat{s})$, and is *indifferent* between s and $\hat{s}$ if $u_i(s) = u_i(\hat{s})$. If $u_i(s) \geq u_i(\hat{s})$ then player i *weakly prefers* s to $\hat{s}$. Player i is only interested in maximizing his own payoff, and not interested in the payoffs to other players (other than in trying to anticipate their actions); any "social" concern a player has about a particular outcome has to be (and could be) built into his own payoff. All players know the available strategies and payoffs of the other players, and know that they know them, etc. (that is, the game is "common knowledge").

The game is played as follows: The players choose their strategies *simultaneously* (without knowing what the other players choose), and receive their respective payoffs for the resulting strategy profile. They cannot enter into any binding agreements about what they should play (which is why the game is called "non-cooperative"). Furthermore, the game is assumed to be played only *once*, and therefore also called a *one-shot* game. Playing the same game (or a varying game) many times leads to the much more advanced theory of *repeated games*.

Much of this book is concerned with two-player games. The players are typically named 1 and 2 or I and II. Then the strategic form is conveniently represented by a table. The rows of the table represent the strategies of player I (also called the *row player*), and the columns represent the strategies of player II (the *column player*). A strategy profile is a strategy pair, that is, a row and a column, with a corresponding cell of the table that contains two payoffs, one for player I and the other for player II.

If m and n are positive integers, then an $m \times n$ game is a two-player game in strategic form with m strategies for player I (the rows of the table) and n strategies for player II (the columns of the table).

Many interesting game-theoretic observations apply already to 2×2 games. Figure 3.1(a) shows the famous *Prisoner's Dilemma* game. Each player has two strategies, called C and D, which stand for "cooperate" and "defect". The payoffs are as follows: If both players choose C, then both get payoff 2. If both players choose D, then both get payoff 1. If the two players choose different strategies, then the player who chooses C gets payoff 0 and the player who chooses D gets the highest possible payoff 3.

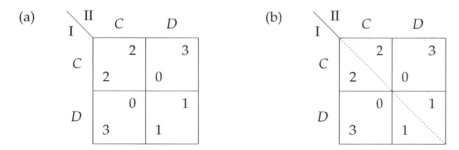

Figure 3.1 Prisoner's Dilemma game (a), its symmetry shown by reflection along the dotted diagonal line in (b).

In each cell of the payoff table, the payoff to the row player is shown in the *lower left* corner and the payoff to the column player in the *upper right* corner. There are several very good reasons for this display:

- There is no ambiguity to whom the payoff belongs. The payoff close to the left is near the strategy labels for the rows, the payoff close to the top is near the labels for the columns; this subdivision is also shown at the top left of the table that shows the two player names, here I and II.

- It is immediately apparent if the game is *symmetric* (that is, it stays the same game when exchanging the players), by reflecting the table along the diagonal line which exchanges the players, their strategies (rows and columns), and the payoffs in each cell, as shown in Figure 3.1(b). This reflection works because the rows are listed from top to bottom and the columns from left to right.

- An $m \times n$ game is specified by two $m \times n$ matrices A and B which contain the payoffs to players I and II. Because the payoffs are *staggered* in each cell, these two matrices are easy to identify separately so that one can focus on the payoffs of a single player.

The only unfamiliar aspect of putting the row player's payoff at the bottom left of a cell is that it is below the column player's payoff, where the usual convention is to consider rows first and columns second. However, the above reasons do not depend on this convention and treat rows and columns symmetrically. Another common display (usually for typesetting reasons) that uses the convention "rows first, columns second" is

$$
\begin{array}{c|c|c|}
 & \text{II} \quad C & D \\
\hline
C & 2,2 & 0,3 \\
\hline
D & 3,0 & 1,1 \\
\hline
\end{array}
\tag{3.1}
$$

where the payoffs are not staggered and therefore harder to separate, and where the symmetry of a game is less apparent. We will always use the staggered payoff display as in Figure 3.1.

The story behind the name "Prisoner's Dilemma" is that of two prisoners held suspect of a serious crime. There is no judicial evidence for this crime except if one of the prisoners testifies against the other. If one of them testifies (choosing "D"), he will be rewarded with immunity from prosecution (payoff 3), whereas the other will serve a long prison sentence (payoff 0). If both testify, their punishment will be less severe (payoff 1 for each). If they both "cooperate" with each other by not testifying at all (choosing "C"), they will only be imprisoned briefly for some minor charge that can be held against them (payoff 2 for each). The "defection" from that mutually beneficial outcome is to testify, which gives a higher payoff no matter what the other prisoner does. However, the resulting payoff is lower to both. This constitutes their "dilemma".

In the Prisoner's Dilemma, both players always have an incentive to defect, so that the strategy pair (D, D) will be the predicted outcome of playing the game, even though both players are worse off than if they played (C, C). The reason is that the non-cooperative model assumes that the players cannot enforce an agreement to play (C, C) (if they could, for example by signing a contract of some sort, this would have to be a separate strategy in the game). The Prisoner's Dilemma shows that individual incentives may jeopardize a common good. For example, in an arms race between two nations, C and D may represent the choices between restricted and unlimited arms build-up. The Prisoner's Dilemma can also be extended to many players. For example, it may be in every fisher's interest not to overfish so that fish stocks do not decline, by introducing a fishing quota that limits each fisher's catch. If adopted collectively, this would benefit all. However, without an enforcement of fishing quotas, each single fisher who exceeds his quota will barely contribute to overfishing but strongly benefit for himself. Collectively, they destroy their fish stock this way. This is also called the "tragedy of the commons". It applies similarly to many types of actions that collectively harm the environment (and each participant) but are individually beneficial.

3.3   Best Responses and Equilibrium

Consider the game in Figure 3.2(a). This is the Prisoner's Dilemma game, with further graphical information added to help the analysis of the game. Namely, certain payoffs are surrounded by boxes that indicate that they are *best-response payoffs*.

A best response is a strategy of a player that is optimal for that player assuming it is known what all other players do. In a two-player game, there is only one other player. Hence, for the row player, a best response is found for each column. In the Prisoner's Dilemma game, the best response of the row player against column C is row D because his payoff when playing D is 3 rather than 2 when playing C (so in the cell for row D and column C, the payoff 3 to player I is put in a box), and the best response against column D is also row D (with payoff 1, in a box, which is larger than payoff 0 for row C). Similarly, the best response of the column player against row C is column D, and against row D is column D, as shown by the boxes around the payoffs for the column player. Because the game is symmetric, the best-response payoffs are also symmetric.

The game in Figure 3.2(b) has payoffs that are nearly identical to those of the Prisoner's Dilemma game, except for the payoff to player II for the top right cell, which is changed from 3 to 1. Because the game is not symmetric, we name the strategies of the players differently, here T and B for the row player (for "top" and "bottom") and l and r for the column player ("left" and "right"). To distinguish them more easily, we often write the strategies of player I in upper case and those of player II in lower case.

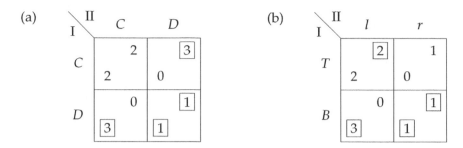

Figure 3.2 Prisoner's Dilemma game (a) with best-response payoffs shown by boxes, and the same done for the Quality game (b).

Figure 3.2(b) is called the *Quality* game. It may describe the situation of player I as a restaurant owner, who can provide food of good quality (strategy T) or bad quality (B), and a potential customer, player II, who may decide to eat there (l) or not (r). The customer prefers l to r only if the quality is good. However, whatever player II does, player I is better off by choosing B. These preferences are shown by the best-response boxes. The best response of player II to row T is different from her best response to row B. The best response of player I to either column is always row B.

In both games in Figure 3.2, consider the bottom right cell of the table, which corresponds to the strategy pair (D, D) in the Prisoner's Dilemma game and to (B, r) in the Quality game. In each game, this is the only cell that has both payoffs surrounded by a box. This means that the two strategies that define that cell are mutual best responses. Such a strategy pair is called an *equilibrium*, which is the central solution concept for non-cooperative games.

We define the concepts of best response and equilibrium for a general N-player game.

Definition 3.1. Consider a game in strategic form with N players where player $i = 1, \ldots, N$ has the strategy set S_i and receives payoff $u_i(s)$ for each strategy profile $s = (s_1, \ldots, s_N)$ in $S_1 \times \cdots \times S_N$. For player i, consider a *partial profile*

$$s_{-i} = (s_1, \ldots, s_{i-1}, s_{i+1}, \ldots, s_N) \tag{3.2}$$

of strategies of the other players, given by their strategies s_j for $j \neq i$. For any strategy s_i in S_i, this partial profile is extended to a full strategy profile denoted by $s = (s_i, s_{-i})$. Then s_i is a *best response* to s_{-i} if

$$u_i(s) = u_i(s_i, s_{-i}) \geq u_i(\bar{s}_i, s_{-i}) \qquad \text{for all } \bar{s}_i \in S_i. \tag{3.3}$$

An *equilibrium* of the game is a strategy profile s where each strategy is a best response to the other strategies in that profile, that is, (3.3) holds for all players $i = 1, \ldots, N$. □

The notation (3.2) for a partial profile s_{-i} of strategies of the players other than player i is very common. For a two-player game ($N = 2$), it is just a strategy of the other player. The notation (s_i, s_{-i}) creates a profile s for all players, where s_i is assumed to inserted at the ith position (as the strategy of player i) to define the strategy profile s. The condition (3.3) states that s_i is a best response to s_{-i} because player i cannot get a better payoff by changing to a different strategy $\bar{s}_i$. That payoff may stay the same (in which case $\bar{s}_i$ is also a best response to s_{-i}). Moreover, only *unilateral* changes by a single player are considered, not joint changes by several players, as for example, a change from (D, D) to (C, C) in the Prisoner's Dilemma. An equilibrium, as a strategy profile of mutual best responses, is therefore a profile where no player can profitably gain by unilaterally changing his strategy, that is, with the remaining strategies kept fixed.

Some notes on terminology: An equilibrium is often called *Nash equilibrium*, named after John Nash (1951) who proved that an "equilibrium point" always exists for finite strategy sets that are extended to include randomized or "mixed" strategies (see Chapter 6). In distinction to such a "mixed equilibrium", an equilibrium according to Definition 3.1 is often called a *pure* equilibrium where players choose their strategies deterministically. The plural of "equilibrium" is *equilibria*.

The congestion games studied in Chapter 2 are N-player games, where the strategy set of each player i is the set of paths from her origin o_i to her destination d_i, and a player's payoff is the negative of her cost as it results from the cost of using the edges on such a path, depending on their usage by all players. Exercise 3.2 considers a three-player game where the strategic form is given directly by a suitable "three-dimensional" table.

Why is equilibrium such an important concept? The analysis of a game should describe what players will (or should) do when following their preferences, taking into account that the other players do the same. A recommendation of a strategy to each player defines a strategy profile. This profile should be an equilibrium because otherwise this recommendation would be self-defeating, that is, at least one player would have an incentive to deviate from her recommended strategy to some other strategy. In that sense, equilibrium is a *necessary* condition of "stability" for a game-theoretic "solution". This does not mean that an equilibrium will always exist, or that the players will always reach an equilibrium when they adapt their strategies to be best responses. In practice, people may not play an equilibrium because the game is only a simplified model of the interactive situation, and even in a game with clear rules (such as Chess) it may be too difficult to play perfectly.

Much of this book is about determining the equilibria of a game, as the first step in a game-theoretic analysis. A separate argument is then which of these equilibria, if any, make sense as a practical recommendation of how to play. Another insight may be that an equilibrium has undesirable properties, like in the Prisoner's Dilemma or in the Braess paradox in Figure 2.3.

3.4 Games with Multiple Equilibria

We consider several well-known 2×2 games that have more than one equilibrium. The game of *Chicken* is shown in Figure 3.3. It is symmetric, as shown by the dotted line, and we have directly marked the best-response payoffs with boxes; recall that neither markup is part of the game description but already part of its analysis. The two strategies A and C stand for "aggressive" and "cautious" behavior, for example of two car drivers that drive towards each other on a narrow road. The aggressive strategy is only advantageous (with payoff 2) if the other player is cautious but leads to a crash (payoff 0) if the other player is aggressive, whereas a cautious strategy always gives payoff 1 to the player who uses it. This game has two equilibria, (C, A) and (A, C), because the best response to aggressive behavior is to be cautious and vice versa.

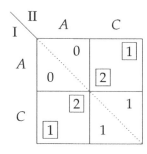

Figure 3.3 The symmetric game of Chicken.

The game known as the *Battle of the Sexes* is shown in Figure 3.4(a); its rather antiquated gender stereotypes should not be taken too seriously. In this scenario, player I and player II are a couple who each decide (simultaneously and independently, which is rather unrealistic) whether to go to a concert (C) or to a sports event (S). The players have different payoffs arising from which event they go to, but that payoff is zero if they have to attend the event alone. The game has two equilibria: (C, C) where they both go to the concert, or (S, S) where they both go to the sports event. Their preferences between these events differ, however.

The Battle of the Sexes game is not symmetric when written as in Figure 3.4(a), because the payoffs for the strategy pairs (C, C) and (S, S) on the diagonal are not the same for both players, which is clearly necessary for symmetry. However, changing the order of the strategies of one player, for example of player I as shown in Figure 3.4(b), makes this a game with a symmetric payoff structure; for a true symmetric game, one would also have to exchange the strategy names C and S of the strategies of player I, which would not represent the actions that player I takes. The game in Figure 3.4(b) is very similar to Chicken. However, because of the

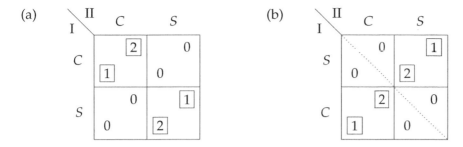

Figure 3.4 The Battle of the Sexes game (a), whose payoff structure is symmetric if the strategies of one player are exchanged, as in (b).

meaning of the strategies, the Battle of the Sexes is a *coordination game* (both players benefit from choosing the same action), whereas Chicken is an "anti-coordination" game.

The *Stag Hunt* game in Figure 3.5 models a conflict between cooperation and safety. Two hunters individually decide between S, to hunt a stag (a male deer), or H, to hunt a hare. Each can catch the hare, without the help of the other player, and then receives payoff 1. In order to catch the stag they have to cooperate – if the other hunter goes for the hare then the stag hunter's payoff is zero. If they both hunt the stag they succeed and each get payoff 3.

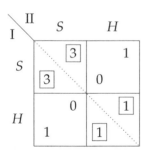

Figure 3.5 The Stag Hunt game.

This symmetric game has two symmetric equilibria, (S, S) and (H, H). The first equilibrium has a much higher payoff to both, but it requires *trusting* that the other player will also play S, because playing H is completely safe no matter what the other player does. The Stag Hunt game is sometimes called the "Trust Dilemma" for that reason.

It would be useful if game theory gave guidance of how to select between multiple equilibria. In the Stag Hunt game, it seems very reasonable to recommend (S, S), because both players get more than when playing (H, H), and it is strictly better to play S if the other player also plays S. However, that reasoning alone is tenuous, because it also applies in the following variation of the game. Suppose each payoff 1 in Figure 3.5 is replaced by 2.9 and these payoffs are in millions of dollars, so a player may get 3 million dollars by playing S (and risk getting nothing if the other player plays H) or safely get 2.9 million dollars by playing H. Here the risk-free choice seems clearly better.

There are a number of different approaches to the problem of equilibrium *selection*. Typically, an equilibrium is meant to fulfill additional *refinement* criteria which are meant to be generally applicable (for example that there should not be another equilibrium with higher payoff to both players, although this may conflict with the inherent "riskiness" of an equilibrium, as the Stag Hunt game demonstrates). This generality is often debated, because it may require that the players use the same theoretical thought process to arrive at the supposedly "right" equilibrium. We leave this problem open, and are content with just identifying when a game has multiple equilibria.

3.5   Dominated Strategies

In the Quality game in Figure 3.2(b), strategy B gives a better payoff to the row player than strategy T, no matter what the other player does. We say B *dominates* T, by which we always mean B *strictly dominates* T. The following definition states the concepts of strict dominance, weak dominance, and payoff equivalence for a general N-player game.

Definition 3.2. Consider an N-player game with strategy sets S_i and payoff functions u_i for each player $i = 1, \ldots, N$, let s_i and t_i be two strategies of some player i, and let $S_{-i} = \times_{j=1, j\neq i}^{N} S_j$ be the set of partial profiles of the other players. Then

- t_i *dominates* (or *strictly dominates*) s_i if

$$u_i(t_i, s_{-i}) > u_i(s_i, s_{-i}) \qquad \text{for all } s_{-i} \in S_{-i} , \tag{3.4}$$

- t_i is *payoff equivalent* to s_i if

$$u_i(t_i, s_{-i}) = u_i(s_i, s_{-i}) \qquad \text{for all } s_{-i} \in S_{-i} ,$$

- t_i *weakly dominates* s_i if t_i and s_i are not payoff equivalent and

$$u_i(t_i, s_{-i}) \geq u_i(s_i, s_{-i}) \qquad \text{for all } s_{-i} \in S_{-i} . \tag{3.5}$$

□

The condition (3.4) of strict dominance is particularly easy to check in a two-player game. For example, if i is the column player, then s_i and t_i are two columns, and s_{-i} is an arbitrary row, and we look at the payoffs of the column player. Then (3.4) states that, row by row, each entry in column t_i is greater than the respective entry in column s_i. For the columns D and C in the Prisoner's Dilemma game in Figure 3.2(a), this can be written as $\binom{3}{1} > \binom{2}{0}$, which is true. However, for the Quality game in Figure 3.2(b), the two payoff columns for r and l are $\binom{1}{1}$ and $\binom{2}{0}$ where this does not hold, and neither r dominates l nor l dominates r. However, for both games (which have the same payoffs to the row player) the bottom row dominates the top row because (3 1) > (2 0) (we compare two vectors component by component).

Note that "t_i dominates s_i" does *not* mean that t_i is "always better" than s_i. In the Prisoner's Dilemma, D dominates C, but playing D can result in payoff 1 and playing C in payoff 2, just not for the same strategy of the other player.

Weak dominance is essentially the same as strict dominance where the strict inequality in (3.4) is replaced by a weak inequality in (3.5). However, it should not be equality throughout (which would mean payoff equivalence). That is, t_i weakly dominates s_i if (3.5) holds and $u_i(t_i, s_{-i}) > u_i(s_i, s_{-i})$ for *at least one* partial profile $s_{-i} \in S_{-i}$.

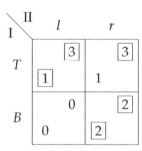

Figure 3.6 A game where column r weakly dominates column l. The weakly dominated strategy l is part of the equilibrium (T, l).

Figure 3.6 shows a game where the strategy T of player I has two best responses, l and r, because they result in the same payoff for player II (in fact for player I as well, but player I does not choose between columns). Against row B, player II's payoff for l is worse than for r, so l is weakly dominated by r. Nevertheless, l is part of an equilibrium, namely (T, l), in fact with a higher payoff to player II than the equilibrium (B, r), so player II arguably has an advantage when "threatening" to play l. Therefore, one should not disregard weakly dominated strategies when analyzing a game, where one is typically interested in finding *all* its equilibria.

In contrast, in an equilibrium a player never plays a strictly dominated strategy, because he can play instead the strategy that dominates it. The following proposition considers what happens when a weakly or strictly dominated strategy is removed from the game.

Proposition 3.3. *Consider an N-player game G in strategic form with strategy sets $S_1, \ldots, S_N$. Let s_i and t_i be strategies of player i. Suppose t_i weakly dominates or is payoff* *equivalent to s_i. Consider the game G' with the same payoff functions as G but where S_i is replaced by $S_i - \{s_i\}$. Then:*

(a) *Any equilibrium of G' is an equilibrium of G.*

(b) *If t_i dominates s_i, then G and G' have the same equilibria.*

Proof. To show (a), consider an equilibrium of G', which is a valid strategy profile in G. If player i could profitably deviate to strategy s_i in G, then, clearly, player i could also profitably deviate to t_i in G', which violates the equilibrium property. The equilibrium property holds for the players other than player i because they have the same strategies and payoffs in G and G'.

To show (b), suppose t_i dominates s_i. Then the dominated strategy s_i is never part of an equilibrium $s = (s_i, s_{-i})$ because of player i's profitable deviation to t_i according to (3.4). Any strategy profile in G that does not contain s_i is also a strategy profile of G'. If that profile is an equilibrium of G, then it is also an equilibrium of G', which has fewer deviation possibilities. Any equilibrium in G' is an equilibrium in G by (a). Hence, G and G' have the same equilibria, as claimed. □

As explained before with Figure 3.6, eliminating a weakly dominated strategy may result in losing equilibria of a game. Proposition 3.3(a) shows that this does not introduce new equilibria. Hence, deleting a weakly dominated strategy may be considered acceptable when trying to find only *one* equilibrium of a game. However, this normally distorts the analysis of the game.

Proposition 3.3(b) shows that removing a (strictly) dominated strategy from a game does not change its equilibria. Because a payoff-maximizing player should not play a dominated strategy, it does make sense to eliminate a dominated strategy from the game by removing it from the player's strategy set. This changes the game, and the smaller game may have new dominated strategies that were not dominated before. Removing such a dominated strategy again, and continuing this process until all remaining strategies are undominated, is called *iterated elimination of dominated strategies*.

In contrast, eliminating *weakly* dominated strategies is problematic because it may lose equilibria, and introduces further issues when iterated. As shown in Exercise 3.1, the *order* in which weakly dominated strategies are eliminated generally matters. The process may get stuck because weak domination turns into payoff equivalence, unless one also removes a payoff equivalent strategy, with a rather arbitrary decision as to which strategy should be eliminated. In addition, even if the elimination of weakly dominated strategies results in a single strategy profile, that profile may depend on the elimination order.

Figure 3.7 demonstrates iterated elimination of dominated strategies for the Quality game. In the first step, the top row T is dominated by B and removed to

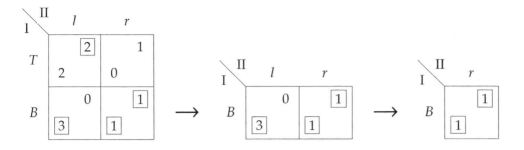

Figure 3.7 Iterated elimination of dominated strategies in the Quality game.

create a 1×2 game. In that game, column r dominates l (which is not true in the original 2×2 game). After eliminating l, a single cell remains that corresponds to the strategy profile (B, r) which is also the unique equilibrium of the original game. The following proposition shows that this holds generally.

Proposition 3.4. *Consider a game G that by a finite iterated elimination of dominated strategies is reduced to a single strategy profile s (that is, every player has only one strategy). Then s is the unique equilibrium of G.*

Proof. As assumed, eliminate a dominated strategy from G, and continue this process in finitely many steps to arrive at a single strategy for each player. The final game only has one strategy profile s, which is trivially an equilibrium because no player can deviate. By Proposition 3.3, the previous game has the same unique equilibrium s, and by going backwards in the produced sequence of games, this eventually holds for the original game G. $\square$

If finite iterated elimination of dominated strategies ends in a unique strategy profile, the game is said to be *dominance solvable*, with the final strategy profile as its solution, which is also the unique equilibrium of the game. The following lemma states the reassuring fact that the order in which dominated strategies are eliminated does not matter, and that several dominated strategies can be removed simultaneously.

Lemma 3.5. *Consider an N-player game G in strategic form with strategy sets $S_1, \ldots,$ S_N, let s_i and $\hat{s}_j$ be two dominated strategies of player i and player j, respectively, possibly the same player ($i = j$). Let G′ be the (otherwise unchanged) game obtained from G by removing s_i from S_i, and let $\hat{G}$ be the game obtained from G by removing $\hat{s}_j$ from S_j. Then $\hat{s}_j$ is dominated in G′, and s_i is dominated in $\hat{G}$, and removing $\hat{s}_j$ from G′ and removing s_i from $\hat{G}$ results in the same game obtained by removing directly s_i and $\hat{s}_j$ from G.*

Proof. Suppose s_i is dominated by t_i and $\hat{s}_j$ is dominated by $\hat{t}_j$. If $\hat{t}_j \neq s_i$, then $\hat{s}_j$ is also dominated in $G′$ by $\hat{t}_j$ because any partial profile s_{-j} in $G′$ is also a partial profile in G where the dominating inequality holds. If $\hat{t}_j = s_i$ (which requires $i = j$), then $\hat{s}_j$ is also dominated in G' (by t_i, because domination is clearly transitive), as claimed. Similarly, if $t_i \neq \hat{s}_j$, then s_i is dominated in $\hat{G}$ by t_i, and if $t_i = \hat{s}_j$, then s_i is dominated in $\hat{G}$ by $\hat{t}_j$. Removing the dominated strategies clearly results in the same game with both s_i and $\hat{s}_j$ removed directly from G. □

Iterated elimination of strictly dominated strategies has two uses. First, it may simplify the game, ideally until only one strategy profile remains, which is the unique equilibrium of the game (which of course only works if the game has only one equilibrium). Second, playing an equilibrium strategy assumes that the other players also play their equilibrium strategies. However, a player will always prefer a dominating strategy over the dominated strategy irrespective of the strategies of the other players. In that sense, choosing a dominating strategy is more "robust" in terms of assumptions about what the other players do. This applies certainly to a strategy that is dominating in the original game; iterated elimination of dominated strategies does require the "rationality" assumption that the other players maximize their payoff (in Figure 3.7, player II chooses r under the assumption that player I will not play T).

3.6   The Cournot Duopoly of Quantity Competition

In this section, we discuss an economic model of competition between two competing firms who choose the *quantity* of their product, where the price decreases with the total quantity on the market. It is also called a *Cournot duopoly* named after Augustin Cournot (1838) who proposed this model, nowadays described as a game with an equilibrium. In this game, the strategy sets of the players are infinite. Finite versions of the game are dominance solvable, and in principle also the infinite version, as we describe at the end of this section.

The players I and II are two firms who choose a nonnegative quantity of producing some good up to some upper bound M, say, so $S_1 = S_2 = [0, M]$. Let x and y be the strategies chosen by player I and II, respectively. For simplicity there are no costs of production, and the total quantity $x + y$ is sold at a price $12 - (x + y)$ per unit, which is also the firm's profit per unit, so the payoffs $a(x, y)$ and $b(x, y)$ to players I and II are given by

$$
\begin{aligned}
\text{payoff to I} \;:\quad & a(x, y) = x \cdot (12 - y - x), \\
\text{payoff to II} \;:\quad & b(x, y) = y \cdot (12 - x - y).
\end{aligned}
\tag{3.6}
$$

$\Rightarrow$ Exercise 3.3 describes an extension that incorporates production costs.

The game (3.6) is clearly symmetric in x and y. Figure 3.8 shows a finite version of this game where the players' strategies are restricted to the four quantities 0,

Figure 3.8 payoff table (player I chooses row, player II chooses column; in each cell the upper-right number is player II's payoff and the lower-left number is player I's payoff; best responses are boxed):

I \ II	0	3	4	6
0	0 / 0	27 / 0	32 / 0	[36] / 0
3	0 / 27	18 / 18	[20] / 15	18 / [9]
4	0 / 32	15 / [20]	[16] / [16]	12 / 8
6	0 / [36]	[9] / 18	8 / 12	0 / 0

Figure 3.8 Symmetric duopoly game between two firms who both have four strategies 0, 3, 4, 6 with payoffs (3.6) if player I chooses x and player II chooses y.

3, 4, or 6. The payoffs are determined by (3.6). For example, for $x = 3$ and $y = 4$ player I gets payoff 15 and player II gets 20. Best-response payoffs, as determined by this payoff table, are marked by boxes. The game has the unique equilibrium $(4, 4)$ where each player gets payoff 16.

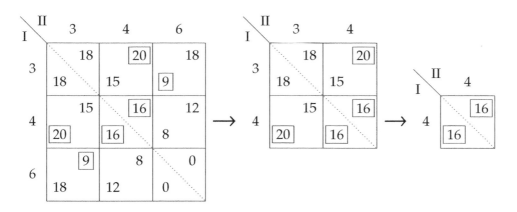

Figure 3.9 The game in Figure 3.8 after eliminating for both players the dominated strategy 0, then strategy 6 (dominated by 4), then strategy 3 (dominated by 4), which shows that the game is dominance solvable.

Figure 3.9 shows that the game can be solved by iterated elimination of dominated strategies. In the 4×4 game, strategy 0 is dominated by 3 or 4, and after eliminating it directly for both players one obtains the 3×3 game shown on the left in Figure 3.9. In that game, strategy 6 is dominated by 4, and after eliminating this strategy for both players the middle 2×2 game is reached. This game has the structure of a Prisoner's Dilemma game because 4 dominates 3 (so that 3 will be eliminated), but the final strategy pair of undominated strategies $(4, 4)$ gives both players a lower payoff than the strategy pair $(3, 3)$. Here the Prisoner's Dilemma arises in an economic context: The two firms could cooperate by equally splitting the optimal "monopoly" quantity 6, but in response the other player would "defect" and choose 4 rather than 3. If both players do this, the total quantity 8 reduces the price to 4 and both players have an overall lower payoff.

The game can also be analyzed when each player has the real interval $[0, M]$ as a strategy set. We can assume that $M = 12$ because if a player chooses a higher quantity than 12 then the price $12 - x - y$ would be negative and the player would be better off by producing nothing. Player I's best response to y is the quantity x that maximizes $a(x, y)$, where

$$a(x, y) = x \cdot (12 - y - x) = -\left(x - \tfrac{12-y}{2}\right)^2 + \left(\tfrac{12-y}{2}\right)^2 \tag{3.7}$$

which clearly has its maximum at $x = \frac{12-y}{2} = 6 - y/2$, where $x \geq 0$ because $y \leq 12$. (When just looking at $x \cdot (12 - y - x)$, this is a parabola with its apex at the midpoint between its zeros at $x = 0$ and $x = 12 - y$. The maximum is also found by solving $\frac{d}{dx} a(x, y) = 0$.) Note that the best response to y is unique. Because the game is symmetric, player II's best response y to x is given by $y = 6 - x/2$.

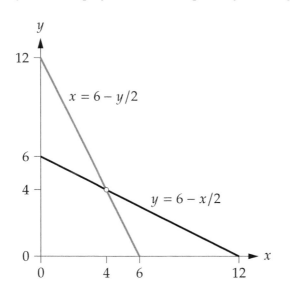

Figure 3.10 Best-response functions $x = 6 - y/2$ and $y = 6 - x/2$ of players I and II for $x, y \in [0, 12]$ in the duopoly game (3.6).

In order to obtain an equilibrium (x, y), the strategies x and y have to be mutual best responses, that is, $x = 6 - y/2$ and $y = 6 - x/2$. This pair of linear equations has the unique solution $x = y = 4$, which is the unique equilibrium of the game, shown as the intersection point $(4, 4)$ of the best-response lines in Figure 3.10. It coincides with the equilibrium of the 4×4 game in Figure 3.8 where 4 was one of the considered four strategies.

In the remainder of this section, we explain how even the game with infinite strategy sets $[0, M]$ is (to an arbitarily close approximation) dominance solvable, by iteratively eliminating whole intervals of dominated strategies from the strategy sets. The first step already happened when we chose $M = 12$: In principle, player I could choose $x' > 12$, but then $a(x', y) = x' \cdot (12 - y - x') < 0$ because $y \geq 0$, which is worse than $a(0, y)$ for any y. That is, if $x' > 12$ then x' is strictly dominated by $x = 0$. The same applies to the other player, so we can assume that both strategy sets are equal to the interval $[0, 12]$.

The following proposition states that when player II is assumed to play y from an interval $[A, B]$, then there is another interval $[A', B'] = [6 - \frac{B}{2}, 6 - \frac{A}{2}]$ that contains all possible best responses of player I (condition (a)), and so that every strategy x' of player I outside the interval $[A', B']$ is dominated by either B' or A' (conditions (b) and (c)). After eliminating these dominated strategies, we can assume that player I only plays strategies from $[A', B']$. If both players are initially considering strategies from the same interval $[A, B]$, then condition (d) shows when this leads to a sequence of successively narrower intervals, which we will discuss afterwards.

Proposition 3.6. *Consider the game with payoffs (3.6) where $y \in [A, B]$ with $0 \leq A \leq B \leq 12$. Let $A' = 6 - \frac{B}{2}$ and $B' = 6 - \frac{A}{2}$. Then*

(a) *Each x in $[A', B']$ is player I's unique best response to some y in $[A, B]$.*

(b) *If $x' > B'$ then strategy x' of player I is strictly dominated by $x = B'$.*

(c) *If $x' < A'$ then strategy x' of player I is strictly dominated by $x = A'$.*

(d) *The interval $[A', B']$ has half the length of the interval $[A, B]$, and $[A', B'] \subseteq [A, B]$ if and only if*

$$4 \in [A + \tfrac{1}{3}(B - A), A + \tfrac{2}{3}(B - A)], \tag{3.8}$$

and if (3.8) holds then it also holds with A', B' instead of A, B.

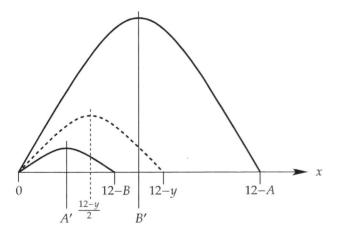

Figure 3.11 The dashed curve shows the payoff $x \cdot (12 - y - x)$ to player I in Proposition 3.6 (negative payoffs are not shown). Its maximum is at $\frac{12-y}{2}$. Here $A' = \frac{12-B}{2}$ and $B' = \frac{12-A}{2}$.

Proof. Assume that player II plays some y in $[A, B]$. According to (3.7), the payoff to player I as a function of x is a parabola. Figure 3.11 shows the two extreme cases

$y = A$ and $y = B$ (solid curves) and an intermediate case (dashed curve). Player I's best response to y is $x = \frac{12-y}{2}$ and therefore between $\frac{12-B}{2}$ and $\frac{12-A}{2}$, that is, in the interval $[A', B']$, shown by the long solid vertical lines in Figure 3.11. The mapping $[A, B] \to [6 - \frac{B}{2}, 6 - \frac{A}{2}]$, $y \mapsto 6 - \frac{y}{2}$ is a bijection, which shows (a).

The case $y = A$ is best possible for player I, which is the largest parabola in Figure 3.11. The figure shows that if $x' > B' = \frac{12-A}{2}$ then player I is certainly better off playing $x = B'$ because for each $y \geq A$ the dashed parabola slopes downwards (possibly even with negative values) from B' to x'. This intuition for (b) is formally shown as follows. Let $x' > B'$. We want to show that $a(B', y) > a(x', y)$, which by (3.7) is equivalent to

$$-\left(B' - \frac{12-y}{2}\right)^2 > -\left(x' - \frac{12-y}{2}\right)^2. \tag{3.9}$$

With

$$x' = B' + c \qquad (c > 0), \qquad y = A + d \qquad (d \geq 0)$$

and therefore $\frac{12-y}{2} = 6 - \frac{A+d}{2} = B' - \frac{d}{2}$ the inequality (3.9) is equivalent to

$$\left(B' - \left(B' - \frac{d}{2}\right)\right)^2 < \left(B' + c - \left(B' - \frac{d}{2}\right)\right)^2,$$

that is, to $(d/2)^2 < (c + d/2)^2$ which is true. This shows (b).

Similarly, the case $y = B$ is worst possible for player I, which is the smallest parabola in Figure 3.11. Then if $x' < A' = \frac{12-B}{2}$ then player I is better off playing $x = A'$ because for each $y \leq B$ the parabola slopes upwards from x' to A'. Formally, let $x' < A'$. We want to show that $a(A', y) > a(x', y)$, that is,

$$-\left(A' - \frac{12-y}{2}\right)^2 > -\left(x' - \frac{12-y}{2}\right)^2. \tag{3.10}$$

Now we set

$$x' = A' - c \qquad (c > 0), \qquad y = B - d \qquad (d \geq 0)$$

and therefore $\frac{12-y}{2} = 6 - \frac{B-d}{2} = A' + \frac{d}{2}$. Then (3.10) is equivalent to

$$\left(A' - \left(A' + \frac{d}{2}\right)\right)^2 < \left(A' - c - \left(A' + \frac{d}{2}\right)\right)^2,$$

that is, to $(-d/2)^2 < (-c - d/2)^2$ or again $(d/2)^2 < (c + d/2)^2$ which is true. This shows (c).

To show (d), clearly $B' - A' = (B - A)/2$. The condition $[A', B'] \subseteq [A, B]$ is equivalent to $A \leq A' = 6 - \frac{B}{2}$ and $B' = 6 - \frac{A}{2} \leq B$, which has the following equivalent statements:

$$A + \frac{B}{2} \leq 6 \leq B + \frac{A}{2},$$

$$A + \frac{A+(B-A)}{2} \leq 6 \leq A + (B - A) + \frac{A}{2},$$

$$\tfrac{3}{2}A + \frac{(B-A)}{2} \leq 6 \leq \tfrac{3}{2}A + (B - A),$$

that is,

$$A + \tfrac{1}{3}(B - A) \leq 4 \leq A + \tfrac{2}{3}(B - A) \tag{3.11}$$

which is stated in (3.8). Assuming that (3.11) holds, we want to show the same inequalities with A', B' instead of A, B, that is,

$$
\begin{aligned}
A' + \tfrac{1}{3}(B' - A') &\leq 4 \leq A' + \tfrac{2}{3}(B' - A'), \\
B' - \tfrac{2}{3}(B' - A') &\leq 4 \leq B' - \tfrac{1}{3}(B' - A'), \\
6 - \tfrac{A}{2} - \tfrac{2}{3}(B' - A') &\leq 4 \leq 6 - \tfrac{A}{2} - \tfrac{1}{3}(B' - A')
\end{aligned}
$$

or

$$2 \leq \tfrac{A}{2} + \tfrac{2}{3}(B' - A'), \qquad \tfrac{A}{2} + \tfrac{1}{3}(B' - A') \leq 2$$

which when multiplied by 2 and using $2(B' - A') = B - A$ is equivalent to (3.11) which we assumed to be true. This completes the proof of (d). $\qquad\square$

By Proposition 3.6, we can apply iterated elimination of dominated strategies to the game (3.6) as follows. Initially, assume that both players consider any strategy in the interval $[A, B] = [0, 12]$. Then $[A', B'] = [0, 6]$. Any strategy x' of player I outside $[A', B']$ (which here means only $x' > 6$) is dominated by $x = B'$. (However, as stated in (a), any strategy in $[A', B']$ is not dominated because it is a best response to some y in $[A, B]$, which shows that the elimination at this stage is best possible.) Similarly, because the game is symmetric, any strategy y' of player II outside $[A', B']$ is dominated by $y = B'$. We eliminate these dominated strategies, so that the new strategy sets will be $[A', B']$. Condition (3.8) states that 4 is in the middle third of the interval $[A, B]$, which is (just about) true if $[A, B] = [0, 12]$ where that middle third is $[4, 8]$. Furthermore, (d) states that then the same holds for $[A', B']$ (indeed, 4 is in the middle third $[2, 4]$ of the interval $[0, 6]$). We can now repeat the elimination of dominated strategies with A', B' instead of A, B, and each time the length $B - A$ of the considered interval is reduced by half. The strategy $x = 4$ and $y = 4$ belongs to all these intervals. In fact, any other strategy will sooner or later be outside a sufficiently small interval and will therefore be eliminated, although there is no bound on the number of elimination rounds for strategies that are very close to 4. In the end, only the strategy pair $(4, 4)$ survives, which is the unique equilibrium of the game.

This shows that the quantity duopoly game (3.6) is dominance solvable (if one allows for an unlimited number of elimination rounds, otherwise up to any desired degree of accuracy). Dominance solvability is considered a stronger solution concept than equilibrium. The equilibrium itself, of course, can be computed directly as the solution (x, y) to the two equations $x = 6 - y/2$ and $y = 6 - x/2$.

3.7   Games without a Pure-Strategy Equilibrium

Not every game has an equilibrium in pure strategies. Figure 3.12 shows two well-known examples. In *Matching Pennies*, the two players reveal a penny which can show Heads (H) or Tails (T). If the pennies match, then player I wins the other player's penny, otherwise player II. No strategy pair can be stable because the losing player would always deviate.

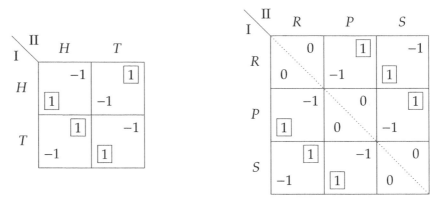

Figure 3.12   Matching Pennies (left) and Rock-Paper-Scissors (right).

Rock-Paper-Scissors is a 3×3 game where both players choose simultaneously one of their three strategies Rock (R), Paper (P), or Scissors (S). Rock loses to Paper, Paper loses to Scissors, and Scissors lose to Rock, and it is a draw otherwise. No two strategies are best responses to each other. Like Matching Pennies, this is a *zero-sum* game because the payoffs in any cell of the table sum to zero. Unlike Matching Pennies, Rock-Paper-Scissors is symmetric. Hence, when both players play the same strategy (the cells on the diagonal), they get the same payoff, which is zero because the game is zero-sum.

The game-theoretic recommendation is to play *randomly* in games like Matching Pennies or Rock-Paper-Scissors that have no equilibrium, according to certain probabilities that depend on the payoffs. As we will see in Chapter 6, any finite game has an equilibrium when players are allowed to use randomized strategies.

3.8   Symmetric Games with Two Strategies per Player

In this section, we consider N-player games that, somewhat surprisingly, always have a pure equilibrium, namely symmetric games where each player has only two strategies.

An N-player game is *symmetric* if each player has the same set of strategies, and if the game stays the same after any permutation (shuffling) of the players and, correspondingly, their payoffs. For two players, this means that the game stays the same when exchanging the players and their payoffs, visualized by reflection along the diagonal.

We now consider symmetric N-player games where each player has two strategies, which we call 0 and 1. Normally, any combination of these strategies defines a separate strategy profile, so there are 2^N profiles, and each of them specifies a payoff to each player, so the game is defined by $N \cdot 2^N$ payoffs. If the game is symmetric, vastly fewer payoffs are needed. Then a strategy profile is determined by how many players choose 1, say k players (where $0 \le k \le N$), and then the remaining $N - k$ players choose 0, so the profile can be written as

$$\underbrace{(1,\ldots,1,}_{k}\underbrace{0,\ldots,0)}_{N-k}. \tag{3.12}$$

Because the game is symmetric, any profile where k players choose 1 has to give the same payoff as (3.12) to any player who chooses 1, and a second payoff to any player who chooses 0. Hence, we need only two payoffs for these profiles (3.12) when $1 \le k \le N - 1$. When $k = 0$ then the profile is $(0,\ldots,0)$ and all players play the same strategy 0 and only one payoff is needed, and similarly when $k = N$ where all players play 1. Therefore, a symmetric N-player game with two strategies per player is specified by only $2N$ payoffs.

Proposition 3.7. *Consider a symmetric N-player game where each player has two strategies, 0 and 1. Then this game has a pure equilibrium. The strategy profile $(1,1,\ldots,1)$ is the unique equilibrium of the game if and only if 1 dominates 0.*

Proof. If 1 dominates 0, then $(1,1,\ldots,1)$ is clearly the unique equilibrium of the game, so suppose this not the case. Because 0 is not dominated and there are only two strategies, for some profile the payoff when playing 0 is greater than or equal to the payoff when playing 1, that is, 0 is a best response to the remaining strategies. Consider such a profile (3.12) where 0 is a best response, with the *smallest* number k of players who play 1, where $0 \le k < N$. Then this profile is an equilibrium: By assumption and symmetry, 0 is a best response to the strategies of the other players for every player who plays 0. If 1 was not a best response for every player who plays 1, then such a player would obtain a higher payoff when changing his strategy to 0. This change would result in a profile where $k - 1$ players play 1, the remaining players play 0, and 0 is a best response, which contradicts the smallest of choice of k. This shows that the game has an equilibrium. $\qquad\square$

We consider this proposition for symmetric games when $N = 2$. The symmetric Prisoner's Dilemma (Figure 3.1) has a single equilibrium, and it indeed uses a dominating strategy. The equilibrium is symmetric, that is, both players play the same strategy. The symmetric Stag Hunt game (Figure 3.5) has two symmetric equilibria. The symmetric Chicken game (Figure 3.3) has two equilibria which are not symmetric. Clearly, in any symmetric two-player game, non-symmetric equilibria come in pairs, here (A, C) and (C, A), obtained by exchanging the players and their strategies. Chicken does not have a symmetric pure equilibrium. The 3×3 game of Rock-Paper-Scissors (Figure 3.12) is symmetric but does not have a pure equilibrium. This shows that Proposition 3.7 cannot be generalized to symmetric games with more than two strategies per player.

⇒ Exercise 3.4 describes a game related to Proposition 3.7.

3.9   Further Reading

The display of staggered payoffs in the lower left and upper right of each cell in the payoff table is due to Thomas Schelling. In 2005, he received, together with Robert Aumann, the Nobel memorial prize in Economic Sciences (officially: The Sveriges Riksbank Prize in Economic Sciences in Memory of Alfred Nobel) "for having enhanced our understanding of conflict and cooperation through game-theory analysis." According to Dixit and Nalebuff (1991, p. 90), he said with excessive modesty: "If I am ever asked whether I ever made a contribution to game theory, I shall answer yes. Asked what it was, I shall say the invention of staggered payoffs in a matrix."

The strategic form is taught in every course on non-cooperative game theory (it is sometimes called the "normal form", now less used in game theory because "normal" is an overworked term in mathematics). Osborne (2004) gives careful historical explanations of the games considered in this chapter, including the original duopoly model of Cournot (1838), and many others. Our Exercise 3.4 is taken from that book. Gibbons (1992) shows that the Cournot game is dominance solvable, with less detail than our proof of Proposition 3.6. Both Osborne and Gibbons disregard Schelling and use comma-separated payoffs as in (3.1).

The Cournot game in Section 3.6 is also a potential game with a strictly concave potential function, which has a unique maximum and therefore a unique equilibrium. Neyman (1997) showed that it is also a unique correlated equilibrium (see Chapter 12). Potential games (Monderer and Shapley, 1996) generalize games with a potential function such as the congestion games considered in Section 2.5.

A classic survey of equilibrium refinements is van Damme (1987). Proposition 3.7 seems to have been shown first by Cheng, Reeves, Vorobeychik, and Wellman (2004, theorem 1).

3.10 Exercises for Chapter 3

Exercise 3.1. Consider the following 3×3 game.

I \ II	l	c	r
T	0 ; 1	1 ; 3	1 ; 1
M	1 ; 1	0 ; 3	1 ; 0
B	2 ; 2	3 ; 3	2 ; 0

(Entries shown as II's payoff ; I's payoff, with II's payoff in the top-right and I's payoff in the bottom-left of each cell.)

(a) Identify all pairs of strategies where one strategy weakly dominates the other.

(b) Assume you are allowed to remove a weakly dominated strategy of some player. Do so, and repeat this process (of iterated elimination of weakly dominated strategies) until you find a single strategy pair of the original game. (Remember that two payoff equivalent strategies do *not* weakly dominate each other!)

(c) Find such an iterated elimination of weakly dominated strategies that results in a strategy pair other than the one found in (b), where *both* strategies, and the payoffs to the players, are different.

(d) What are the equilibria (in pure strategies) of the game?

Exercise 3.2. Consider the following three-player game in strategic form.

I \ II	l	r
T	3,4,4	1,3,3
B	8,1,4	2,0,6

III: L

I \ II	l	r
T	4,0,5	0,1,6
B	5,1,3	1,2,5

III: R

Each player has two strategies: Player I has the strategies T and B (the top and bottom row), player II has the strategies l and r (the left and right column in each 2×2 panel), and player III has the strategies L and R (the right or left panel). The payoffs to the players in each cell are given as triples of numbers to players I, II, III.

(a) Identify all pairs of strategies where one strategy strictly, or weakly, dominates the other. [*Hint: Make sure you understand what dominance means for more than two players. Be careful to consider the correct payoffs.*]

(b) Apply iterated elimination of strictly dominated strategies to this game. What are the equilibria of the game?

Exercise 3.3. Consider the duopoly game as studied in Section 3.6 where player I and II produce nonnegative quantities x and y and the unit price on the market is $12 - x - y$. In addition, the players now have a *cost of production*, which is a cost of 1 per unit for player I and a cost of 2 per unit for player II. Write down the payoff to the players as a modification of (3.6), find their best response functions, and determine the equilibrium of this game. What are the players' profits in equilibrium?

Exercise 3.4. (Osborne, 2004, Exercise 33.1, Contributing to a public good.)

Each of N people chooses whether or not to contribute a fixed amount towards the provision of a public good. The good is provided if and only if at least K people contribute, where $2 \leq K \leq N$. If the good is not provided, contributions are not refunded. Each person strictly ranks outcomes from best to worst as follows:

(i)   any outcome in which the good is provided and she does not contribute,

(ii) any outcome in which the good is provided and she contributes,

(iii) any outcome in which the good is not provided and she does not contribute,

(iv) any outcome in which the good is not provided and she contributes.

Formulate this situation as a strategic-form game and find its equilibria. You need to know the number of players, the strategies that a player has available, and the players' preferences for strategy profiles. Is there a symmetry here? Is there an equilibrium in which more than K people contribute? One in which K people contribute? One in which fewer than K people contribute?

4

Game Trees with Perfect Information

This chapter considers game trees, the second main way for defining a non-cooperative game in addition to the strategic form. In a game tree, players move *sequentially* and (in the case of perfect information studied in this chapter) are aware of the previous moves of the other players. In contrast, in a strategic-form game players move *simultaneously*. In this "dynamic" setting, a *play* means a specific run of the game given by a sequence of actions of the players.

A game tree is described by a tree whose nodes represent possible states of play. Some nodes are *chance nodes* where the next node is determined randomly according to a known probability distribution. A *decision node* belongs to a player who makes a move to determine the next node. Play ends at a terminal node or *leaf* of the game tree where each player receives a payoff.

Game trees can be solved by *backward induction*, where one finds optimal moves for each player, given that all future moves have already been determined. Backward induction starts with the decision nodes closest to the leaves, and assumes that a player always makes a payoff-maximizing move. Such a move is in general not unique, so that the players' moves determined by backward induction may be interdependent (see Figure 4.4).

A *strategy* in a game tree is a derived concept. A strategy defines a move at every decision node of the player and therefore a *plan of action* for every possible state of play. Any strategy *profile* therefore defines a payoff to each player (which may be an expected payoff if there are chance moves), which defines the strategic form of the game.

Because every strategy is a combination of moves, the number of strategies in a game tree is very large. *Reduced strategies* reduce this number to some extent. In a reduced strategy, moves at decision nodes that cannot be reached due to an earlier own of the player are left *unspecified*.

Backward induction defines a strategy profile which is an equilibrium of the game (Theorem 4.6). This is called a *subgame-perfect equilibrium* (SPE) because it defines an equilibrium in every subgame. (In a game tree with perfect information, a subgame is just a subtree of the game. In games with imperfect information, treated in Chapter 10, an SPE can in general *not* be found by backward induction.)

In Section 4.7, we consider *commitment games*, which are two-player games where a given strategic-form game is changed to a game tree where one player moves first and the other player is informed about the first player's move and can react to that move. It is assumed that the second player always chooses a best response, as in an SPE. This changes the game, typically to the advantage of the first mover. We study the commitment game for the Quality game and for the Cournot duopoly game of quantity competition from Section 3.6, where the commitment game is known as a *Stackelberg leadership* game.

Every game tree can be converted to strategic form. A general game in strategic form can only be represented by a tree with extra components that model imperfect information. Game trees with imperfect information will be treated in Chapter 10.

4.1 Prerequisites and Learning Outcomes

Chapter 3 is a prerequisite for this chapter. Chapter 1 is not, although combinatorial games are sequential games with perfect information as considered here, and therefore illustrative examples. We will recall the mathematical concepts of directed graphs and trees, and of the expected value of a random variable, as background material.

After studying this chapter, you should be able to

- interpret game trees with their moves and payoffs;
- create the strategies of a player as combinations of moves, and know how to count them;
- combine strategies into *reduced* strategies, which identify moves at decision nodes that are unreachable due to an earlier *own* move of the player, and understand where the corresponding "wildcard" symbol " ∗ " is placed in the move list that represents the strategy;
- write down the strategic form of a game tree, with players' payoffs in the correct place in each cell, and with computed expected payoffs if needed;
- explain backward induction and why it requires full (unreduced) strategies, which may not be unique;
- construct any equilibrium, even if not subgame-perfect, directly from the game tree, also for more than two players, as in Exercise 4.5;
- create the commitment game from a game in strategic form, and compare its subgame-perfect equilibria with the equilibria in the strategic form.

4.2   Definition of Game Trees

Figure 4.1 shows an example of a game tree. We always draw trees downwards, with the starting node, called the *root*, at the top. (Conventions on drawing game trees vary. Sometimes trees are drawn from the bottom upwards, sometimes from left to right, and sometimes with the root at the center with edges in any direction.)

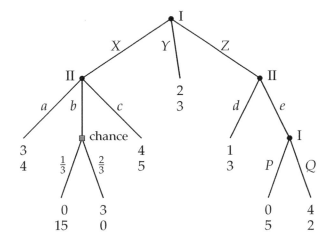

Figure 4.1 Example of a game tree. The square node indicates a chance move. At a leaf of the tree, the top payoff is to player I, the bottom payoff to player II.

The *nodes* of the tree denote states of play (which have been called "positions" in the combinatorial games considered in Chapter 1). Nodes are connected by lines, called *edges*. An edge from a node u to a *child* node v (where v is drawn below u) indicates a possible *move* in the game. This may be a move of a "personal" player, for example move X of player I in Figure 4.1. Then u is also called a *decision node*. Alternatively, u is a *chance node*, like the node u that follows move b of player II in Figure 4.1. We draw decision nodes as small filled circles and chance nodes as squares. After a chance node u, the next node v is determined by a random choice made with the probability associated with the edge that leads from u to v. In Figure 4.1, these probabilities are $\frac{1}{3}$ for the left move and $\frac{2}{3}$ for the right move.

At a terminal node or *leaf* of the game tree, every player gets a payoff. In Figure 4.1, leaves are not explicitly drawn, but the payoffs given instead, with the top payoff to player I and the bottom payoff to player II.

It does not matter how the tree is drawn, only how the nodes are connected by edges, as summarized in the background material on directed graphs and trees. The following is the formal definition of a game tree.

Definition 4.2. A *game tree* is a tree with the following additional information:

- A finite set of N players.
- Each non-terminal node of the tree is either a *decision node* and belongs to one of the N players, or is a *chance node*.

Background material: Directed graphs and trees

A *directed graph* (or *digraph*) is given by a finite set of *nodes* and a finite set of *edges*. Every edge is an ordered pair of nodes u and v, written as uv. A *walk from u to v* is a sequence of nodes $u_0, u_1, \ldots, u_m$ for $m \geq 0$ where $u_k u_{k+1}$ is an edge for $0 \leq k < m$, and $u = u_0$ and $v = u_m$. If these nodes are all distinct, the walk is called a *path*.

A (directed) *tree* is a digraph with a distinguished node, the *root*, so that for every node u there is a unique walk from the root to u. If uv is an edge of the tree, then v is called a *child* of u and u is called the *parent* of v. In a tree, a node v is a *descendant* of a node u if there is a nonempty walk from u to v. A node without a child is called a *leaf* or *terminal node* of the tree.

We represent trees graphically with the root at the top and edges drawn downwards from a parent to a child.

Lemma 4.1. *In a tree, the unique walk from the root to a node is a path. Every node has a unique parent, except for the root which has no parent.*

Proof. If the walk $u_0, u_1, \ldots, u_m$ from the root u_0 to the node u_m had a repeated node $u_i = u_j$ for some $i < j$, we could remove the nodes $u_{i+1}, \ldots, u_j$ from the walk and obtain a second walk from u_0 to u_m, which is not allowed. Hence, the walk has no repeated nodes and is therefore a path.

Consider a tree node u which is not the root, and let vu be the last edge on the unique path from the root to u. If there was a second edge wu with $w \neq v$, then the path from the root to w extended by u would give a second path from the root to u, which is not allowed. Hence, v is the unique parent of u. If the root had a parent v, the path from the root to v could be continued to a walk from the root to itself, which would be a second walk in addition to the empty path that consists of a single node. Again, this is not allowed. □

- For a decision node u, every edge uv to a child v of u is assigned a different *move*. These moves are called the *moves at u*.

- For a chance node u, the edges uv to the children v of u are assigned *probabilities* (that is, nonnegative reals that sum to 1).

- Every leaf of the game tree has an N-tuple of real-valued payoffs, one for each player. □

The game tree, with its decision nodes, moves, chance probabilities and payoffs, is known to the players, and defines the game completely. The game is played by starting at the root. At a decision node, the respective player chooses a move, which determines the next node. At a chance node, the move is made randomly according to the given probabilities. Play ends when a leaf is reached, where all players receive their payoffs.

Players are interested in maximizing their own payoffs. If the outcome of playing the game is random, then the players are assumed to be interested in maximizing their *expected payoffs*.

Background material: Expected value of a finite random variable

A *random variable* U takes real values that depend on a random event. Let U take finitely many values $u_1, \ldots, u_k$ that occur with probabilities $p_1, \ldots, p_k$, respectively. Then the *expected value* or *expectation* of U is defined as

$$\mathbb{E}[U] = p_1 u_1 + \cdots + p_k u_k . \tag{4.1}$$

Because the probabilities $p_1, \ldots, p_k$ sum to 1, the expected value is the weighted average of the values u_i with their probabilities p_i as weights.

As a special case, if the random variable is deterministic and takes only a single value u_1 (with probability $p_1 = 1$), then that value is the expected value, $\mathbb{E}[U] = u_1$.

In Figure 4.1, the expected payoff to player I after the chance move is $\frac{1}{3}0 + \frac{2}{3}3 = 2$, and to player II it is $\frac{1}{3}15 + \frac{2}{3}0 = 5$. In this game, the chance node could therefore be replaced by a leaf with payoff 2 for player I and payoff 5 for player II. These expected payoffs reflect the players' preferences for random outcomes; Chapter 5 discusses in detail the assumptions that underly the use of such payoffs.

In a two-player game, we normally name the players I and II and refer to player I as "he" and to player II as "she". Moves of player I are typically in upper case and of player II in lower case.

Figure 4.2 demonstrates the tree property that every node is reached by a unique path from the root. This fails to hold in the left picture (a) for the middle leaf. Such a structure may make sense in a game. For example, the figure could represent two people in a pub where first player I chooses to pay for the drinks (move P) or to accept (move A) that the first round is paid by player II. In the second round, player II may then decide to either accept not to pay (move a) or to pay (p). Then the players may only care how often, but not when, they have paid a round, with the middle payoff pair $(1, 1)$ for two possible "histories" of moves. The tree property prohibits states of play with more than one history, because the history is represented by the unique path from the root. Even if certain differences in the game history do not matter, these states of play are distinguished, mostly as a matter of mathematical convenience. The correct representation of the above game is shown in the right picture, Figure 4.2(b).

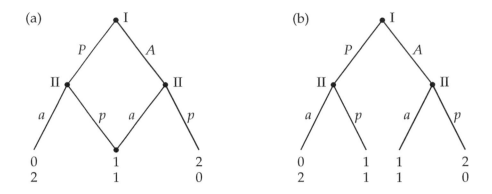

Figure 4.2 The left picture (a) is not a tree because the leaf where both players receive payoff 1 is reachable by two different paths. The correct game tree representation is shown on the right in (b).

Game trees are also called *extensive games* with *perfect information*. Perfect information means that a player always knows the state of play and therefore the complete history of play up to then. Game trees can be enhanced with an additional structure that represents "imperfect information", which is the topic of Chapter 10. Imperfect information is required to represent simultaneous moves. In the game trees considered in the present chapter, players move *sequentially* and observe each other's moves.

Note how game trees differ from the combinatorial games in Chapter 1:

- A combinatorial game can be described very compactly, in particular when it is given as a sum of games. For general game trees, such game sums are usually not considered.

- The "rules" in a game tree are much more flexible: more than two players and chance moves are allowed, players do not have to alternate, and payoffs do not just stand for "win" or "lose".

- The flexibility of game trees comes at a price, though: The game description is much longer than a combinatorial game. For example, a simple instance of Nim may require a huge game tree. Regularities like the mex rule do not apply to general game trees.

4.3   Backward Induction

Which moves should the players choose in a game tree? "Optimal" play should maximize a player's payoff. This can be decided irrespective of other players' actions when the player is the last player to move. In the game in Figure 4.3(a), player II maximizes her payoff by move r. Going backward in time, player I makes his move T or B at the root of the game tree, where he will receive either 1 or 2, assuming the described future behavior of player II. Consequently, he will choose B, and play ends with payoff 2 to each player. These chosen moves are shown in Figure 4.3(b) with arrows on the edges; similar to the boxes that we put around best-response payoffs in a strategic-form game, this is additional information to help the analysis of the game and not part of the game specification.

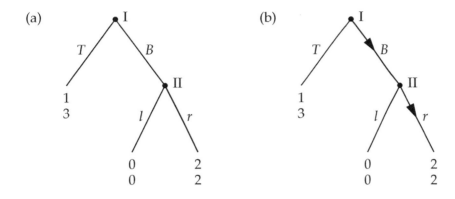

Figure 4.3 A two-player game tree (a) with backward induction moves (b).

This process is called *backward induction*: Starting with the decision nodes closest to the leaves, a player's move is chosen that maximizes that player's payoff at the node. In general, a move is chosen in this way for each decision node provided all subsequent moves have already been decided. Eventually, this will determine a move for every decision node, and hence for the entire game.

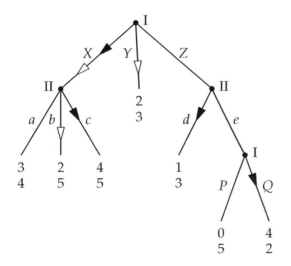

Figure 4.4 The game tree in Figure 4.1 with the chance move replaced by its expected payoffs, and backward induction moves that depend on player II's choice between b (white arrow, then two possible choices X or Y by player I) and c (black arrow, unique choice X by player I).

The move selected by backward induction is not necessarily unique, if there is more than one move that gives maximal payoff to the player. This may influence the choice of moves earlier in the tree. In the game in Figure 4.4, backward induction chooses either move b or move c for player II, both of which give her payoff 5 (which is an expected payoff for move b in the original game in Figure 4.1) that exceeds her payoff 4 for move a. At the rightmost node, player I chooses Q. This determines the preceding move d by player II which gives her the higher payoff 3 as opposed to 2 (via move Q). In turn, this means that player I, when choosing between X, Y, or Z at the root of the game tree, will get payoff 2 for Y and payoff 1 for Z. The payoff when he chooses X depends on the choice of player II: if that is b, then player I gets 2, and he can choose either X or Y, both of which give him maximal payoff 2. These choices are shown with white arrows in Figure 4.4. If player II chooses c, however, then the payoff to player I is 4 when choosing X, so this is the unique optimal move.

To summarize, the possible combinations of moves that can arise in Figure 4.1 by backward induction are, by listing the moves for each player: (XQ, bd), (XQ, cd), and (YQ, bd). Note that Q and d are always chosen, but that Y can only be chosen in combination with the move b by player II.

We do *not* try to argue that player II "should not" choose b because of the "risk" that player I might choose Y. This would be a "refinement" of backward induction that we do not pursue.

The moves determined by backward induction are therefore, in general, not unique, and possibly interdependent.

Backward induction gives a *unique* recommendation to each player if there is always only one move that gives maximal payoff. This applies to *generic* games. A generic game is a game where the payoff parameters are real numbers with no special dependency between them (like two payoffs being equal). In particular, it should be allowed to replace them with values nearby. For example, the payoffs may be given in some practical scenario which has some "noise" that effects the precise value of each payoff. Then two such payoffs are equal with probability zero, and so this case can be disregarded. In generic games, the optimal move is always unique, so that backward induction gives a unique result.

4.4   Strategies in Game Trees

In a game in strategic form, each player's strategies are assumed as given. For a game tree, a strategy is a derived concept. The best way to think of a strategy in a game tree is as a *plan of action* for every decision node.

Definition 4.3. In a game tree, a *strategy* of a player specifies a move for every decision node of that player.   □

Hence, if a player has k decision nodes, a strategy of that player is a k-tuple of moves. For simplicity of notation, we often write down a strategy as a list of moves, one for each decision node of the player (and not as a comma-separated tuple surrounded by parentheses). In the game tree in Figure 4.1, the strategies of player I are XP, XQ, YP, YQ, ZP, ZQ. The strategies of player II are ad, ae, bd, be, cd, ce. When specifying a strategy in that way, this must be done with respect to a fixed order of the decision nodes in the tree, in order to identify each move uniquely. This matters when a move name appears more than once, for example in Figure 4.2(b). In that tree, the strategies of player II are aa, ap, pa, pp, with the understanding that the first move refers to the left decision node and the second move to the right decision node of player II.

Backward induction defines a move for every decision node of the game tree, and therefore for every decision node of each player, which in turn gives a strategy for each player. The result of backward induction is therefore a strategy profile.

With a strategy for each player, we can define the strategic form of the game tree, which is determined as follows. Consider a strategy profile, and assume that players move according to their strategies. If there are no chance moves, their play leads to a unique leaf. If there are chance moves, a strategy profile may lead to a probability distribution on the leaves of the game tree, with resulting *expected* payoffs. For chance moves that occur earlier in the tree, the computation of expected payoffs is more complicated than in Figure 4.1; see Figure 4.8 for an example. In general, any strategy profile defines an expected payoff to each player; this includes the deterministic case where payoffs and expected payoffs conincide, as mentioned after (4.1).

Definition 4.4. The *strategic form* of a game tree is defined by the set of strategies for each player according to Definition 4.3, and the expected payoff to each player resulting from each strategy profile. □

We first consider the simple game tree in Figure 4.3, which is shown in Figure 4.5 together with its strategic form (already encountered in Figure 3.6), which is a 2×2 game. In this game, each player has only one decision node, so a strategy is the same as a move. (In general, a strategy specifies several moves, so do not use the terms "strategy" and "move" interchangeably!) Note also that we write the payoff to player I at the top at each leaf of the game tree, but at the bottom left of a cell in the payoff table.

The game tree in Figure 4.3 has only three leaves (and no chance moves) but the strategic form has four cells. Therefore, the payoffs for one of the leaves appear twice, namely for the strategy pairs (T, l) and (T, r). The reason is that the leftmost leaf is reached immediately after move T of player I, and then the move l or r of player II does not matter.

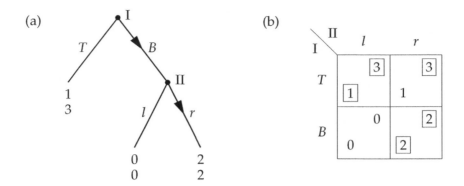

Figure 4.5 Game tree (a) of the Threat game with backward induction moves, and its strategic form (b) with best responses. Note: In the game tree, the payoffs to player I are shown at the *top*, but in the strategic form they appear at the *bottom left* of each cell.

The game tree and the strategic form specify the same game, but are played differently. In the game tree, play proceeds sequentially. In the strategic form, the players choose their strategies simultaneously, and then they play according to these "plans of action".

In the game in Figure 4.5, backward induction specifies the unique strategy profile (B, r), which is also an equilibrium (which holds generally, see Theorem 4.6 below). The game has a second equilibrium (T, l), which can also be seen from the game tree: If player II chooses l, then player I's best response is T. Player II can *threaten* this suboptimal move because she never has to make it when player I chooses T, as he is incentivized to do. The game is also known as the *Threat* game for that reason. Player II gets a higher payoff in this equilibrium (T, r). We will distinguish equilibria that are obtained by backward induction and those that are not, as discussed in Section 4.6.

Figure 4.6 shows the strategic form of the game tree in Figure 4.1. In that game, players I and II both have two decision nodes with three and two moves, respectively, and therefore 3×2 strategies each. The strategic form has 36 cells, which have many repetitions because the game tree has only eight (or, after contracting the two chance outcomes as in Figure 4.4, seven) leaves.

4.5 Reduced Strategies

In Figure 4.6, the two rows XP and XQ have identical payoffs for both players, and similarly the two rows YP and YQ. The reason is that the strategies XP and XQ specify that player I chooses X at his first decision node and either P or Q at his second decision node. However, after choosing X, the second decision node cannot

I \ II	ad		ae		bd		be		cd		ce	
XP		4		4		5		5		5		5
	3		3		2		2		4		4	
XQ		4		4		5		5		5		5
	3		3		2		2		4		4	
YP		3		3		3		3		3		3
	2		2		2		2		2		2	
YQ		3		3		3		3		3		3
	2		2		2		2		2		2	
ZP		3		5		3		5		3		5
	1		0		1		0		1		0	
ZQ		3		2		3		2		3		2
	1		4		1		4		1		4	

Figure 4.6 Strategic form of the extensive game in Figure 4.1. The less redundant reduced strategic form is shown in Figure 4.7 below.

be reached. The second decision node can only be reached if player I chooses Z as his first move, and then the two strategies ZP and ZQ may lead to different leaves (if player II chooses e, as in her three strategies ae, be, or ce). Because player I himself chooses between X, Y, or Z, it makes sense to replace both strategies XP and XQ by a less specific "plan of action" $X*$ that precribes only move X. Similarly, the two strategies YP and YQ are collapsed into $Y*$. We *always* use the star " $*$ " as a placeholder. It stands for any unspecified move at the respective unreachable decision node, to identify the node where this move is left unspecified.

Leaving moves at unreachable nodes unspecified in this manner defines a *reduced strategy* according to the following definition. Because the resulting expected payoffs remain uniquely defined, tabulating these reduced strategies and the payoff for the resulting reduced strategy profiles gives the *reduced strategic form* of the game. The reduced strategic form of the game tree of Figure 4.1 is shown in Figure 4.7, with indicated best-response payoffs.

Definition 4.5. In a game tree, a *reduced strategy* of a player specifies a move for every decision node of that player, except for those decision nodes that are unreachable due to an earlier own move, where the move is replaced by " $*$ ". A *reduced strategy profile* is a tuple of reduced strategies, one for each player of the game. The *reduced strategic form* of a game tree lists all reduced strategies for each

I \ II	ad	ae	bd	be	cd	ce
X*	4 / [3]	4 / 3	[5] / [2]	[5] / 2	[5] / [4]	[5] / [4]
Y*	[3] / 2	[3] / 2	[3] / [2]	[3] / 2	[3] / 2	[3] / 2
ZP	3 / 1	[5] / 0	3 / 1	[5] / 0	3 / 1	[5] / 0
ZQ	[3] / 1	2 / [4]	[3] / 1	2 / [4]	[3] / 1	2 / [4]

Figure 4.7 Reduced strategic form of the extensive game in Figure 4.1. The star *
stands for an arbitrary move at the second decision node of player I, which is not
reachable after move X or Y.

player, and tabulates the expected payoff to each player for each reduced strategy
profile.   □

The preceding definition generalizes Definitions 4.3 and 4.4. It is important
that the only moves that are left unspecified are at decision nodes which are
unreachable due to an earlier *own* move of the player. A reduced strategy must
not disregard a move because another player may not move there, because that
possibility cannot be excluded by looking only at the player's own moves. In the
game tree in Figure 4.1, for example, no reduction is possible for the strategies of
player II, because neither of her moves at one decision node precludes a move at
her other decision node. Therefore, the reduced strategies of player II in that game
are the *same* as her (unreduced, original) strategies.

In Definition 4.5, reduced strategies are defined without any reference to
payoffs, only by the structure of the game tree. One could derive the reduced
strategic form in Figure 4.7 from the strategic form in Figure 4.6 by identifying the
strategies that have identical payoffs for both players and replacing them by a single
reduced strategy. Some authors define the reduced strategic form in terms of this
elimination of duplicate strategies in the strategic form. (Sometimes, dominated
strategies are also eliminated when defining the reduced strategic form.) We define
the reduced strategic form without any reference to payoffs because strategies may
have identical payoffs by accident, due to special payoffs at the leaves of the game
(this does not occur in generic games). Moreover, we prefer that reduced strategies
refer to the structure of the game tree, not to some relationship of the payoffs,
which is a different aspect of the game.

The following is a recipe for filling, reasonably quickly, the payoff table for the reduced strategic form (or, if one wishes, the full strategic form) of a game tree like Figure 4.1:

- For each player, choose a fixed order of the decision nodes, where nodes that come earlier in the tree should come first. If the player has k decision nodes, a (reduced) strategy has k "slots" to define a move, one for each decision node.

- Generate all (reduced) strategies. A reduced strategy defines for each slot a move, or an unspecified move " $*$ " if the respective decision node is unreachable due to an earlier move of the same player. For two players, list the (reduced) strategies of player I as rows and of player II as columns.

- For each *leaf* of the game tree, consider the path from the root to the leaf. Assume first that there are no chance moves on that path. For each player, consider the moves of that player on that path. For example, the leftmost leaf in Figure 4.1 with payoffs $(3, 4)$ has move X for player I and move a for player II.

- For each player, identify *all strategies* that agree with the moves on the path to the leaf. For the leftmost leaf in Figure 4.1 and the reduced strategic form in Figure 4.7, these strategies are given by the reduced strategy $X*$ for player I (or the unreduced strategies XP and XQ in Figure 4.6), and the strategies ad and ae for player II. For all these strategy pairs, enter the payoff pair, here $(3, 4)$, in the respective cells. Repeat this for each leaf.

  As another example, the leaf with payoffs $(2, 3)$ that directly follows move Y is reached by move Y of player I (which defines row $Y*$ in the reduced strategic form in Figure 4.7), and *no moves* of player II, so *all strategies* of player II agree with these moves; that is, the entire row $Y*$ is filled with the payoff pair $(2, 3)$.

- If there are chance moves on the path from the root to the leaf, let p be the product of chance probabilities on that path. Multiply each payoff at the leaf with p, and *add* the result to the current payoffs in the respective cells that are determined as before (assuming the payoffs in each cell are initially zero). For an example see Figure 4.8.

This recipe for creating the strategic form of a game tree is much less laborious than considering each strategy pair and identifying to which leaf it leads (or leaves, in case there are chance moves).

However, it should be clear that the strategic form is typically substantially larger than the game tree. It is therefore often easier to directly analyze the game tree rather than the strategic form.

Strategies are combinations of moves, so for every additional decision node of a player, each move at that node can be combined with move combinations already considered. The number of move combinations therefore grows *exponentially* with

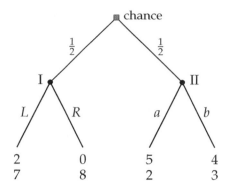
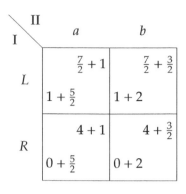

Figure 4.8 Creating the strategic form by iterating through tree leaves when there are chance moves, which lead to expected payoffs.

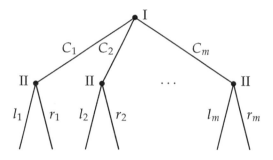

Figure 4.9 Game tree (without payoffs) with m moves $C_1, \ldots, C_m$ of player I at the root of the game tree. After each move C_i, player II has two moves l_i and r_i. Player II has 2^m strategies.

the number of decision nodes of the player, because it is the product of the numbers of moves at each node. In the game in Figure 4.9, for example, where m possible initial moves of player I are followed each time by two possible moves of player II, the number of strategies of player II is 2^m. Moreover, this is also the number of reduced strategies of player II because no reduction is possible. This shows that even the reduced strategic form can be exponentially large in the size of the game tree. If a player's move is preceded by an own earlier move, the reduced strategic form is smaller because then that move can be left unspecified in any reduced strategy that does not choose the preceding own move.

⇒ Exercise 4.4 is about reduced strategies.

In the reduced strategic form, an equilibrium is defined, as in Definition 3.1, as a profile of reduced strategies, each of which is a best response to the others. In this context, we will for brevity sometimes refer to reduced strategies simply as "strategies". This is justified because, when looking at the reduced strategic form, the concepts of dominance and equilibrium can be applied directly to the reduced strategic form, for example to the table that defines a two-player game. Then "strategy" means simply a row or column of that table. The term "reduced strategy" is only relevant when referring to the game tree.

The equilibria in Figure 4.7 are identified as those pairs of (reduced) strategies that are best responses to each other, with both payoffs surrounded by a box in the respective cell of the table. These equilibria in reduced strategies are $(X*, bd)$, $(X*, cd)$, $(X*, ce)$, and $(Y*, bd)$.

4.6 Subgame-Perfect Equilibrium (SPE)

In this section, we consider the relationship between backward induction and equilibrium. The concept of equilibrium is based on the strategic form because it applies to a strategy profile. In a game tree, the strategies in that profile are assumed to be chosen by the players *before* play starts, and the concept of best response assumes that the other players play the moves in the strategy profile.

With the game tree as the given specification of the game, it is often desirable to keep its sequential interpretation. The strategies chosen by the players should therefore also express some "sequential rationality" as expressed by backward induction. That is, the moves in a strategy profile should be optimal for any part of the game, including subtrees that cannot be reached due to earlier moves, possibly by other players, like the little tree that starts with the decision node of player II in the Threat game in Figure 4.3(a).

In a game tree, a *subgame* is any subtree of the game, given by a node of the tree as the root of the subtree and all its descendants. (Note: This definition of a subgame applies only to games with perfect information, which are the game trees considered so far. In extensive games with imperfect information, which we consider in Chapter 10, a subtree is a subgame only if all players *know* that they are in that subtree; see Section 10.10.) A strategy profile that defines an equilibrium for every subgame is called a *subgame-perfect equilibrium* or SPE.

Theorem 4.6. *Backward induction defines an SPE.*

Proof. Recall the process of backward induction: Starting with the nodes closest to the leaves, consider a decision node u, say, with the assumption that all moves at descendants of u have already been selected. Among the moves at u, select a move that *maximizes* the expected payoff to the player who moves at u. (Expected payoffs must be regarded if there are chance moves after u.) In that manner, a move is selected for every decision node, which determines an entire strategy profile.

We prove the theorem inductively (see Figure 4.10): Consider a non-terminal node u of the game tree, which may be a decision node (as in the backward induction process), or a chance node. Suppose that the moves at u are $c_1, c_2, \ldots, c_k$, which lead to subtrees $T_1, T_2, \ldots, T_k$ of the game tree. If u is a chance node, then instead of the moves we have probabilities $p_1, p_2, \ldots, p_k$. Assume, as inductive hypothesis, that the moves selected so far define an SPE in each of these trees. (As

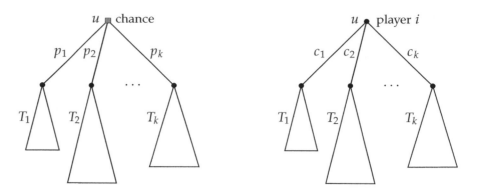

Figure 4.10 Induction step for proving Theorem 4.6 when the next node u belongs to chance (left) or to player i (right).

the "base case" of the induction, this is certainly true for each subtree $T_1, T_2, \ldots, T_k$ that is just a leaf of the tree; if all subtrees are leaves, then u is a "last" decision node considered first in the backward induction process.) The induction step is completed if one shows that, by selecting the move at u, one obtains an SPE for the subgame with root u.

First, suppose that u is a chance node, so that the next node is chosen according to the fixed probabilities specified in the game tree. Then backward induction does not select a move for u, and the inductive step holds trivially: For every player, the payoff in the subgame that starts at u is the expectation of the payoffs for the subgames T_l weighted with the probabilities p_l for $1 \leq l \leq k$, and if a player could improve on that payoff, she would have to do so by changing her moves within at least one subtree T_l, which, by inductive hypothesis, she cannot.

Second, suppose that u is a decision node, let i be the player to move at u, and consider a player j *other* than player i. Again, for player j, the moves in the subgame that starts at u are completely specified. Irrespective of what move player i selects at u, player j cannot improve her payoff because that would mean she could improve her payoff already in some subtree T_l.

Third, consider player i who moves at u. Backward induction selects a move for u that is best for player i, given the remaining moves. Suppose player i could improve his payoff by choosing some move c_l and additionally change his moves in the subtree T_l. But the resulting improved payoff would *only* be the improved payoff in T_l, that is, the player could already get a better payoff in T_l itself. This contradicts the inductive assumption that the moves selected so far define an SPE for T_l. This completes the induction.   □

Theorem 4.6 has two important consequences. In backward induction, each move can be chosen deterministically, so that backward induction determines a profile of pure strategies. Theorem 4.6 therefore implies that game trees have equilibria, and it is not necessary to consider randomized strategies. Secondly, subgame-perfect equilibria exist. For game trees, we can use "SPE" synonymously with "strategy profile obtained by backward induction".

Corollary 4.7. *Every game tree with perfect information has a pure-strategy equilibrium.*

Corollary 4.8. *Every game tree with perfect information has an SPE.*

An SPE can only be described as a profile of *unreduced* (that is, fully specified) strategies. Recall that in the game in Figure 4.1, the SPE are (XQ, bd), (XQ, cd), and (YQ, bd). The reduced strategic form of the game in Figure 4.7 shows the equilibria $(X*, bd)$, $(X*, cd)$, and $(Y*, bd)$. However, we cannot call any of these an SPE because they leave the second move of player I unspecified, as indicated by the $*$ symbol. In this case, replacing $*$ by Q results in all cases in an SPE. The full strategies are necessary to determine if they define an equilibrium in every subgame. As seen in Figure 4.7, the game has, in addition, the equilibrium $(X*, ce)$. This is not subgame-perfect because it prescribes the move e, which is not part of an equilibrium in the subgame that starts with the decision node where player II chooses between d and e: Replacing $*$ by P would not be the best response by player I to e in that subgame, and when replacing $*$ by Q, player II's best response would not be e. This property can only be seen in the game tree, not even in the unreduced strategic form in Figure 4.6, because it concerns an unreached part of the game tree due to the earlier move X of player I.

> $\Rightarrow$ Exercise 4.3 is very instructive for understanding subgame-perfect equilibria. Exercise 4.5 studies a three-player game.

4.7 Commitment Games

The main difference between game trees and games in strategic form is that in a game tree the players move sequentially and are aware of any previous moves of the other players. In contrast, in a strategic-form game the players move simultaneously. The difference between these two descriptions becomes striking when *changing* a two-player game in strategic form to a *commitment game*.

The commitment game is a new game obtained from a two-player game in strategic form. One of the players (we always choose player I) moves *first* by choosing one of his strategies. Then, unlike in the strategic form, player II is *informed* about the move of player I and moves second. Player II chooses one of her original strategies, which can depend on the move of player I. The resulting payoffs to both players are as specified in the original game.

An example is shown in Figure 4.11. The game in strategic form in (a) is the Quality game in Figure 3.2(b). The commitment game, shown in (b) and (c), is a

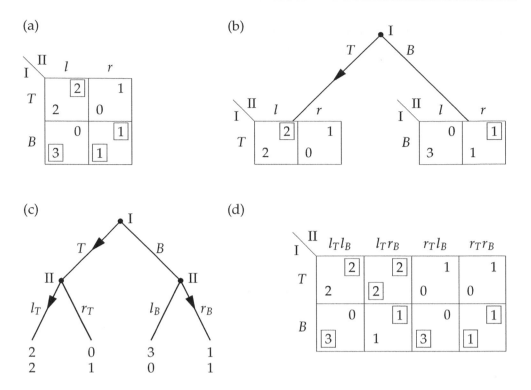

Figure 4.11   The simultaneous Quality game (a) from Figure 3.2(b), converted to its sequential *commitment* game (b) where player I moves first and player II can *react* to player I's move. The conventional representation of (b) is the extensive game (c). The arrows in (b) and (c) show the backward induction outcome. The strategic form of (c) with best-response payoffs is shown in (d).

game tree with perfect information, where player I moves first and player II moves second. The original strategies T and B of player I become his moves T and B. In Figure 4.11(b), we have shown the subsequent choices of player II as two 1×2 games which are the two rows of the strategic form in (b), which depend on the move T or B of player I, with the same payoffs as in (a). Player II can then separately choose between l and r. We assume that player II chooses her best move, shown by the best-response payoffs, which is l in response to T and r in response to B. We assume that player I anticipates that move of player II and therefore chooses T, as shown by the backward induction arrow.

The graphical description in Figure 4.11(b) just illustrates how the commitment game is obtained from the original game in strategic form. It represents a game tree, which is shown in (c), where each move of player I leads to a separate decision node for player II, whose moves are the original strategies of player II. In order to emphasize that these moves l and r are *separate* and depend on the decision node, we have in (c) given them subscripts, and call them l_T and r_T after move T, and l_B and r_B after move B. Figure 4.11(d) shows the resulting strategic form of the commitment game. It shows that the original strategic-form game and the commitment game are very different.

In the commitment game, we then look for subgame-perfect equilibria, that is, apply backward induction. Figure 4.11(b) and (c) show that the resulting unique SPE is $(T, l_T r_B)$, which means that (going backwards) player II when offered high quality T chooses l_T and when offered low quality B chooses r_B. In consequence, player I will then choose T because that gives him the higher payoff 2 rather than just 1 with B.

Compared to the equilibrium in the original Quality game, the commitment game has a higher payoff (in the SPE) to both players. This holds because the game has changed. It is important that player I has the power to *commit* to T and cannot change later back to B after seeing that player II chose l. Without that commitment power of player I, the players would not act as described and the game would not be accurately reflected by the game tree.

The commitment game in Figure 4.11 also demonstrates the difference between an equilibrium, which is a strategy profile, and the *equilibrium path*, which is described by the moves that are actually played in the game when players use their equilibrium strategies. (If the game has no chance moves, the equilibrium path is just a path in the game tree from the root to a leaf; if the game has chance moves, the path may branch.) In contrast, a strategy profile defines a move in every part of the game tree. Here, the SPE $(T, l_T r_B)$ specifies moves l_T and r_B for the two decision nodes of player II. On the equilibrium path, the moves are T of player I followed by move l_T of player II. However, it is *not sufficient* to simply call this equilibrium (T, l_T) or (T, l), because player II's move r_B must be specified to know that T is player I's best response.

We only consider SPE when analysing commitment games. For example, the equilibrium (B, r) of the original Quality game in Figure 4.11(a) can also be found in (d) where player II *always* chooses the move r that corresponds to her equilibrium strategy in (a), which defines the strategy $r_T r_B$ in the commitment game. This determines the equilibrium $(B, r_T r_B)$ in (d), because B is a best response to $r_T r_B$, and $r_T r_B$ is a best response to B. However, this equilibrium of the commitment game is not subgame-perfect, because it prescribes the suboptimal move r_T off the equilibrium path; this does not affect the equilibrium property because T is not chosen by player I. (This holds as a general argument: consider any equilibrium (x, y) of the original game, and in the commitment game let player I commit to x and let player II always respond with y; then this is an equilibrium of the commitment game, which, however, is in general not subgame-perfect.)

In order to compare a strategic-form game with its commitment game in an interesting way, we consider only subgame-perfect equilibria of the commitment game.

In practical applications, it is often of interest to "change the rules of the game" and see what happens (with a proper game-theoretic analysis). Changing the Quality game to its commitment game shows how both players benefit.

Commitment games are also called *leadership games* where the first player to move is called the *leader* and the second player the *follower*. The leader's ability to commit irrevocably to his strategy is important, and is also called *commitment power*. The ability to commit typically confers a *first-mover advantage*, which states that the leader is not worse off than in the original game where the players act simultaneously (as shown in Exercise 4.7, this holds definitely when best responses are always unique).

This statement must be interpreted carefully. For example, if either player has the power to commit, then it is not necessarily best to go first. For example, consider changing the game in Figure 4.11(a) so that player II can commit to her strategy, and player I moves second. Then player I will always respond by choosing B because this is his dominant strategy in (a). Backward induction would then amount to player II choosing r, and player I choosing $B_l B_r$, with the low payoff 1 to both. Then player II is not worse off than in the simultaneous-choice game, as asserted by the first-mover advantage, but does not gain anything either. In contrast, making player I the first mover as in Figure 4.11(b) is beneficial to both.

Commitment games are also known as *Stackelberg games*, after the economist Heinrich von Stackelberg who applied them to the Cournot duopoly of quantity competition such as the game that we considered in Section 3.6. We study this commitment game first for its finite version in Figure 3.8 and then its infinite version (3.6) for arbitrary nonnegative x and y. We will see that in this game, the leader is better off than in the simultaneous game, but the follower is worse off.

The commitment game for the game in Figure 3.8 is shown in Figure 4.12, just as Figure 4.11(b) is obtained from Figure 4.11(a). The best responses of player II are given by her (original) best responses to the first move $0, 3, 4, 6$ of player I, which are $6, 4, 4, 3$, respectively, with payoff pairs $(0, 36)$, $(15, 20)$, $(16, 16)$, and $(18, 9)$ to the two players. Among these, 18 is the best for player I. Hence, by backward induction, player I will commit to his rightmost move 6. The corresponding payoffs are 18 for player I and 9 for player II, compared to 16 and 16 in the simultaneous game in Figure 3.8. In this symmetric game, the ability of one player to credibly commit himself gives a clear first-mover advantage to the leader which, unlike in the Quality game, is here at the expense of the follower.

We now consider the commitment game for the game with payoffs $a(x, y)$ and $b(x, y)$ in (3.6) to players I and II when their strategies x and y are arbitrary reals

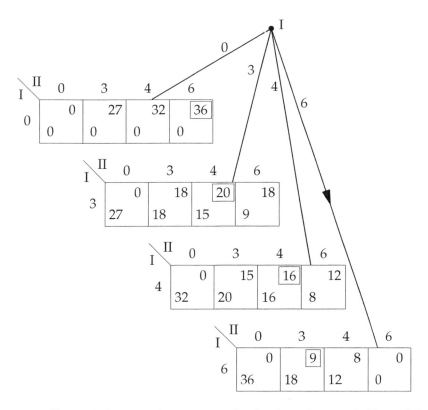

Figure 4.12 Commitment game for the duopoly game in Figure 3.8.

in $[0, 12]$ (they will not choose $x > 12$ or $y > 12$ which would give them negative payoffs). We have seen that the best response of player II to x is $y(x) = 6 - x/2$. In the commitment game, it is indeed useful to see player II's choice as such a *function* $y(x)$ of the commitment x of player I.

As always in a commitment game, we look for an SPE and assume that the follower chooses her best response, even off the equilibrium path. The backward induction outcome is therefore the solution x to the problem of maximizing $a(x, y(x))$, that is, of maximizing $x \cdot (12 - y(x) - x) = x \cdot (6 - x/2) = 6x - x^2/2$. The derivative of this function with respect to x is $6 - x$, which is zero for $x = 6$. That is, player I commits to 6 and player II responds with $y(x) = 6 - x/2$ for any commitment x (which defines the full strategy of player II in this SPE), in particular with $y(6) = 3$ on the equilibrium path. The resulting payoffs are 18 for player I and 9 for player II. The strategies 6 and 3 are part of the 4×4 game in Figure 3.8, along with the equilibrium strategy 4 of both players in that simultaneous game.

⇒ Exercise 4.7 studies a game with infinite strategy sets like (3.6), and makes interesting observations, in particular in (d), on the payoffs to the two players in the commitment game compared to the original simultaneous game.

4.8 Further Reading

Backward induction is also known as *Zermelo's algorithm*, attributed to an article of Zermelo (1913) on Chess. However, Zermelo was not concerned with backward induction but instead proved that if the white player can force a win, then he can do so in a bounded number of moves (Schwalbe and Walker, 2001). For that reason, "backward induction" is the better term.

Reduced strategies are occasionally defined in terms of the strategic form by identifying strategies that have identical payoffs for all players. We define this reduction independently of payoffs only via the structure of the game tree, which is clearer. We also give the ∗ symbol (which is sometimes omitted) for each unspecified move and each decision node (there may be several such decision nodes) in order to ensure that even a reduced strategy has k "slots" of moves when a player has k decision nodes.

The economist von Stackelberg (1934) analyzed various ways how firms can act in a duopoly situation, one of which is the leadership game where one firm commits to a certain output and the other firm reacts with a best response. We have described it with concepts of modern game theory (see also Gibbons, 1992), which were not around in 1934.

The commitment game of the game (3.6) treated at the end of Section 4.7 is a symmetric game where players choose their strategies from an interval, which has a unique symmetric equilibrium, and where the players' best response functions are unique and monotonic in the other player's choice. Because the game is symmetric, it suffices to analyze the game when player I is the leader in the corresponding commmitment game, which has an SPE payoff to the leader and to the follower. The leader's payoff is at least as high as in the simultaneous game (see Exercise 4.6). Curiously, in such games the payoff to the follower is either worse than in the simultanenous game, or even higher than the payoff to the leader (as in Exercise 4.7). That is, the seemingly natural case that both players profit from sequential play compared to simultaneous play, but the leader more so than the follower, does not occur. The proof is short and uses only some monotonicity conditions (von Stengel, 2010, theorem 1).

4.9 Exercises for Chapter 4

Exercise 4.1. Consider the following game tree. As always, the top payoffs at a leaf are for player I and bottom payoffs for player II.

(a) What is the number of strategies of player I and of player II?

(b) How many reduced strategies do they have?

(c) Give the reduced strategic form of the game.

(d) What are the equilibria of the game in reduced strategies?

(e) What are the subgame-perfect equilibria of the game?

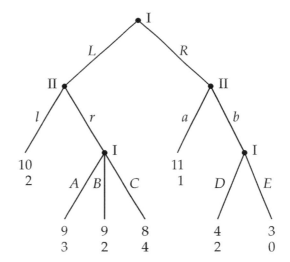

Exercise 4.2. Consider the following game tree.

(a) What is the number of strategies of player I and of player II? How many reduced strategies does each of the players have?

(b) Give the reduced strategic form of the game.

(c) What are the equilibria of the game in reduced strategies? What are the subgame-perfect equilibria of the game?

(d) Identify every pair of reduced strategies where one strategy weakly or strictly dominates the other, and indicate if the dominance is weak or strict.

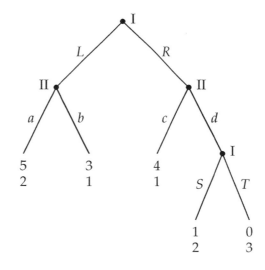

Exercise 4.3. Consider the following game trees.

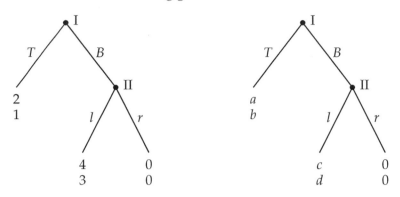

(a) Find all equilibria for the game tree on the left. Which of these are subgame-perfect?

(b) In the game tree on the right, the payoffs a, b, c, d are positive real numbers. For each of the following statements (i), (ii), (iii), decide if it is true or false, justifying your answer with an argument or counterexample; you may refer to any standard results. For any $a, b, c, d > 0$,

   (i)   the game always has a subgame-perfect equilibrium (SPE);

   (ii) the payoff to player II in any SPE is always at least as high as her payoff in any equilibrium;

   (iii) the payoff to player I in any SPE is always at least as high as his payoff in any equilibrium.

Exercise 4.4. Consider the following two game trees (a) and (b). Payoffs have been omitted because they are not relevant for the question. In each case, how many strategies does each player have? How many reduced strategies?

(a)   (b)

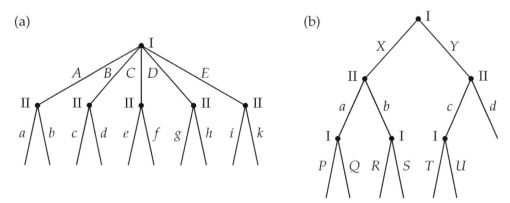

Exercise 4.5. Consider the following three-player game tree. At a leaf, the topmost payoff is to player I, the middle payoff is to player II, and the bottom payoff is to player III.

(a) How many strategy *profiles* does this game have?

(b) Identify all pairs of strategies where one strategy strictly, or weakly, dominates the other.

(c) Find all equilibria (in pure strategies). Which of these are subgame-perfect?

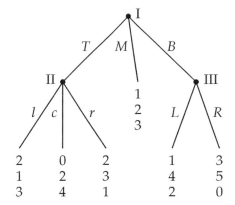

Exercise 4.6. Consider a game G in strategic form. Recall that the *commitment game* derived from G is defined by letting player I choose one of his strategies x, which is then *announced* to player II, who can then in each case choose one of her strategies in G as a response to x. The resulting payoffs are as in the original game G.

(a) If G is an $m \times n$ game, how many strategies do player I and player II have, respectively, in the commitment game?

For each of the following statements, determine whether they are true or false; justify your answer by a short argument or counterexample.

(b) In an SPE of the commitment game, player I never commits to a strategy that is strictly dominated in G.

(c) In an SPE of the commitment game, player II never chooses a move that is a strictly dominated strategy in G.

Exercise 4.7. Let G be the following game: Player I chooses a non-negative real number x, and simultaneously player II chooses a non-negative real number y. The resulting (symmetric) payoffs are

$$x \cdot (4 + y - x) \quad \text{for player I,}$$
$$y \cdot (4 + x - y) \quad \text{for player II.}$$

(a) Given x, determine player II's best response $y(x)$ (which is a function of x), and player I's best response $x(y)$ to y. Find an equilibrium, and give the payoffs to the two players.

(b) Find an SPE of the commitment game (where player I moves first), and give the payoffs to the two players.

(c) Are the equilibria in (a) and (b) unique?

(d) Let G be a game where the best response $y(x)$ of player II to any strategy x of player I is always unique. Show that in any SPE of the commitment game (where player I moves first), the payoff to player I is at least as large as his payoff in any equilibrium of the original game G.

5

Expected Utility

In a non-cooperative game, every outcome is associated with a payoff to each player that the player wants to maximize. The payoff represents the player's preference for the outcome. The games considered so far often have a pure-strategy equilibrium, where the preference applies to deterministic outcomes and is usually straightforward. From Chapter 6 onwards we will consider mixed strategies, which are actively randomized actions. The game is then analyzed under the assumption that players faced with random outcomes maximize their *expected payoffs*. The present chapter is about justifications for this assumption.

A player's payoff is often called *utility*, with a *utility function* $u : X \to \mathbb{R}$ where X is the set of outcomes of a game. Suppose the outcomes $x_1, \ldots, x_n$ have probabilities $p_1, \ldots, p_n$, respectively, which is called a *lottery* over the set of outcomes. The expected utility for this lottery P is $\mathbb{E}(u, P) = p_1 u(x_1) + \cdots + p_n u(x_n)$. An expected-utility function represents a preference of the player for lotteries in the sense that the player strictly prefers one lottery over another if it has a higher expected utility, and is indifferent between two lotteries if their expected utility is the same.

One of the "rationality" assumptions of game theory is that players are expected-utility maximizers. This seems like a rather strong assumption in the sense that the player is "risk neutral" with respect to the received utility or payoff. However, as we explain in this chapter, this is a misconception. The payoff does not mean money, but is an abstract and much more general representation of a player's preference for lotteries, as long as this preference fulfills certain *consistency* conditions.

The consistency conditions about the player's preference between lotteries are known as the axioms (basic assumptions) of von Neumann and Morgenstern (1947). The two main axioms are

- the preference is total (any two lotteries are comparable) and transitive;

- any preference or indifference between two lotteries is preserved when one of the lotteries is substituted for the other in an otherwise identical larger lottery (*independence* axiom).

The theorem of von Neumann and Morgenstern (1947) states that these axioms (plus a more technical axiom) allow the construction of an expected-utility function.

We derive this theorem in this chapter. The main idea is relatively straightforward and explained in the first summary Section 5.2, under the assumption (which we relax later) that there is a least preferred outcome x_0 and a most preferred outcome x_1. With these "reference outcomes", every other outcome z is replaced by an equally preferred lottery between x_0 and x_1, where x_1 is chosen with probability α. The utility for z can then be defined as $u(z) = \alpha$. With this representation "utility = probability", the expected-utility theorem holds because an expected probability is again a probability (see Figure 5.1).

The remaining sections of this "foundational" chapter give in essence a brief introduction to *decision theory*, which studies how an individual or a group can make a decision between several given alternatives (see Section 5.3). We focus on a single player who faces decisions under *risk*, that is, the alternatives are lotteries with known probabilities for outcomes (Section 5.4). As part of a decision analysis, a utility function is often used as a tool for simplifying complex decisions, by eliciting the decision maker's preferences from artificial simpler scenarios.

A lottery with only one possible outcome means to get that outcome with certainty. Any expected-utility function u therefore also represents preferences under certainty, via the values $u(x)$ for outcomes x. If the decision is only between alternatives that have certain outcomes, the utility function is called *ordinal* and can be changed without changing the preference by applying any strictly increasing transformation (Section 5.5).

For decisions between risky alternatives, the expected-utility function is *cardinal*, which means that the preference is only preserved when applying positive-affine transformations like for a temperature scale (Section 5.6). Then the utility function u is uniquely determined by fixing the "utility scale" by choosing an *origin* x_0 so that $u(x_0) = 0$, and a *unit*, that is, a more preferred outcome x_1 so that $u(x_1) = 1$. All other values of u are then already determined by the player's preference for *simple lotteries*, which have only two outcomes x and y, say, where x occurs with probability $1 - \alpha$ and y with probability α. We use the notation $x \wedge_\alpha y$ (see (5.16), unique to this book) for this simple lottery where $\wedge_\alpha$ is meant to represent a little "probability tree" where the right branch has probability α.

Simple lotteries specify fully how the values $u(z)$ of the utility function result from probabilities, namely as $u(z) = \alpha$ if the player is indifferent between z and the simple lottery $x \wedge_\alpha y$, where $u(x) = 0$ and $u(y) = 1$. This also means that the player's preference for lotteries can be *constructed* from these simple lotteries as soon as one accepts the consistency axioms. In this way, constructing a utility function is a useful decision aid, because lotteries with many outcomes and associated probabilities are rather complex prospects. In practice, finding a probability α so that the player is indifferent between z and $x \wedge_\alpha y$ is conceptually easier if x and y are not extreme like the worst and best possible outcomes that can occur. For that reason, we do not even assume (as in Section 5.2) that such worst and best possible outcomes exist for our main Theorem 5.4. The proof of the theorem then requires adapting the origin and unit of the utility function to the lottery under consideration, which leads to somewhat lengthy but straightforward manipulations with positive-affine transformations.

In Section 5.7 we explain the consistency axioms of *independence* and *continuity* that imply the existence of an expected-utility function, which is proved in full in Section 5.8. In Section 5.9, outcomes are themselves real numbers (such as amounts of money) where a lottery has an expected value. *Risk aversion* means that getting this expected value for certain is always preferred to the (risky) lottery, which is equivalent to a *concave* utility function. In the final Section 5.10 we discuss the "rationality" of the utility axioms (as a short "philosophical" discussion that we generally avoid in this book), and give further references.

5.1   Prerequisites and Learning Outcomes

We need the concept of a binary relation, like that of an order (see Definition 1.2). Expected values have been defined in (4.1).

After studying this chapter, you should be able to

- understand a player's preference as a binary relation on the set of lotteries over outcomes, and its expected-utility representation;

- distinguish ordinal functions for decisions under certainty and cardinal utility functions for decisions under risk;

- understand the role of a utility function for creating a total and transitive preference, for example an additive function for outcomes with multiple attributes (Exercise 5.1);

- explain the lexicographic order and why it is problematic both in practice, see the example (5.10), and theory, see Figure 5.3;

- apply positive-affine transformations to payoffs and know why they represent the same preference for lotteries;

- use simple lotteries and their notation $x \wedge_\alpha y$;

- state the von Neumann–Morgenstern consistency axioms and their use;

- construct an expected-utility function via probabilities;

- apply concepts of risk aversion and risk neutrality to outcomes that are themselves numbers.

5.2   Summary

This section is a kind of "executive summary" (with more details later) of the main theorem of this chapter: If the player's preference for lotteries fulfills certain consistency axioms, then this preference can be represented by an expected-utility function.

We study *decisions under risk* of a single decision maker (here still called *player* for brevity). Risk means the player faces *lotteries* with outcomes $x_1, \ldots, x_n$ from a set X, which depend on the lottery and occur with *known* probabilities $p_1, \ldots, p_n$. For simplicity, we consider only lotteries with finitely many outcomes, but the set X of possible outcomes may be infinite, for example a real interval.

The set of lotteries over X is denoted by $\Delta(X)$. We start with a binary *preference relation* $\precsim$ on $\Delta(X)$, where $P \precsim Q$ means that the player weakly prefers lottery Q to lottery P. If $P \precsim Q$ and $Q \precsim P$ hold then this is written as $P \sim Q$ and means that the player is *indifferent* between P and Q. If $P \precsim Q$ but not $Q \precsim P$ then this defines the *strict preference* $P \prec Q$.

The goal is to study conditions on the player's preference relation $\precsim$ on $\Delta(X)$ so that it can be represented by an *expected-utility* function. This is a function $u : X \to \mathbb{R}$ so that for $P, Q \in \Delta(X)$,

$$P \precsim Q \qquad \Leftrightarrow \qquad \mathbb{E}(u, P) \leq \mathbb{E}(u, Q) \tag{5.1}$$

where $\mathbb{E}(u, P)$ is the expected value of u under P. In a decision situation (and later in a game) the player therefore wants to maximize his expected utility.

If a utility function exists, then it imposes a number of conditions on the preference $\precsim$ of the player. Suitable conditions called *consistency axioms* will be sufficient for the existence of an expected-utility function.

The first axiom states that $\precsim$ is *total* and *transitive*. The only difference to a total order (see Definition 1.2) is that $P \sim Q$ is allowed to hold if $P \neq Q$.

The lotteries in $\Delta(X)$ include deterministic lotteries that choose a single outcome x with probability 1, which are denoted by $\mathbf{1}_x$. The preference relation $\precsim$ therefore applies to deterministic outcomes x and y where $x \precsim y$ stands for $\mathbf{1}_x \precsim \mathbf{1}_y$, which holds if and only if $u(x) \leq u(y)$ for an expected-utility function u. This preference for deterministic outcomes is preserved if $u(x)$ is replaced by $\phi(u(x))$ for a strictly increasing function ϕ, for example $\phi(u) = e^u$. That is, the utility function for deterministic outcomes is *ordinal* in the sense that more preferred outcomes have higher values of the utility function, and this is all that matters.

Whereas a utility function for decisions under certainty is ordinal, an expected-utility function u for decisions under risk is *cardinal*. This means that the only admitted transformations of u are *positive-affine*, given by $au + b$ for constants a and b with $a > 0$, for example as between the temperature scales Celsius and Fahrenheit. This uniqueness holds because an expected-utility function evaluates lotteries via their expected value. It already applies when considering only *simple* lotteries that have only two outcomes (Proposition 5.2). The "cardinal" property of an expected-utility function refers to its uniqueness up to positive-affine transformations, not to an intrinsic "strength" of preferences for outcomes.

By applying a positive-affine transformation, for two arbitrary outcomes x_0 and x_1 so that $x_0 \prec x_1$, one can choose the expected-utility function u so that $u(x_0) = 0$ and $u(x_1) = 1$, which fixes the *origin* and *unit* of the utility scale. Consider now any outcome z so that $x_0 \precsim z$ and $z \precsim x_1$ (which we abbreviate as $x_0 \precsim z \precsim x_1$ using that $\precsim$ is transitive). Then $0 \leq u(z) \leq 1$, and the precise value of $u(z)$ is given by the probability α so that the player is indifferent between z and the simple lottery that chooses x_0 with probability $1 - \alpha$ and x_1 with probability α (which we write as $x_0 \wedge_\alpha x_1$, see (5.16), where $\wedge_\alpha$ is meant to suggest a small probability tree). This holds because the expected utility $(1 - \alpha)u(x_0) + \alpha u(x_1)$ for this simple lottery is α.

With this scaling $u(x_0) = 0$ and $u(x_1) = 1$, each value $u(z)$ of the expected-utility function is therefore just the *probability* α of choosing x_1 (and x_0 otherwise) that makes the player indifferent between this lottery $x_0 \wedge_\alpha x_1$ and getting z with certainty. This insight has profound consequences. Assume for simplicity that x_0 is least preferred and x_1 is most preferred among all outcomes, so that $x_0 \precsim z \precsim x_1$ for all outcomes z. Define $u(z)$ in $[0, 1]$ as the probability α so that $z \sim x_0 \wedge_\alpha x_1$. Then α is unique, and u is therefore unique up to positive-affine transformations.

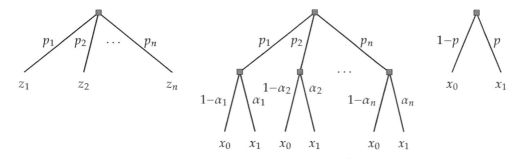

Figure 5.1 Equivalent probability trees for the lottery P with probabilities $p_1, \ldots, p_n$ for outcomes $z_1, \ldots, z_n$ and probability $p = \sum_{i=1}^{n} p_i \alpha_i$ for outcome x_1.

Most importantly, the preference for an *arbitrary* lottery P is represented by its *expectation* $\mathbb{E}(u, P)$. Namely, suppose that P chooses the outcomes $z_1, \ldots, z_n$ with probabilities $p_1, \ldots, p_n$, respectively. Let $z_i \sim x_0 \wedge_{\alpha_i} x_1$ where $\alpha_i = u(z_i)$

for $1 \leq i \leq n$. Figure 5.1 shows on the left the probability tree for P. In the middle tree, each outcome z_i is replaced by its equally preferred simple lottery $x_0 \wedge_{\alpha_i} x_1$ between x_0 and x_1. In that tree, the only outcomes are x_0 and x_1. Along the ith edge from the root, outcome x_1 is reached with probability $p_i \alpha_i$ because independent probabilities are multiplied. Summed over $i = 1, \ldots, n$, this gives the overall probability

$$p = \sum_{i=1}^{n} p_i \alpha_i \tag{5.2}$$

for x_1 and $1 - p$ for x_0. This is shown in the right tree in Figure 5.1 that aggregates the probabilities for x_0 and x_1. Hence, the player is indifferent between P and $x_0 \wedge_p x_1$ where $p = \mathbb{E}(u, P)$ by (5.2) because $\alpha_i = u(z_i)$ for $1 \leq i \leq n$. The comparison between different lotteries P and Q therefore becomes a comparison between their corresponding probabilities for x_1: If $p = \mathbb{E}(u, P)$ as described and in a similar way $q = \mathbb{E}(u, Q)$, then $P \precsim Q$ if and only if $p \leq q$, which is exactly the property (5.1) of an expected-utility function.

In summary, assume "reference outcomes" x_0 and x_1 with $u(x_0) = 0$ and $u(x_1) = 1$ so that $x_0 \precsim z_i \precsim x_1$ for the outcomes $z_1, \ldots, z_n$ of the considered lottery P. Because the values $u(z_i)$ of the utility function are probabilities α_i for simple lotteries $x_0 \wedge_{\alpha_i} x_1$ between these reference outcomes, the expected utility of P as in Figure 5.1 is just the aggregated probability p for x_1 as in (5.2), because probabilities can be multiplied.

Figure 5.1 is therefore a derivation of an expected-utility function. Crucially, this argument can be made in terms of the preference relation $\precsim$ alone, provided the following hold:

- If $x_0 \prec z \prec x_1$ for outcomes x_0, x_1, z, then $z \sim x_0 \wedge_\alpha x_1$ for some $\alpha \in (0, 1)$.

- If $z \sim x_0 \wedge_\alpha x_1$, then substituting in any lottery the outcome z with $x_0 \wedge_\alpha x_1$ creates an equally preferred lottery.

- In combined lotteries, only the probability for an outcome matters, not the structure of the lottery.

The first of these conditions is called the *continuity* axiom. The second condition, together with the corresponding condition with $\sim$ replaced by $\prec$, is called the *independence* axiom. These axioms are assumed to hold for general lotteries and not just outcomes. The third condition is not considered a constraint on the preference relation because decisions under risk deal with "objective" known probabilities.

Hence, any preference relation $\precsim$ on lotteries that is total, transitive, and fulfills the continuity and independence axioms can be represented by an expected-utility function. This is the main Theorem 5.4. Its full proof does not assume that "extreme" outcomes x_0 and x_1 exist so that $x_0 \precsim z \precsim x_1$ for all outcomes z to define $u(z) = \alpha$ via $z \sim x_0 \wedge_\alpha x_1$. Instead, x_0 and x_1 can be chosen depending on the considered lottery to define the utility function $u(z)$ "locally", which is then extended to all outcomes z using related lotteries and positive-affine transformations.

The remainder of this chapter explains in more detail what decision theory is about. For the development of game theory it can be skipped, except for the background material on line segments in Figure 5.4, which is needed for the geometric concept of convexity in the following chapters. Section 5.9 defines concave utility functions, which are used in Chapter 11 on bargaining.

5.3   Decisions Under Risk

The following situation illustrates a decision problem under risk. Suppose you have booked a rental car for a week. Your liability for damage to the car is 1,000 Euros. At the rental desk you are offered additional insurance at the cost of 20 Euros per day that would reduce this excess from 1,000 Euros to zero, which would cost you an extra 140 Euros. Should you buy this insurance or not? The choice is between being 140 Euros out of pocket for sure, or risking with some probability p that you pay up to 1,000 Euros in case you damage the car.

There are a number of further complications to this decision:

- This is a one-off decision – you are not renting cars all the time. If you did, you could think for yourself whether during 50 days of car rental you are likely to create a damage of 1,000 Euros, because that is what you would pay for sure if you took out this insurance all the time.

- If there is damage, maybe it is just a scratch that costs less than 1,000 Euros. This means that the risk is defined by multiple outcomes with different probabilities.

- Buying the insurance will give you peace of mind so you do not have to worry about damaging the car, which is an extra benefit. (On the other hand, would you drive less safely?)

- Maybe the car rental itself costs only 200 Euros, which is is raised significantly by an extra 140 Euros. (The "independence axiom" says that this should not matter.)

- If you decline the insurance and then have to pay for damage, you will *regret* your decision. Here the decision theorist will tell you: The risky choice means by definition that you do not know the future. If you took the risk, accept your decision that you did so.

- Paying 1,000 Euros might put you in severe financial difficulties. Hopefully, it doesn't. However, this consideration may be a good guide in other situations, for example whether you should pay another 80 GBP for insuring your new 400-GBP refrigerator against "accidental breakdown" for three years after purchase. (Such insurances are popular in the UK but tend to be over-priced.)

- You do not know the probability p that your car gets a 1,000-Euro damage. Thinking of whether you will create such a damage during 50 days of driving a rental car may help estimating that probability. This also depends on how safely you can drive in the place that you are visiting.

Concerning the probability p, the *expected cost* if you do not buy insurance is $1,000 \cdot p$ Euros, which exceeds 140 Euros if $p > 0.14$. However, there is no rule that says "buy the insurance if $p > 0.14$" because you may well prefer to buy the insurance if $p = 0.1$, for example. (The actual p is probably much lower than suggested by the prices of the rental car companies – separate "car rental excess" insurers cover the excess for 60 Euros per *year* for multiple rentals, so (a) this is an indication that the risk is low, (b) this is the kind of insurance you might prefer to buy in advance.)

Assume that in this car rental problem, you have an estimate of the probability p that you will damage the car (perhaps with a rather high p to be on the safe side). Then the problem whether you should buy the extra insurance for 140 Euros is called a *decision problem under risk*, which is one of the problems of decision theory and which is the one we study.

Decision theory deals with situations that are distinguished in the following respects:

- decisions of an *individual* or of a *group*;
- decisions under *certainty*, under *risk*, or under *uncertainty*;
- outcomes with single or multiple *attributes*.

The car rental problem has only a single decision maker. The problem of group decisions is to aggregate the preferences of multiple individuals. For example, there may be three candidates x, y, z for a job that is decided by a committee of three people A, B, C who have the following preferences:

$$
\begin{array}{ll}
A : & x \prec y \prec z \\
B : & y \prec z \prec x \\
C : & z \prec x \prec y
\end{array}
\tag{5.3}
$$

which means that A prefers y over x and z over both x and y, and correspondingly for B and C. One possible decision rule could be a *pairwise comparison by majority vote*. In this case, A and B (the majority) prefer z over y, and A and C prefer y over x, and B and C prefer x over z. Hence, even though each committee member has a strict order of the candidates, this pairwise comparison yields a *cycle* and therefore no clear winner. This is known as the *Condorcet Paradox* and one of the problems of group decisions, as part of the area of *Social Choice*. In this chapter, we only consider an individual decision maker and his preferences, whom we still call *player* for brevity.

In decisions under *certainty*, the outcome of the decision is known for sure. Decisions under *risk* deal with situations where the outcomes may be random but have probabilities that (as accurately as necessary) are known to the player. Decisions under *uncertainty* concern situations where the actual probability of an outcome is unknown. One approach of decision theory (which we will not cover) is to find conditions on preferences that amount to a player's (possibly implicitly defined) *subjective probabilities* for such outcomes that can be used like ordinary probabilities. We consider the situation of *risk* where probabilities are known.

Some decision problems involve only a single *attribute* like money, for which the preference is typically easy to decide when the outcome is certain (the preference is to pay less rather than more). Many decisions involve multiple attributes where the decision is no longer easy, such as whether going to a good but expensive restaurant or a less good but cheaper restaurant, with the two attributes of quality and price. The attribute structure of outcomes will not be relevant for us, but is useful for demonstrating complications that can arise with seemingly natural decision rules, as in the example (5.10) below.

5.4 Preferences for Lotteries

We consider the following setup of a single player in a decision situation. A set X contains the possible *outcomes*, and the set $\Delta(X)$ of (finite) lotteries over X is formally defined as follows.

Definition 5.1. Let X be a nonempty set of *outcomes*. A *lottery over* X is given by finitely many elements $x_1, \ldots, x_n$ of X (also called the outcomes of P) with corresponding probabilities $p_1, \ldots, p_n$, which are nonnegative numbers that sum to 1. The set of lotteries over X is denoted by $\Delta(X)$. $\qquad\square$

If we compare two lotteries P and Q we can assume that they assign probabilities $p_1, \ldots, p_n$ and $q_1, \ldots, q_n$ to the *same* outcomes $x_1, \ldots, x_n$, if necessary by letting some probabilities be zero if an outcome has positive probability for only one of the lotteries. This is simplest if X is itself finite, in which case we can take $n = |X|$, but we may also consider cases where X is infinite.

We assume that the player has a *preference* for any two lotteries P and Q. If, when faced with the choice between P and Q, the player is willing to accept Q rather than P, we write this as

$$P \precsim Q$$

and also say the player *(weakly) prefers* Q to P. We then define the *indifference* relation $\sim$ by

$$P \sim Q \quad \Leftrightarrow \quad P \precsim Q \quad \text{and} \quad Q \precsim P \tag{5.4}$$

where $P \sim Q$ means the player is *indifferent* between P and Q. The *strict preference* relation $\prec$ is defined by

$$P \prec Q \qquad \Leftrightarrow \qquad P \precsim Q \quad \text{and} \quad \text{not} \ Q \precsim P \tag{5.5}$$

where $P \prec Q$ means the player *strictly prefers* Q to P; it is clearly equivalent to

$$P \prec Q \qquad \Leftrightarrow \qquad P \precsim Q \quad \text{and} \quad \text{not} \ P \sim Q.$$

We will derive a number of conditions on the weak preference relation $\precsim$ on $\Delta(X)$ that makes this relation mathematically tractable for its use in decision theory and game theory. These will be restrictions on $\precsim$ that may not be *descriptive* of how people actually behave in certain decision situations. Instead, these restrictions are often called *normative*, that is, they state how people *should* behave, for example in order to come to a conclusive decision. In that way, the theory may be used as a tool in a decision analysis.

The goal will be to find conditions for the existence of a function $u : X \to \mathbb{R}$, which is called a *utility function* for the player, so that the player strictly prefers a lottery Q over lottery P if Q has higher *expected utility* under Q than under P, and is indifferent between two lotteries if they have the same expected utility.

Expected utility is the expected value of the utility function for the given lottery. Suppose the lottery P has probabilities $p_1, \ldots, p_n$ for the outcomes $x_1, \ldots, x_n$, respectively. Then the utility function u applied to these outcomes defines a random variable. In extension of (4.1), we denote its expectation by $\mathbb{E}(u, P)$, given by

$$\mathbb{E}(u, P) = p_1 u(x_1) + \cdots + p_n u(x_n). \tag{5.6}$$

Then the utility function u represents the player's preference $\precsim$ on $\Delta(X)$ if, as stated in (5.1), for any two lotteries P and Q in $\Delta(X)$ we have $P \precsim Q$ if and only if $\mathbb{E}(u, P) \leq \mathbb{E}(u, Q)$.

If a utility function exists, then it imposes a number of conditions on the preference $\precsim$ of the player, which are *necessary* conditions for a utility function. A selected set of these conditions is also *sufficient* for the existence of a utility function. These conditions are considered by many as sufficiently weak and plausible to justify them as a normative assumptions about the player's preference.

A first necessary condition that follows from (5.1) is that $\precsim$ is a *total* relation, because the order $\leq$ on $\mathbb{R}$ is total: For any two lotteries P and Q,

$$P \precsim Q \quad \text{or} \quad Q \precsim P. \tag{5.7}$$

In a sense, this seems already problematic because lotteries represent complex scenarios, with multiple outcomes and their probabilities, so a person may already face difficulties in deciding between one or the other. In fact, the representation (5.1) with a utility u function may help in such a decision, once u has been determined by lotteries that are not very complex. We will see that lotteries that have only *two outcomes* will suffice for that purpose.

5.5 Ordinal Preferences for Decisions Under Certainty

As a first step, we ignore uncertainties and look only at lotteries that have only a single outcome x, which has necessarily probability 1. Let $\mathbf{1}_x$ be the lottery in $\Delta(X)$ that assigns probability 1 to x. Then clearly $\mathbb{E}(u, \mathbf{1}_x) = u(x)$. If x and y are two outcomes, we write $x \precsim y$ if $\mathbf{1}_x \precsim \mathbf{1}_y$ which is simply the player's weak preference of y over x as a deterministic outcome (technically, the preference $\precsim$ is a relation on the set $\Delta(X)$ of lotteries). Therefore, if u is an expected-utility function according to (5.1), then for any $x, y \in X$,

$$x \precsim y \qquad \Leftrightarrow \qquad u(x) \le u(y), \tag{5.8}$$

that is, the utility function represents the player's preference over outcomes that occur with certainty.

In a number of decision situations, in particular when the outcome has only a single attribute, the preference for receiving the outcome for certain is clear. For example, if x is the amount of money saved, then $u(x)$ is strictly increasing in x, or if x is money to be paid, then $u(x)$ is strictly decreasing in x.

An immediate consequence of the representation (5.1) is that the preference relation $\precsim$ is *transitive*,

$$x \precsim y \quad \text{and} \quad y \precsim z \quad \Rightarrow \quad x \precsim z. \tag{5.9}$$

This is typically not problematic for single-attribute decisions, but already for two attributes some decision rules can lead to a non-transitive preference relation, as the following example demonstrates. Suppose there are three candidates x, y, z for a job. A candidate c is described by a pair of attributes (c_1, c_2), where the attribute "intelligence" is represented by an IQ score c_1 from a cognitive test, and the attribute "experience" by c_2 years of previous work in the profession. Their attributes are as follows:

candidate (c_1, c_2)	IQ score c_1	years experience c_2
$x = (x_1, x_2)$	121	5
$y = (y_1, y_2)$	124	4
$z = (z_1, z_2)$	128	2

$$\tag{5.10}$$

The hiring manager considers intelligence as more important, and considers experience as relevant only for candidates with equal intelligence. This amounts to the *lexicographic order* $\leq_{\text{lex}}$ where two attribute pairs (c_1, c_2) and (d_1, d_2) are compared according to

$$(c_1, c_2) \leq_{\text{lex}} (d_1, d_2) \qquad \Leftrightarrow \qquad c_1 \leq d_1 \quad \text{or} \quad (c_1 = d_1 \text{ and } c_2 \leq d_2). \qquad (5.11)$$

Both attributes are real numbers. Because $\leq$ is a total order (see Definition 1.2) on $\mathbb{R}$, it is easy to see that $\leq_{\text{lex}}$ is also a total order on $\mathbb{R}^2$. If $\leq_{\text{lex}}$ defines the preference relation, then z is the most preferred candidate in (5.10), because $x_1 < z_1$ and $y_1 < z_1$.

However, the lexicographic order $\leq_{\text{lex}}$ in (5.11) means that whenever $c_1 < d_1$ then (c_1, c_2) is always strictly preferred to (d_1, d_2) no matter how close c_1 and d_1 are, even if d_2 is vastly better than c_2; the order between c_2 and d_2 matters only when $c_1 = d_1$.

Suppose that, for that reason, the manager adopts the following rule: If the IQ score between two candidates differs by less than 5, then the candidate with higher experience is preferred. Then in (5.10) x is preferred to y because their IQ scores differ by only 3 and x has more experience; similarly, y is preferred to z. However, z is preferred to x because their IQ difference is high enough to count. The rule therefore leads to an intransitivity of the strict preference.

While the rule itself is plausible (for example because the IQ result is considered "noisy"), the intransitivity creates a cycle. A similar cycle resulted from the pairwise comparison rule in the Condorcet paradox (5.3) (where the three rankings by the committee members A, B, C could also be considered as three "attributes" of the candidates x, y, z). However, a cyclic preference, for whatever reason it occurs, cannot result in a satisfactory choice of one of the outcomes in the cycle because there is always a more preferred outcome. If the decision is then made in some arbitrary way, the cycle represents in effect an indifference among the outcomes in the cycle. It is therefore useful to impose the transitivity rule (5.9) as one of the normative conditions on the preference relation $\precsim$.

The rule "if the IQ scores differ by less than 5, go by experience" is intransitive because "differing by less than 5", which replaces the condition $c_1 = d_1$ in (5.11), is not a transitive relation. One way out of this would be to round the IQ score to the nearest multiple of 5 and then apply the lexicographic rule (5.11). In (5.10), the IQ scores of x, y, z would then be rounded to 120, 125, 130 and create the (transitive) ranking $x \prec y \prec z$. However, this rounding process is sensitive to small changes: if z had an IQ score of 127, this would be rounded to 125 and considered equal to the rounded IQ score of y, with the preference switching to $x \prec z \prec y$. Rather than using the lexicographic rule, it is probably better to realize that there may be a *tradeoff* between the two attributes. That tradeoff that is better represented by *weights* in an additive representation, such as "2 IQ points have the same weight as 1 year job experience", which is represented by a utility function such as $u(c_1, c_2) = c_1 + 2c_2$. Such an additive utility function may be in practice a good approximation of a player's preference, and choosing its weights may help the player to clarify the necessary tradeoffs. (Exercise 5.1 relates to such an "additive" utility function.)

A utility function u that represents a preference for deterministic outcomes as in (5.8) has a lot of freedom for its shape. Namely, for the comparison of outcomes that occur with certainty, only the *ordinal comparison* between values of u matters, that is, whether $u(x)$ is less than, greater than, or equal to $u(y)$. This implies that $u(x)$ can also be replaced by a $\phi(u(x))$ for a strictly increasing function ϕ (also called a *transformation* of the utility function). For example, if x is "personal wealth" and always $x > 0$, then the preference $x \lesssim y$ can either be represented by the wealth itself, that is, if and only if $x \le y$, or, for example, by the logarithm of that wealth, namely $\log x \le \log y$, where the strictly increasing transformation is given by $\phi = \log$. The represented preference (a higher personal wealth is strictly preferred) is exactly the same. A utility function for decisions under certainty is therefore also called an *ordinal* utility function.

5.6 Cardinal Utility Functions and Simple Lotteries

As soon as a utility function represents a preference for risky choices, its values cannot be transformed by an arbitrary strictly increasing transformation. Suppose a player can choose between the following two lotteries P and Q. Lottery P gives with probability $\frac{1}{2}$ either 100 dollars or 400 dollars, whereas lottery Q gives 240 dollars for certain. If the utility for receiving x dollars is just x, then the expected utility for P is $\frac{1}{2}100 + \frac{1}{2}400 = 250$ whereas for Q it is 240, so the player strictly prefers P to Q. However, if the utility for receiving x dollars is $\sqrt{x}$, then the expected utility for P is $\frac{1}{2}\sqrt{100} + \frac{1}{2}\sqrt{400} = \frac{1}{2}10 + \frac{1}{2}20 = 15$ whereas for Q it is $\sqrt{240} \approx 15.5$, so the player strictly prefers Q to P. There is no particular reason to assume a special utility function like $u(x) = \sqrt{x}$, but the player should be able to express a preference between the lotteries P and Q, which should be either $Q \prec P$, or $P \prec Q$, or the indifference $P \sim Q$. That is, an expected-utility function that represents a preference for lotteries is no longer just ordinal but *cardinal*.

By definition, a cardinal utility function is *unique up to positive affine transformations*. That is, if $u : X \to \mathbb{R}$ is a utility function, then the function $v : X \to \mathbb{R}$ defined by

$$v(x) = a \cdot u(x) + b \tag{5.12}$$

for real constants a and b where $a > 0$ is also a utility function, and these are the only allowed transformations. More formally, call two functions $u, v : X \to \mathbb{R}$ *scale-equivalent* if (5.12) holds for some reals a and b with $a > 0$, for all x in X. This is clearly an equivalence relation because it is reflexive (take $a = 1$, $b = 0$), symmetric because (5.12) is equivalent to $u(x) = \frac{1}{a}v(x) - \frac{b}{a}$, and transitive because if $w(x) = cv(x) + d$ with $c > 0$ then $w(x) = (ca)u(x) + (cb + d)$. We will show that any two expected-utility functions are scale-equivalent.

When replacing $u(x)$ with $v(x) = a\,u(x) + b$, a different factor a changes the *unit* of the utility function (as if changing a monetary payoff from dollars to cents), and the additive constant b changes its *origin* (where the utility function takes value zero). This change of origin in a positive-affine transformation is familiar from the *temperature scales* Celcius and Fahrenheit, which are related by the positive-affine transformation $F = 1.8\,C + 32$ to convert a temperature of C degrees Celsius to F degrees Fahrenheit; for example, a temperature of 20 degrees Celsius equals 68 degrees Fahrenheit. The temperature conversion from Celsius to *Kelvin* keeps the unit and only changes the origin, where degrees in Kelvin are degrees in Celsius plus 273.15 (so that the physically lowest possible temperature of "absolute zero" is represented by zero degrees Kelvin).

We first show that any expected-utility function u that fulfills (5.1) can be replaced by the scale-equivalent function v in (5.12), which we also write as $au + b$, for any fixed $a, b \in \mathbb{R}$ with $a > 0$. Namely, by (5.6),

$$
\begin{aligned}
\mathbb{E}(v, P) &= \mathbb{E}(au + b, P) \\
&= p_1(au(x_1) + b) + \cdots + p_n(au(x_n) + b) \\
&= a(p_1 u(x_1) + \cdots + p_n u(x_n)) + (p_1 + \cdots + p_n)b \\
&= a\mathbb{E}(u, P) + b
\end{aligned}
\tag{5.13}
$$

so that for any lotteries P and Q we have the equivalent statements

$$
\begin{aligned}
P \precsim Q \quad &\Leftrightarrow \quad \mathbb{E}(u, P) && \le && \mathbb{E}(u, Q) \\
&\Leftrightarrow \quad a\mathbb{E}(u, P) + b && \le && a\mathbb{E}(u, Q) + b \\
&\Leftrightarrow \quad \mathbb{E}(v, P) && \le && \mathbb{E}(v, Q)
\end{aligned}
\tag{5.14}
$$

which proves the claim. While (5.14) is immediate from the simple observation (5.13), it emphasizes that "utility" is an abstract scale to represent a player's preference with an origin that can be arbitrarily shifted (by adding b) and whose units can be scaled by multiplication by some positive factor a. Despite its name, "utility" is not an intrinsic "usefulness" of an outcome, but instead represents this preference (which is why we use the more neutral term "payoff" for the games in the other chapters of this book).

Our second claim is that an expected-utility function *only* admits the positive-affine transformations in (5.12) in order to represent a given preference for lotteries. For that purpose, it will suffice to look at lotteries that have only one or two outcomes.

The following operation combines two existing lotteries. Suppose $\alpha \in [0, 1]$ and P and Q are two lotteries over X. Then the *combined lottery* $(1 - \alpha)P + \alpha Q$ is a new lottery that can be thought of a two-stage process as in the following picture, which we call a *probability tree* (a game tree that has only chance moves),

(5.15)

where at the first stage P is chosen with probability $1 - \alpha$ and Q with probability α, and then the chosen lottery is played with its respective probabilities for outcomes (as always, different random chance moves are independent). If these outcomes are x_i with probabilities p_i under P and q_i under Q for $1 \le i \le n$, then the combined lottery $(1 - \alpha)P + \alpha Q$ can also be considered as a single-stage process that chooses x_i with probability $(1 - \alpha)p_i + \alpha q_i$ for $1 \le i \le n$, and is therefore a lottery in the usual sense.

We always write a combined lottery as $(1 - \alpha)P + \alpha Q$ so that the probability α chooses the *second* lottery Q, because for $\alpha \in [0, 1]$ the case $\alpha = 0$ as the left endpoint of the interval $[0, 1]$ naturally corresponds to the left branch in (5.15) which chooses P, and the case $\alpha = 1$ as the right endpoint of the interval $[0, 1]$ corresponds to the right branch which chooses Q. This correspondence would be lost if (as it is often done) we wrote such a combination as $\beta P + (1 - \beta)Q$ for $\beta \in [0, 1]$ (which with $\beta = 1 - \alpha$ is of course mathematically equivalent).

A *simple lottery* is a lottery that chooses outcome x with probability $1 - \alpha$ and outcome y with probability α, which we denote by $x \wedge_\alpha y$ (as in a little probability tree). Recall that the lottery that chooses the certain outcome x is written as $\mathbf{1}_x$, so that by definition

$$x \wedge_\alpha y = (1 - \alpha)\mathbf{1}_x + \alpha \mathbf{1}_y.$$ (5.16)

Note that the simple lottery $x \wedge_\alpha y$ is a lottery with one or two outcomes (it has only one outcome if $x = y$ or if $\alpha = 0$ or $\alpha = 1$). Clearly,

$$\mathbb{E}(u, x \wedge_\alpha y) = (1 - \alpha)u(x) + \alpha u(y).$$ (5.17)

In the following proposition, we assume that the player's preference for simple lotteries is represented by their expected value of a function $u : X \to \mathbb{R}$ as in (5.1) and (5.17), which we call an "expected-utility function for simple lotteries". The proposition states that u is then unique up to scale-equivalence. Once the player's preference $\precsim$ for simple lotteries is determined in this way, his preference for general lotteries is automatically determined (provided it fulfills certain assumptions which we will consider later).

Proposition 5.2. *Let S be a subset of $\Delta(X)$ that includes all simple lotteries. Let $u : X \to \mathbb{R}$ be a function that fulfills (5.1) for all $P, Q \in S$. Then if v is another such* *function, then u and v are scale-equivalent, that is, (5.12) holds for some a, b with a > 0 for all x ∈ X.*

Proof. In this proof, let "utility function" be a function u that fulfills (5.1) for all simple lotteries $P, Q \in S$. Consider a given utility function $\hat{u} : X \to \mathbb{R}$. If the decision maker is indifferent between all outcomes, and hence between all lotteries, then any utility function is constant and all utility functions are scale-equivalent (by adding suitable constants). Hence, we can assume there are two outcomes x_0 and x_1 with $x_0 < x_1$ so that $\hat{u}(x_0) < \hat{u}(x_1)$. Define the function $u : X \to \mathbb{R}$ by

$$u(x) = \frac{1}{\hat{u}(x_1) - \hat{u}(x_0)} \hat{u}(x) - \frac{\hat{u}(x_0)}{\hat{u}(x_1) - \hat{u}(x_0)}$$

which is a utility function that is scale-equivalent to $\hat{u}$ and which fulfills

$$u(x_0) = 0, \qquad u(x_1) = 1. \tag{5.18}$$

We show that any other utility function $v : X \to \mathbb{R}$ is scale-equivalent to u, via

$$v(x) = a\, u(x) + b \qquad \text{with} \qquad a = v(x_1) - v(x_0), \quad b = v(x_0). \tag{5.19}$$

Here, a and b are uniquely determined by (5.18) and

$$v(x_0) = a\, u(x_0) + b, \qquad v(x_1) = a\, u(x_1) + b. \tag{5.20}$$

Consider any outcome z. Then one of the following three cases applies:

$$z < x_0, \qquad x_0 \lesssim z \lesssim x_1, \qquad \text{or} \qquad x_1 < z. \tag{5.21}$$

We first consider the middle case $x_0 \lesssim z \lesssim x_1$ in (5.21). Then $0 = u(x_0) \leq u(z) \leq u(x_1) = 1$. Let $\alpha = u(z)$, and consider the simple lottery $P = x_0 \wedge_\alpha x_1$, so that by (5.17)

$$\mathbb{E}(u, P) = \mathbb{E}(u, x_0 \wedge_\alpha x_1) = (1 - \alpha)u(x_0) + \alpha u(x_1) = \alpha = u(z)$$

and therefore $z \sim P$. The utility function v represents the same indifference, so that by (5.20) and (5.18)

$$
\begin{aligned}
v(z) &= \mathbb{E}(v, P) = \mathbb{E}(v, x_0 \wedge_\alpha x_1) \\
&= (1 - \alpha)v(x_0) + \alpha v(x_1) \\
&= (1 - \alpha)(a\, u(x_0) + b) + \alpha(a\, u(x_1) + b) \\
&= \alpha \cdot a + b = a\, u(z) + b
\end{aligned}
$$

which shows (5.19) for $x = z$.

Next, consider the case $z < x_0$, which is similar but slightly more involved. Because $z < x_0 < x_1$, we now choose a simple lottery $Q = z \wedge_\beta x_1$ so that $Q \sim x_0$, that is,

$$(1 - \beta)u(z) + \beta u(x_1) = u(x_0)$$

which holds if $\beta(u(x_1) - u(z)) = u(x_0) - u(z)$ or $\beta = \frac{-u(z)}{1-u(z)} = 1 - \frac{1}{1-u(z)}$ where $0 < \beta < 1$ because $u(z) < u(x_0) = 0$. Furthermore,

$$u(z) = \frac{-\beta}{1-\beta}. \tag{5.22}$$

Because $Q \sim x_0$, we have

$$
\begin{aligned}
\mathbb{E}(v, Q) &= v(x_0) \\
(1-\beta)v(z) + \beta v(x_1) &= v(x_0) \\
(1-\beta)v(z) + \beta(a\,u(x_1) + b) &= a\,u(x_0) + b \\
(1-\beta)v(z) + \beta(a + b) &= b \\
(1-\beta)v(z) &= -\beta\,a + (1-\beta)b \\
v(z) &= \tfrac{-\beta}{1-\beta} a + b = a\,u(z) + b \tag{5.23}
\end{aligned}
$$

which shows (5.19) for $x = z$ if $z < x_0$. The third case $x_1 < z$ in (5.21) is similar, by choosing a simple lottery $x_0 \wedge_\gamma z$ so that the player is indifferent between that lottery and x_1, and again using (5.20) (see Exercise 5.2). This proves the claim. $\square$

The proof of Proposition 5.2 highlights the meaning of the values of the utility function: Consider two outcomes x_0 and x_1 so that $x_0 < x_1$. First of all, the utility function u can always be scaled so that, as in (5.18), the origin of the utility scale is $u(x_0) = 0$, with $u(x_1) = u(x_1) - u(x_0) = 1$ as its unit. Second, consider some outcome z so that $x_0 < z < x_1$. This condition indicates that there is some probability α so that the player is indifferent between z and the simple lottery $x_0 \wedge_\alpha x_1$. That is, we have the equivalent statements

$$z \sim x_0 \wedge_\alpha x_1, \qquad u(z) = (1-\alpha)u(x_0) + \alpha u(x_1),$$

and

$$\alpha = \frac{u(z) - u(x_0)}{u(x_1) - u(x_0)} \tag{5.24}$$

where by (5.18) $\alpha = u(z)$. In other words, the utility $u(z)$ for the outcome z that is between the less and more preferred outcomes x_0 and x_1 is just the *probability* for getting the more preferred outcome x_1 in the simple lottery $x_0 \wedge_\alpha x_1$, assuming the scaling (5.18). Without that scaling, α is the more general quotient (5.24), which follows from the observation that for any real numbers A, B, C with $A < C$,

$$A < B < C, \quad B = (1-\alpha)A + \alpha C \qquad \Leftrightarrow \qquad \alpha = \frac{B - A}{C - A}, \quad 0 < \alpha < 1. \tag{5.25}$$

Taking two outcomes x_0 and x_1 and identifying the probability α so that $z \sim x_0 \wedge_\alpha x_1$ is therefore a possible way to *construct* the values $u(z)$ of the utility function u for outcomes z that fulfill $x_0 < z < x_1$. If x_0 is least preferred and

x_1 most preferred among all outcomes, then this determines the utility function uniquely (where if $z \sim x_0$ then $u(z) = u(x_0)$ and if $z \sim x_1$ then $u(z) = u(x_1)$). However, even if such extreme outcomes exist (for example if X is finite), in practice they might be not as useful for constructing the utility function because the player may be uncomfortable to consider simple lotteries between the worst and best possible outcomes. More realistic outcomes x_0 and x_1 are typically more useful instead. Then this leads to the other cases $z \prec x_0$ and $x_1 \prec z$ in (5.21), where one can construct other simple lotteries as in the proof of Proposition 5.2. The probability in that lottery is then used to find the unknown utility value $u(z)$ using the general expression (5.24), but with permuted roles of the outcomes z, x_0, x_1 according to the player's preference for z relative to x_0 and x_1.

In summary, an expected-utility function is unique up to positive-affine transformations already if it represents the player's preference for simple lotteries, that is, lotteries that have two outcomes, with an arbitrary probability α for the second outcome (and probability $1 - \alpha$ for the first outcome). Finding a suitable α so that the player is indifferent between this lottery and another outcome can be used to construct the utility function. The utility function is then useful for deciding the player's preference for more complicated lotteries.

Next, we will consider *only* the preference between lotteries and certain *consistency axioms* for this preference. We then show that a preference relation that fulfills these axioms can be represented by an expected-utility function.

5.7 Consistency Axioms

Consider a preference relation $\precsim$ on the set $\Delta(X)$ of lotteries over the outcome set X, and its derived relations $\sim$ and $\prec$ in (5.4) and (5.5). In order to prove that $\precsim$ can be represented by an expected-utility function according to (5.1), we consider a number of sufficient assumptions, called *axioms*. These are reasonable, normative conditions about how a player's preference should look like.

The first condition, already stated in (5.7) and (5.9), is that $\precsim$ is total and transitive. As argued above, this is a useful normative assumption because otherwise the player would sometimes not be able to make up his mind; this may result in an arbitrary choice and factual indifference, which is typically not what is intended.

The following conditions on $\precsim$ affect the combination of lotteries as in (5.15). First of all, it is assumed that only the eventual probabilities for outcomes matter and not whether these probabilities result from a multi-stage process.

Let P, Q, R be lotteries and α be a probability. The following condition states that indifference between lotteries is preserved when they are substituted into other lotteries:

$$P \sim Q \quad \Rightarrow \quad (1-\alpha)R + \alpha P \sim (1-\alpha)R + \alpha Q. \qquad (5.26)$$

Here the player is indifferent between P and Q, and stays indifferent if these lotteries occur with some probability α in a combined lottery that otherwise chooses some other lottery R. A similar condition states the same for strict preference, where $\alpha > 0$ is required:

$$P \prec Q, \ 0 < \alpha \leq 1 \quad \Rightarrow \quad (1-\alpha)R + \alpha P \prec (1-\alpha)R + \alpha Q. \qquad (5.27)$$

The two conditions (5.26) and (5.27) are also known as *independence of irrelevant alternatives* or just *independence* axioms. The "irrelevant alternative" is the lottery R which should not affect the indifference $P \sim Q$ or the strict preference $P \prec Q$.

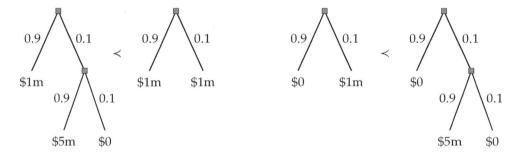

Figure 5.2 The Allais paradox, which is an empirical violation of (5.27) for three hypothetical outcomes x, y, z of receiving 1, 5, or 0 million dollars, $P = y \wedge_{0.1} z$ and $Q = \mathbf{1}_x$, and where R is either $\mathbf{1}_x$ or $\mathbf{1}_z$.

Figure 5.2 shows an example, known as the *Allais paradox*, where in certain hypothetical situations people may violate condition (5.27). Consider three outcomes x, y, z where x means "receive one million dollars", y means "receive five million dollars", and z means "receive zero dollars", let P be the simple lottery $y \wedge_{0.1} z$ and let $Q = \mathbf{1}_x$. The preference $P \prec Q$ means that the player would rather receive one million dollars for sure than five million dollars with 90 percent probability and otherwise nothing, which may be the player's choice. Consider now the lottery $R = \mathbf{1}_z$ and let $\alpha = 0.1$. Then the combined lottery $(1-\alpha)R + \alpha P$, shown on the right in Figure 5.2, means that the player receives nothing with 91 percent probability and five million dollars with 9 percent probability, and $(1-\alpha)R + \alpha Q$ that he receives nothing with 90 percent probability and one million dollars with 10 percent probability. Given that the player very likely receives nothing anyhow, his preference may now be reversed, that is, $(1-\alpha)R + \alpha Q \prec (1-\alpha)R + \alpha P$, which contradicts condition (5.27). On the other hand, if $R = \mathbf{1}_x$, shown on the left in Figure 5.2, then $(1-\alpha)R + \alpha P$ means that the player receives one million dollars with 90 percent probability, nothing with 1 percent probability, and five million dollars with 9 percent probability, whereas $(1-\alpha)R + \alpha Q$ means that he receives one million with certainty, which may well preserve his original preference $P \prec Q$ also in combination with R, in agreement with (5.27).

The suggested remedy for resolving the inconsistent preference in Figure 5.2, which violates (5.27), is that the player should (a) decide whether his preference is $P \prec Q$ or $Q \prec P$ (or $P \sim Q$), which ignores the irrelevant alternative $\mathbf{1}_z$ or $\mathbf{1}_x$, and (b) consider how much weight he should give his *regret* when the underdesirable outcome $\mathbf{1}_z$ occurs in the lottery $P = \mathbf{1}_z \wedge_{0.9} \mathbf{1}_y$ and whether it matters that this regret is "invisible" in the combined lottery $(1 - \alpha)\mathbf{1}_z + \alpha P$ that he prefers in Figure 5.2. If the regret matters, it should be made part of the outcome z which then takes two forms "receiving zero dollars when I could have gotten one million", and "receiving zero dollars but that was the likely outcome anyhow". In summary, the inconsistency in Figure 5.2 may well be observed in practice, but the independence axiom (5.27) may help resolve it by highlighting that the basic decision should be between P and Q.

One consequence of (5.27) is the following, rather natural, *monotonicity* property, which states that increasing the probability for a more preferred part Q of a combined lottery makes the combined lottery more preferable.

Lemma 5.3. *Consider a preference relation $\precsim$ on $\Delta(X)$ that fulfills (5.27). Then for P, Q in $\Delta(X)$,*

$$P \prec Q, \quad 0 \le \alpha < \beta \le 1 \quad \Rightarrow \quad (1 - \alpha)P + \alpha Q \prec (1 - \beta)P + \beta Q. \quad (5.28)$$

Proof. Let $P \prec Q$ and $0 \le \alpha < \beta \le 1$, which implies that $0 \le \frac{\alpha}{\beta} < 1$ and therefore $0 < 1 - \frac{\alpha}{\beta} \le 1$. The following probability trees convey the proof:

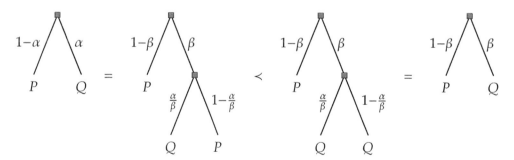

We use (5.27) twice, first to show that $\frac{\alpha}{\beta}Q + (1 - \frac{\alpha}{\beta})P \prec \frac{\alpha}{\beta}Q + (1 - \frac{\alpha}{\beta})Q$, and then by substituting either combined lottery for Q' in $(1 - \beta)P + \beta Q'$, which shows (5.28). □

The next condition is called the *continuity* axiom, where P, Q, R are lotteries and α is a probability:

$$P \prec Q \prec R \quad \Rightarrow \quad \exists \alpha \in (0, 1) : Q \sim (1 - \alpha)P + \alpha R. \quad (5.29)$$

To see why condition (5.29) may be considered problematic, let P be "die in the next minute", Q be "receive nothing", and R "receive one dollar". Then $Q \sim (1-\alpha)P+\alpha R$

means that the player would risk his life with positive probability $1 - \alpha$ just to receive one dollar with probability α, which seems unreasonable. The counter-argument to this is that $1 - \alpha$ can be arbitrarily small as long as it is positive, for example 2^{-500} as when 500 flips of a fair coin come out "heads" 500 times in a row. This is a minimal risk, and people do risk their lives with small probabilities for small gains, for example when cycling without a helmet. Condition (5.29) is a technical, not a practical restriction.

The continuity axiom (5.29) holds necessarily when the preference $\prec$ is represented by an expected-utility function u, because $P \prec Q \prec R$ implies $\mathbb{E}(u, P) < \mathbb{E}(u, Q) < \mathbb{E}(u, R)$, so that by (5.25) with $\alpha = \frac{\mathbb{E}(u,Q) - \mathbb{E}(u,P)}{\mathbb{E}(u,R) - \mathbb{E}(u,P)}$ we have $0 < \alpha < 1$ and

$$\mathbb{E}(u, Q) = (1 - \alpha)\mathbb{E}(u, P) + \alpha\mathbb{E}(u, R) = \mathbb{E}(u, (1 - \alpha)P + \alpha R) \tag{5.30}$$

and therefore $Q \sim (1 - \alpha)P + \alpha R$.

Because a utility function $u : X \to \mathbb{R}$ takes real values, there are sets X with a preference relation $\precsim$, already restricted to deterministic outcomes, that cannot be represented with a utility function according to (5.8). The classic example for this is $X = [0, 1]^2$ with $\precsim$ as the lexicographic order $\leq_{\text{lex}}$ defined in (5.11), illustrated in Figure 5.3.

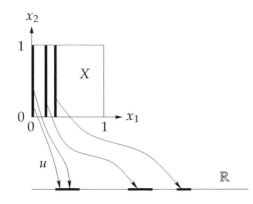

Figure 5.3 The set $X = [0, 1]^2$ with $\precsim$ as the lexicographic order $\leq_{\text{lex}}$. A utility function $u(x_1, x_2)$ would have to map the uncountably many vertical strips of X to nonempty disjoint intervals in $\mathbb{R}$, which is not possible.

Suppose there was a utility function $u : [0, 1]^2 \to \mathbb{R}$ so that, according to (5.11),

$$(x_1, x_2) \prec (y_1, y_2) \Leftrightarrow (x_1, x_2) \leq_{\text{lex}} (y_1, y_2) \quad \text{and} \quad (x_1, x_2) \neq (y_1, y_2)$$
$$\Leftrightarrow x_1 < y_1 \quad \text{or} \quad (x_1 = y_1 \text{ and } x_2 < y_2)$$
$$\Leftrightarrow u(x_1, x_2) < u(y_1, y_2). \tag{5.31}$$

Consider now a "vertical strip" of the form $\{x_1\} \times [0, 1]$, a subset of X, of which three are shown as bold lines in Figure 5.3. For any two points (x_1, x_2) and (x_1, y_2) on such a strip so that $x_2 < y_2$ (for example, the endpoints $(x_1, 0)$ and $(x_1, 1)$) we have $u(x_1, x_2) < u(x_1, y_2)$. For two such strips $\{x_1\} \times [0, 1]$ and $\{y_1\} \times [0, 1]$ with

$x_1 < y_1$ the lexicographic order dictates $u(x_1, 1) < u(y_1, 0)$. Therefore, the images under u of these strips must not overlap, and they are subsets of pairwise disjoint intervals in the range $\mathbb{R}$ of u. This would work fine if there were only finitely many such strips (as when they have a certain width as drawn in the diagram), but mathematically they have no width, and therefore the images of these strips "cannot fit" disjointly into $\mathbb{R}$. Technically, for any $x_1 \in [0, 1]$ we can choose a *rational number* $r(x_1)$ so that $u(x_1, 0) < r(x_1) < u(x_1, 1)$. If $x_1 < y_1$ then $u(x_1, 1) < u(y_1, 0)$ and therefore $r(x_1) < r(y_1)$. Hence, the mapping $x_1 \mapsto r(x_1)$ defines an injection from the set $[0, 1]$ to the set $\mathbb{Q}$ of rationals, which implies that $[0, 1]$ has at most the set-theoretic cardinality of $\mathbb{Q}$. This is a contradiction because $\mathbb{Q}$ is a countable set but $[0, 1]$ is not. Hence, a real-valued utility function $u : [0, 1]^2 \to \mathbb{R}$ cannot represent the lexicographic order $\leq_{\text{lex}}$ as in (5.31).

The extreme "sensitivity" to changes in the first component is a mathematical reason not to use a lexicographic rule as a preference relation. For the same reason, the rule is also behaviorally questionable, as we discussed earlier after (5.11) and after (5.29) for the lexicographic preference of "safety to your life" over "monetary gain".

As shown in (5.30), the existence of an expected-utility function implies the continuity axiom (5.29). If the preference $\precsim$ restricted to deterministic outcomes is the lexicographic order $\leq_{\text{lex}}$ on $X = [0, 1]^2$, then it violates the continuity axiom, and an expected-utility function does not exist. In the next section, we show how the continuity axiom (5.29) is used in the construction of an expected-utility function.

5.8 Existence of an Expected-Utility Function

The following is the central theorem of this chapter. The main idea of the proof has been given in Section 5.2. The proof given here is longer because it does not assume that there exists a least preferred outcome x_0 and most preferred outcome x_1. Instead, any two outcomes x_0 and x_1 can be chosen to define the origin and unit of the utility function. Then the utility function can be defined for further outcomes z (for example, with $z < x_0$) as in Proposition 5.2. The expected-utility property is shown to hold for arbitrary lotteries P for a function v that is scale-equivalent to u and scaled with the least and most preferred outcomes of P (where we use that P has only finitely many outcomes).

Theorem 5.4. *Let $\Delta(X)$ be the set of lotteries on the set of outcomes X and let $\precsim$ be a preference relation on $\Delta(X)$. Assume that $\precsim$ is total and transitive, and fulfills the independence axioms (5.26) and (5.27) and the continuity axiom (5.29). Then there exists an expected-utility function $u : X \to \mathbb{R}$ that represents $\precsim$ according to (5.1).*

Proof. The fact that $\precsim$ is total and transitive implies that $\sim$ and $\prec$ as defined in (5.4) and (5.5) are also total and transitive, which we use all the time, for example when writing $x_0 \precsim z \precsim x_1$ as short for $x_0 \precsim z$ and $z \precsim x_1$. We will make repeated use of the continuity axiom. Note that if $P \prec Q \prec R$ and $Q \sim (1 - \alpha)P + \alpha R$ as in (5.29), then α is unique, because if $\alpha < \beta$ then $(1 - \alpha)P + \alpha R \prec (1 - \beta)P + \beta R$ by Lemma 5.3, so we cannot also have $Q \sim (1 - \beta)P + \beta R$.

If the player is indifferent between all outcomes, then he is also indifferent between all lotteries over outcomes, and a constant utility function suffices. Hence, assume there are two outcomes x_0 and x_1 so that $x_0 \prec x_1$. As in the proof of Proposition 5.2, we use x_0 and x_1 to fix the origin and unit of the utility function u as in (5.18), that is, $u(x_0) = 0$ and $u(x_1) = 1$. For any outcome z we then choose the value $u(z)$ of the utility function with the help of a suitable simple lottery:

- If $z \sim x_0$ then let $u(z) = 0$, and if $z \sim x_1$ then let $u(z) = 1$.

- If $x_0 \prec z \prec x_1$ then find α so that $z \sim (1 - \alpha)\mathbf{1}_{x_0} + \alpha \mathbf{1}_{x_1} = x_0 \wedge_\alpha x_1$ according to (5.29) and let $u(z) = \alpha$.

- If $z \prec x_0 \prec x_1$ then find β so that $x_0 \sim z \wedge_\beta x_1$, again by (5.29), and let $u(z) = \frac{-\beta}{1-\beta}$ as in (5.22).

- If $x_0 \prec x_1 \prec z$ then find γ so that $x_1 \sim x_0 \wedge_\gamma z$ by (5.29) and let $u(z) = \frac{1}{\gamma}$.

In all cases, $u(z)$ is determined by the player's indifference between the certain outcome and the simple lottery. For example, in the last case, $u(x_1) = 1 = \mathbb{E}(u, x_0 \wedge_\gamma z) = (1 - \gamma)u(x_0) + \gamma u(z) = \gamma u(z)$, so that $u(z) = \frac{1}{\gamma}$.

With $u : X \to \mathbb{R}$ determined in this way, we now show that u fulfills (5.1), that is, u represents the player's preference between any two lotteries P and Q. Let z_i for $1 \le i \le n$ be the outcomes of P or Q where z_i has probability p_i under P and q_i under Q. We first assume that all these outcomes fulfill $x_0 \precsim z_i \precsim x_1$. Let $\alpha_i = u(z_i)$, where by the construction of u we have $z_i \sim x_0 \wedge_{\alpha_i} x_1$. We now replace, by the successive substitution of equivalent lotteries according to (5.26), each outcome z_i by its equivalent lottery $x_0 \wedge_{\alpha_i} x_1$. That is (see Figure 5.1),

$$
\begin{aligned}
P &= \sum_{i=1}^n p_i \mathbf{1}_{z_i} \sim \sum_{i=1}^n p_i(x_0 \wedge_{\alpha_i} x_1) \\
&= \sum_{i=1}^n p_i\big((1 - \alpha_i)\mathbf{1}_{x_0} + \alpha_i \mathbf{1}_{x_1}\big) \\
&= \Big(1 - \sum_{i=1}^n p_i \alpha_i\Big)\mathbf{1}_{x_0} + \Big(\sum_{i=1}^n p_i \alpha_i\Big)\mathbf{1}_{x_1} \\
&= (1 - p)\mathbf{1}_{x_0} + p\mathbf{1}_{x_1} = x_0 \wedge_p x_1
\end{aligned}
\tag{5.32}
$$

with

$$
p = \sum_{i=1}^n p_i \alpha_i = \sum_{i=1}^n p_i\, u(z_i) = \mathbb{E}(u, P).
\tag{5.33}
$$

Similarly,

$$Q \sim x_0 \wedge_q x_1 \qquad \text{with} \qquad q = \sum_{i=1}^{n} q_i \alpha_i = \sum_{i=1}^{n} q_i\, u(z_i) = \mathbb{E}(u, Q). \qquad (5.34)$$

The derivation (5.32) is the core of the argument:

- By construction, the utility $u(z_i)$ of the outcome z_i is the *probability* α_i so that the player is indifferent between z_i and the simple lottery $x_0 \wedge_{\alpha_i} x_1$ that chooses the better outcome x_1 with probability α_i.

- Each outcome z_i is replaced by its simple lottery $x_0 \wedge_{\alpha_i} x_1$ in P, which by repeated application of (5.26) gives an equally preferred lottery.

- The resulting overall lottery has only the two outcomes x_0 and x_1, with probability $p = \sum_{i=1}^{n} p_i \alpha_i$ for x_1 because independent *probabilities can be multiplied*.

- The sum $\sum_{i=1}^{n} p_i \alpha_i$ is equal to the expectation $\mathbb{E}(u, P)$ by (5.33). The corresponding statement for Q is (5.34).

With the help of the monotonicity Lemma 5.3, we show that $P \precsim Q$ holds if and only if $\mathbb{E}(u, P) \le \mathbb{E}(u, Q)$, that is, if and only if $p \le q$. Suppose $P \precsim Q$. If $q < p$ then, because $x_0 \prec x_1$, (5.28) (with x_0, x_1, q, p for P, Q, α, β) implies $Q \sim x_0 \wedge_q x_1 \prec x_0 \wedge_p x_1 \sim P$, that is, $Q \prec P$, a contradiction. This shows that $P \precsim Q$ implies $p \le q$. To show the converse, let $p \le q$. If $p = q$ then clearly $P \sim Q$, and if $p < q$ then $P \sim x_0 \wedge_p x_1 \prec x_0 \wedge_q x_1 \sim Q$, again by (5.28), that is, $P \prec Q$. This shows that $p \le q$ implies $P \precsim Q$, as required.

It remains to show (5.1) for lotteries P and Q that have outcomes $z_1, \ldots, z_n$ with $z_i \prec x_0$ or $x_1 \prec z_i$ for some i. Let z_1 be the least preferred and z_n be the most preferred of all outcomes $z_1, \ldots, z_n$. We adopt the same approach as before, but use two new "reference outcomes" y_0 and y_1 instead of x_0 and x_1:

- If $z_1 \prec x_0$ then let $y_0 = z_1$, otherwise let $y_0 = x_0$. If $x_1 \prec z_n$ then let $y_1 = z_n$, otherwise let $y_1 = x_1$. Then $y_0 \precsim z_i \precsim y_1$ for all outcomes z_i of the lotteries P and Q.

- Define the function $v : X \to \mathbb{R}$ by

$$v(x) = \frac{u(x) - u(y_0)}{u(y_1) - u(y_0)},$$

which is scale-equivalent to u and fulfills

$$v(y_0) = 0, \qquad v(y_1) = 1.$$

- Replace every outcome z_i with an equally preferred simple lottery between y_0 and y_1 for some probability β_i (which is possible because $y_0 \precsim z_i \precsim y_1$),

$$z_i \sim y_0 \wedge_{\beta_i} y_1 \qquad (1 \le i \le n). \qquad (5.35)$$

- Substitute in P and Q the simple lotteries in (5.35) for the outcomes z_i, using repeatedly (5.26). The player is then indifferent between P and a suitable simple lottery with the two outcomes y_0 and y_1, and similarly for Q,

$$P \sim y_0 \wedge_{\hat{p}} y_1 , \qquad \hat{p} = \sum_{i=1}^{n} p_i \beta_i , \qquad Q \sim y_0 \wedge_{\hat{q}} y_1 , \qquad \hat{q} = \sum_{i=1}^{n} q_i \beta_i .$$

Then we are done if we can show that

$$\beta_i = v(z_i) \qquad (0 \le i \le n) \tag{5.36}$$

which implies that $\hat{p} = \mathbb{E}(v, P)$ and $\hat{q} = \mathbb{E}(v, Q)$ as in (5.33) and (5.34), where we argue as before to prove that $P \precsim Q$ if and only if $\hat{p} \le \hat{q}$. This shows that v is an expected-utility function, and so is the scale-equivalent function u.

The probabilities β_i are uniquely determined by the indifference conditions (5.35), but we cannot assume that (5.36) holds because v, although scale-equivalent to u, is not yet proved to be an expected-utility function because we have not yet shown this for u (for the general case that $z_i < x_0$ or $x_1 < z_i$ for some i).

Hence, it remains to prove (5.36). We do this, using (5.26), with the help of suitable indifferences between simple lotteries that involve y_0, x_0, x_1, y_1 and z_i for $0 \le i \le n$, using the definition of $u(z_i)$. The computations are straightforward applications of affine transformations but somewhat lengthy.

The goal is to represent all of $z_1, \dots, z_n, x_0, x_1$ as simple lotteries between y_0 and y_1. We first do this for x_0 and x_1, with

$$x_0 \sim y_0 \wedge_{\eta_0} y_1 , \qquad x_1 \sim y_0 \wedge_{\eta_1} y_1 , \tag{5.37}$$

where we claim that

$$\eta_0 = v(x_0) = \frac{-u(y_0)}{u(y_1) - u(y_0)} , \qquad \eta_1 = v(x_1) = \frac{1 - u(y_0)}{u(y_1) - u(y_0)} . \tag{5.38}$$

To show (5.38), we have $y_0 \precsim x_0 < x_1$ and therefore

$$x_0 \sim y_0 \wedge_{\beta} x_1 , \qquad u(y_0) = \frac{-\beta}{1 - \beta} \tag{5.39}$$

according to the construction of $u(y_0)$ (which applies also if $y_0 = x_0$ where $\beta = 0$). With an obvious extension of the notation (5.16) to combinations of lotteries, we therefore have

$$x_0 \sim y_0 \wedge_{\beta} x_1 \sim y_0 \wedge_{\beta} (y_0 \wedge_{\eta_1} y_1) = y_0 \wedge_{\eta_0} y_1$$

which shows that

$$\eta_0 = \beta \eta_1 . \tag{5.40}$$

Similarly, using that $x_0 < x_1 \lesssim y_1$ and the construction of $u(y_1)$,

$$x_1 \sim x_0 \wedge_\gamma y_1, \qquad u(y_1) = \frac{1}{\gamma} \tag{5.41}$$

(also if $y_1 = x_1$ where $\gamma = 1$). Therefore,

$$x_1 \sim x_0 \wedge_\gamma y_1 \sim (y_0 \wedge_{\eta_0} y_1) \wedge_\gamma y_1 = y_0 \wedge_{\eta_1} y_1$$

which shows that

$$\eta_1 = (1 - \gamma)\eta_0 + \gamma. \tag{5.42}$$

We solve (5.40) and (5.42) for η_0 and η_1 in terms of β and γ, namely

$$\eta_1 = (1 - \gamma)\beta\eta_1 + \gamma, \qquad \eta_1 = \frac{\gamma}{1 - (1 - \gamma)\beta},$$

which agrees with the expression for η_1 in (5.38) because by (5.39) and (5.41)

$$\frac{1 - u(y_0)}{u(y_1) - u(y_0)} = \frac{1 + \frac{\beta}{1-\beta}}{\frac{1}{\gamma} + \frac{\beta}{1-\beta}} = \frac{\gamma(1 - \beta) + \gamma\beta}{1 - \beta + \gamma\beta} = \frac{\gamma}{1 - (1 - \gamma)\beta}.$$

Furthermore, by (5.40)

$$\frac{-u(y_0)}{u(y_1) - u(y_0)} = \frac{\frac{\beta}{1-\beta}}{\frac{1}{\gamma} + \frac{\beta}{1-\beta}} = \frac{\gamma\beta}{1 - \beta + \gamma\beta} = \beta\eta_1 = \eta_0$$

which shows (5.38).

To show (5.36), we make the usual case distinctions for z_i. First, if $x_0 \lesssim z_i \lesssim x_1$ then $z_i \sim x_0 \wedge_{\alpha_i} x_1$ with $u(z_i) = \alpha_i$. We substitute (5.37) and obtain

$$z_i \sim x_0 \wedge_{\alpha_i} x_1 \sim (y_0 \wedge_{\eta_0} y_1) \wedge_{\alpha_i} (y_0 \wedge_{\eta_1} y_1) = y_0 \wedge_{\beta_i} y_1$$

which shows

$$\begin{aligned}
\beta_i &= (1 - \alpha_i)\eta_0 + \alpha_i\eta_1 = \eta_0 + \alpha_i(\eta_1 - \eta_0) = \eta_0 + u(z_i)(\eta_1 - \eta_0) \\
&= \frac{-u(y_0)}{u(y_1) - u(y_0)} + u(z_i)\frac{1}{u(y_1) - u(y_0)} = v(z_i),
\end{aligned} \tag{5.43}$$

which again shows (5.36).

If $y_0 \lesssim z_i < x_0$, then by construction of $u(z_i)$ we have $x_0 \sim z_i \wedge_{\delta_i} x_1$ with

$$u(z_i) = \frac{-\delta_i}{1 - \delta_i} = 1 - \frac{1}{1 - \delta_i}, \qquad \frac{1}{1 - \delta_i} = 1 - u(z_i). \tag{5.44}$$

Then $z_i \sim y_0 \wedge_{\beta_i} y_1$ and thus

$$x_0 \sim z_i \wedge_{\delta_i} x_1 \sim (y_0 \wedge_{\beta_i} y_1) \wedge_{\delta_i} (y_0 \wedge_{\eta_1} y_1) = y_0 \wedge_{\eta_0} y_1,$$

that is, $\eta_0 = (1 - \delta_i)\beta_i + \delta_i\eta_1$ and therefore by (5.44)

$$\beta_i = \frac{\eta_0 - \delta_i\eta_1}{1 - \delta_i} = \eta_0(1 - u(z_i)) + u(z_i)\eta_1 = \eta_0 + u(z_i)(\eta_1 - \eta_0) = v(z_i)$$

as in (5.43).

Finally, if $x_1 < z_i \lesssim y_1$, then by construction of $u(z_i)$ we have $x_1 \sim x_0 \wedge_{\gamma_i} z_i$ with $u(z_i) = 1/\gamma_i$. Then $z_i \sim y_0 \wedge_{\beta_i} y_1$ and thus

$$x_1 \sim x_0 \wedge_{\gamma_i} z_i \sim (y_0 \wedge_{\eta_0} y_1) \wedge_{\gamma_i} (y_0 \wedge_{\beta_i} y_1) = y_0 \wedge_{\eta_1} y_1 ,$$

that is, $\eta_1 = (1 - \gamma_i)\eta_0 + \gamma_i\beta_i$ and therefore

$$\beta_i = \frac{\eta_1 - (1 - \gamma_i)\eta_0}{\gamma_i} = \frac{\eta_1}{\gamma_i} - \frac{\eta_0}{\gamma_i} + \eta_0 = u(z_i)(\eta_1 - \eta_0) + \eta_0 = v(z_i)$$

again as in (5.43). This shows (5.36) for all z_i as claimed, and therefore the theorem.

$\square$

5.9   Risk Aversion

So far, we have considered an abstract set X of outcomes and lotteries for these outcomes, and have shown that a utility function $u : X \to \mathbb{R}$ exists that represents the player's preference for these lotteries if this preference satisfies certain consistency axioms.

In this section, we assume that X is a *real interval*. Each outcome is therefore a real number. This is typically a payment in *money* which the player wants to maximize. Is there any relationship between x and $u(x)$, apart from u being monotonically increasing? The concept of *risk aversion* means that if P is a lottery over X with an *expected monetary amount* $\hat{x}$, then the player prefers to get the certain amount $\hat{x}$ to the lottery P. As an example, suppose your current wealth is $\hat{x}$ dollars, and you are offered a lottery where you win or lose with equal probability 100 dollars. This lottery has outcomes $\hat{x} - 100$ and $\hat{x} + 100$ with probability $\frac{1}{2}$ each, and its expected value is clearly $\hat{x}$ dollars. If you prefer this safe amount (i.e., not play the lottery), then you are risk averse (for this specific lottery). A *risk neutral* player is always indifferent between $\hat{x}$ and P, and a *risk seeking* player would prefer the lottery. None of these concepts necessarily has to apply in general, because the preference may depend on P. However, in practice a player is typically risk averse. We will see that this is equivalent to the utility function being *concave*. In order to define concavity, consider the background material and Figure 5.4 on line segments, which provide a useful geometric view of probabilities.

A function u is called *concave* if any two points on the graph of the function are connected by a line segment that is *nowhere above* the graph of the function.

Background material: Line segments

Figure 5.4 shows two points x and y, here in the plane, but the picture may also be regarded as a suitable view of the situation in a higher-dimensional space. The (infinite) *line* that goes through the points x and y is obtained by adding to the point x, regarded as a vector, any multiple of the difference vector $y - x$. The resulting vector $x + \lambda \cdot (y - x)$, for $\lambda \in \mathbb{R}$, is x when $\lambda = 0$ and y when $\lambda = 1$. Figure 5.4 shows some examples a, b, c of other points on that line. When $0 \le \lambda \le 1$, like for point a, the resulting points give the *line segment* that joins x and y. If $\lambda > 1$, then one obtains points on the line through x and y on the other side of y relative to x, like the point b in Figure 5.4. For $\lambda < 0$, the corresponding point, like c in Figure 5.4, is on the other side of x relative to y.

The expression $x + \lambda(y - x)$ can be re-written as

$$(1 - \lambda)x + \lambda y \tag{5.45}$$

where the given points x and y appear only once. This expression (5.45) (with $1 - \lambda$ as the coefficient of the first point x and λ as the coefficient of the second point y) shows how the real interval $[0, 1]$ naturally maps to the line segment that joins x and y via the possible values of $\lambda \in [0, 1]$, with $0 \mapsto x$ and $1 \mapsto y$ for the endpoints. See also the right diagram in Figure 5.4.

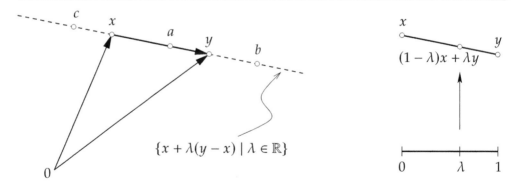

Figure 5.4   Left diagram: The *line* through the points x and y is given by the points $x + \lambda(y - x)$ where $\lambda \in \mathbb{R}$. Examples are point a for $\lambda = 0.6$, point b for $\lambda = 1.5$, and point c for $\lambda = -0.4$. The *line segment* that connects x and y results when λ is restricted to $0 \le \lambda \le 1$. The right diagram shows how the real interval $[0, 1]$ is naturally mapped to that line segment via the map $\lambda \mapsto (1 - \lambda)x + \lambda y$.

That is, for all x and x' in the domain of u,

$$(1 - \lambda)u(x) + \lambda u(x') \le u((1 - \lambda)x + \lambda x') \qquad \text{for all } \lambda \in [0, 1]. \tag{5.46}$$

As an example, Figure 5.5 shows the function $u(x) = \sqrt{x}$ defined on $[0, 1]$. On the graph of the function (as a subset of $\mathbb{R}^2$) we show three white dots that are the points $(x, u(x))$, $((1 - \lambda)x + \lambda x', u((1 - \lambda)x + \lambda x'))$, and $(x', u(x'))$. The black dot is the convex combination $(1 - \lambda) \cdot (x, u(x)) + \lambda \cdot (x', u(x'))$ of the outer white dots, which is below the middle white dot, as asserted in (5.46). This holds generally and shows that u is concave.

If we have equality in (5.46), then that part of the graph of the function is itself a line segment. If we never have equality, that is, (5.46) holds as a strict inequality whenever $x \neq x'$ and $0 < \lambda < 1$, then the function u is said to be *strictly concave*.

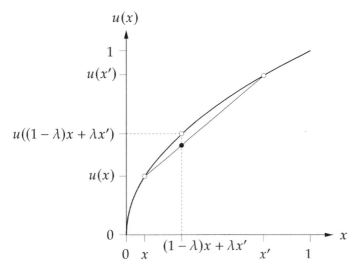

Figure 5.5 Example of a concave function $u(x) = \sqrt{x}$ on the interval $[0, 1]$.

It is easy to see that a function that is *twice differentiable* is concave if and only if its second derivative is always nonnegative, and strictly concave if its second derivative is always positive (as in the case of the square root function $\sqrt{x}$ for $x > 0$). Furthermore, it can be shown that, except at the endpoints of the interval that is the domain of the function, concavity implies continuity.

The following observation states that risk aversion is equivalent to the concavity of the player's utility function. For simple lotteries its proof is almost immediate.

Proposition 5.5. *Let X be a real interval and let $u : X \to \mathbb{R}$ be an expected-utility function. Consider any lottery P over X with expected value $\hat{x} = \mathbb{E}(P)$. Then the player is risk averse, that is, always $P \precsim \hat{x}$, if and only if u is concave.*

Proof. Because X is a real interval, any lottery P over X is a random variable that has an expectation $\hat{x} = \mathbb{E}(P)$, where $\hat{x} \in X$. Consider first (see (5.16)) a simple lottery $P = x \wedge_\lambda x'$ for $\lambda \in [0, 1]$ and $x, x' \in X$, where $\hat{x} = \mathbb{E}(P) = (1 - \lambda)x + \lambda x'$. Then $P \precsim \hat{x}$ is equivalent to $\mathbb{E}(u, P) = (1 - \lambda)u(x) + \lambda u(x') \leq u(\hat{x})$, which is exactly the concavity condition (5.46).

It remains to show that if u is concave then $P \precsim \mathbb{E}(P)$ for any lottery P that assigns positive probabilities $p_1, \ldots, p_n$ to outcomes $x_1, \ldots, x_n$ for $n \geq 3$. As inductive hypothesis, assume this is true for $n - 1$ instead of n (we have just shown it for $n = 2$). Write P as $Q \wedge_{p_n} x_n$ where Q assigns probability $\frac{p_i}{1-p_n}$ to x_i for $1 \le i \le n - 1$. By inductive hypothesis, $\mathbb{E}(Q, u) \le u(\mathbb{E}(Q))$. Therefore (where the second inequality holds because u is concave),

$$
\begin{aligned}
\mathbb{E}(P, u) &= \sum_{i=1}^{n-1} p_i u(x_i) + p_n x_n \\
&= (1 - p_n)\mathbb{E}(Q, u) + p_n x_n \\
&\le (1 - p_n)u(\mathbb{E}(Q)) + p_n x_n \\
&\le u((1 - p_n)\mathbb{E}(Q) + p_n x_n) = u(\mathbb{E}(P)),
\end{aligned}
$$

which completes the proof.   □

We stress again that risk aversion requires real-valued outcomes, such as money, where the expectation of a lottery is itself an outcome which can be compared with the lottery.

Beyond concavity, risk aversion is not easy to quantify. If the player has a wealth of $\hat{x}$ dollars and is risk averse, he will prefer $\hat{x}$ to an equal gamble between $\hat{x} - 100$ and $\hat{x} + 100$ dollars. The player should therefore be willing to pay a "risk premium" π so that he is indifferent between safely receiving $\hat{x} - \pi$ and the gamble. For the utility function $u(x) = -e^{-cx}$ for some $c > 0$, it is easy to see that π does not depend on $\hat{x}$, and that this is up to positive-affine transformations the only concave utility function with this property. If the values of $\hat{x}$ are realistic, and the lottery itself (like winning or losing 100 dollars) is comparable to the risks under consideration, then estimating the risk premium π may be a useful approach to constructing the utility function.

5.10   Discussion and Further Reading

The consistency axioms that imply the existence of an expected-utility function are sometimes called "rationality axioms". A player is called "rational" when he acts in accordance with his own consistent preferences. In game theory, further assumptions are that all players know the entire game, including the preferences of all players, that everybody is rational, and often that this is *common knowledge* (everybody knows that everybody knows, and so on). For more detailed discussions see Aumann and Brandenburger (1995), Aumann and Dreze (2008), and the book by Perea (2012).

Game theory, and economic theory in general, is often criticized for these idealized assumptions. It should be clear that the assumptions are part of the mathematical model, which may not be particularly realistic, and need to be questioned when considering the conclusions of a game-theoretic analysis. These conclusions already pose difficulties when a solution (such as equilibrium) is not unique. But even if unique, the conclusions are normative (stating what players should do) rather than descriptive (stating what players actually do).

A further source of misunderstanding game theory is the use of words for mathematical concepts that have strong connotations elsewhere. It is clear that a game tree is not an actual tree but describes a sequence of decision "branches", so "tree" can be used as a neutral word. However, the word "utility" suggests an intrinsic strength of preference associated with the "usefulness" of an outcome. As argued in this chapter, it should be considered as a mathematical tool for representing a preference for risky outcomes.

Because of the connotations of the word "utility", we instead use the more neutral and active term "payoff" in the context of games in this book. However, it is useful to remember that payoffs represent a preference and can therefore be scaled with positive-affine transformations.

A number of classic books on game theory, such as Luce and Raiffa (1957) and Myerson (1991), start with the rationality assumptions that justify the use of expected payoffs. Unusually, Osborne and Rubinstein (1994) even define games in terms of preference relations rather than payoffs.

The concept of expected utility is generally attributed to Daniel Bernoulli (1738), who proposed the logarithm as a utility function for lotteries that involve money. For an informative and entertaining historical account see Szpiro (2020), and also Szpiro (2010) on approaches to group decision making, the Condorcet Paradox, and social choice in general. The consistency axioms for an expected-utility function are due to, and often named after, von Neumann and Morgenstern (1947). The paradox in Figure 5.2 has been described by Allais (1953). For a short history of the discussion around these axioms, and further references, see Moscati (2016).

A classic book on utility theory, which connects axioms about preferences and their representation with a utility function, is Fishburn (1970). An abstract approach to preferences, due to Herstein and Milnor (1953), is called a "mixture set" where two lotteries (including safe outcomes) P and Q and a probability α are combined into a new lottery $(1 - \alpha)P + \alpha Q$ as in (5.15), with suitable axioms about these mixtures of P, Q, α. Rather than considering such abstract "mixtures", we have kept their meaning as lotteries, so that the algebraic manipulations, for example as used in the proof of Theorem 5.4, stay concrete.

Utility theory is not a good descriptive theory of human decision making, as documented and extensively studied by the psychologists Daniel Kahneman and Amos Tversky. Their collaboration and research is vividly described in the popular-science book by Lewis (2016). The example (5.10) of intransitive preferences is due to Tversky (1969).

An important paper on risk aversion is Pratt (1964).

5.11   Exercises for Chapter 5

Exercise 5.1. Find positive weights w in a utility function of the form $u(c_1, c_2) = c_1 + w \cdot c_2$ to create at least four different strict rankings among the candidates x, y, z in (5.10). [*Hint: Experiment with some values of w and vary them. What are the values of $u(c_1, c_2)$ for the three candidates as a function of w?*]

Exercise 5.2. Complete the proof of Proposition 5.2 for the third case $x_1 \prec z$ in (5.21), to show that $v(z) = a\,u(z) + b$. As stated in the proof, choose a simple lottery $Q = x_0 \wedge_\gamma z$ so that the player is indifferent between that lottery and x_1, and proceed similarly to (5.23) using (5.18) and (5.20).

Exercise 5.3.
(a) Prove the equivalence (5.25), including the inequalities, for real numbers A, B, C that fulfill $A < C$.

(b) Prove $(1 - \alpha)\mathbb{E}(u, P) + \alpha\mathbb{E}(u, R) = \mathbb{E}(u, (1 - \alpha)P + \alpha R)$ for two lotteries P and R, which is the second equation in (5.30).

6

Mixed Equilibrium

A game in strategic form does not always have an equilibrium in which each player chooses her strategy deterministically. As we describe in this chapter, Nash (1951) showed that any finite strategic-form game has an equilibrium if players are allowed to use *mixed* strategies. A mixed strategy is an actively randomized choice of the player's given "pure" strategies according to certain probabilities. A profile of mixed strategies is called a *mixed equilibrium* if no player can improve her expected payoff by unilaterally changing her strategy.

In this chapter, we briefly recall from Chapter 5 how a player's payoffs represent her preference for random outcomes via an *expected-utility* function. This is illustrated with a single-player decision problem whether to comply or not with a legal requirement, which is then turned into a game. The game is known as an *Inspection game* between an inspector as player I and an inspectee as player II. This game has no equilibrium in pure strategies. However, active randomization comes up naturally in this game, because the inspector can and will choose to inspect only some of the time. The mixed equilibrium in this game demonstrates many aspects of general mixed equilibria, in particular that a player's randomization probabilities depend on the payoffs of the *other* player.

We then turn to general two-player games in strategic form, already studied in Chapter 3, here called bimatrix games. The payoff matrices for the two players are convenient for representing expected payoffs via multiplication with vectors of probabilities that are the players' mixed strategies. These mixed strategies, as vectors of probabilities, have a geometric interpretation, which we study in detail (and even more so in Chapter 9). Mixing with probabilities is geometrically equivalent to taking *convex combinations*.

In a two-player game we usually name the players I and II, and denote their pure strategies as rows $i = 1, \ldots, m$ and columns $j = 1, \ldots, n$, respectively. In N-player games, which we only consider for general definitions, the players are numbered as players $i = 1, \ldots, N$, which is the most commonly used notation.

The *best-response condition* gives a convenient, finite condition for when a mixed strategy is a best response to the mixed strategy of the other player (or against the mixed strategies of all other players in a game with more than two players). Namely, only the *pure* strategies that are best responses are allowed to be played with positive probability by a best-response mixed strategy. These pure best responses must have maximal, and hence equal, expected payoff. This is stated as a "trivial condition" by Nash (1951, p. 287), and it is indeed not hard to prove (see Proposition 6.1). However, the best-response condition is central for understanding and computing mixed-strategy equilibria.

In Section 6.5, we state and prove Nash's theorem that every game has an equilibrium. We follow Nash's original proof (for two players, to simplify notation), and state and briefly motivate *Brouwer's fixed-point theorem* used in that proof. In Chapter 7 we prove the fixed-point theorem itself.

The important final Sections 6.6–6.8 show how to find equilibria in bimatrix games, in particular when one player has only two strategies.

The special and interesting case of *zero-sum* games is treated in Chapter 8.

6.1 Prerequisites and Learning Objectives

Prerequisites are Chapter 3 and knowing that payoffs are only unique up to positive-affine transformations, as discussed in Chapter 5. Fluency with matrix multiplication is useful, which we will recall as background material. We also need the algebraic representation of line segments explained in Figure 5.4.

After studying this chapter, you should be able to:

- explain what mixed strategies and mixed equilibria are;
- state the main theorems about mixed strategies, namely the best-response condition, and Nash's theorem on the existence of mixed equilibria;
- (as an important skill) find all mixed equilibria of small bimatrix games, using the "difference trick" and, where necessary, the upper-envelope method;
- give the definition of a degenerate game, and find all equilibria of small degenerate games.

6.2 Compliance Inspections

In this section, we consider a two-player *Inspection game* where randomized or *mixed* strategies occur naturally. Against a mixed strategy, the other player faces a lottery, which is evaluated according to its expected payoffs, as studied in detail in the preceding Chapter 5.

To recap this briefly, Figure 6.1 shows a game with a single player who can decide to *comply* with a legal requirement, like buying a train ticket, or to *cheat*

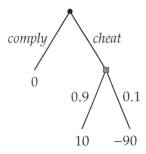

Figure 6.1 One-player decision problem between *comply* and *cheat*. With the stated payoffs, the player is indifferent.

otherwise. The payoff when she chooses to comply is 0. Cheating involves a 10 percent chance of getting caught and having to pay a penalty, stated as the negative payoff −90, and otherwise a 90 percent chance of gaining a payoff of 10. With these numbers, cheating leads to a random outcome with an expected payoff of $0.9 \cdot 10 + 0.1 \cdot (-90)$, which is zero, so that the player is indifferent between her two moves. In fact, assuming that the two outcomes with payoffs 0 and 10 have been set as origin and unit of the utility function, the utility of −90 for "getting caught" is determined by this indifference. If the player is indifferent when the probability of getting caught is only four percent, then −90 has to be replaced by $-u$ so that

$$0 = 0.96 \cdot 10 + 0.04 \cdot (-u),$$

which is equivalent to $u = 9.6/0.04 = 240$. Also, recall that the origin and unit of the utility function are arbitrary and can be changed by a positive-affine transformation. For example, $0, 10, -90$ could be replaced by $0, 1, -9$ or $-1, 0, -10$.

In practice, the preference for risk of a player may not be known. A game-theoretic analysis should be carried out for different choices of the payoff parameters in order to test how much they influence the results.

We now consider the Inspection game. For player II, called *inspectee*, the strategy *comply* means again to fulfill some legal obligation (such as buying a transport ticket, or paying tax). The inspectee has an incentive to violate this obligation by choosing to *cheat*. Player I, the *inspector*, should verify that the inspectee complies, but doing so requires inspections which are costly. If the inspector inspects and catches the inspectee cheating, then the inspectee has to pay a penalty.

Figure 6.2 shows possible payoffs for the Inspection game. The outcome that defines the reference payoff zero to both players is that the inspector chooses *Don't Inspect* and the inspectee chooses to comply. Without inspection, the inspectee prefers to cheat, represented by her payoff 10, with negative payoff −10 to the inspector. The inspector may also decide to *Inspect*. If the inspectee complies, inspection leaves her payoff 0 unchanged, while the inspector incurs a cost with a negative payoff of −1. If the inspectee cheats, however, inspection will result in

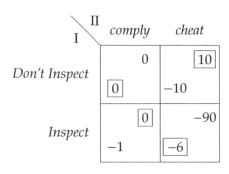

Figure 6.2 Inspection game between the inspector (player I) and inspectee (player II).

a penalty (payoff −90 for player II) and still create a certain amount of hassle for player I (payoff −6).

In all cases, player I would strongly prefer if player II complied, but this is outside of player I's control. However, the inspector prefers to inspect if the inspectee cheats (because −6 is better than −10), as shown by the best-response box for payoff −6 in Figure 6.2. If the inspector always preferred *Don't Inspect*, then this would be a dominating strategy and be part of a (unique) equilibrium where the inspectee cheats.

The best-response payoffs in Figure 6.2 show that this game has no equilibrium in pure strategies. If any player settles on a deterministic choice (like *Don't Inspect* by player I), then the best reponse of the other player is unique (here to cheat by player II), to which the original choice is not a best reponse (player I prefers to inspect when the other player cheats, against which player II in turn prefers to comply). The strategies in an equilibrium must be best responses to each other, so in this game this fails for any pair of pure strategies.

What should the players do in the game of Figure 6.2? One possibility is that they prepare for the worst, that is, choose a so-called max-min strategy. A max-min strategy maximizes the player's worst payoff against all possible choices of the opponent. The max-min strategy (as a pure strategy) for player I is to inspect (where the inspector guarantees himself payoff −6), and for player II it is to comply (which guarantees her a payoff of 0). However, this is not an equilibrium and hence not a stable recommendation to the two players, because player I could switch his strategy to *Don't Inspect* and improve his payoff.

A *mixed strategy* of player I in this game is to inspect only with a certain probability. In the context of inspections, randomizing is also a practical approach that reduces costs. Even if an inspection is not certain, a sufficiently high chance of being caught should deter the inspectee from cheating.

A mixed strategy of player I is given by a probability p for *Inspect*, and thus $1 - p$ for *Don't Inspect*. Instead of his two pure strategies, player I can now choose a probability $p \in [0, 1]$. This includes his pure strategies *Don't Inspect* when $p = 0$ and *Inspect* when $p = 1$. A mixed strategy of player II is given by a probability q

for *cheat* and $1 - q$ for *comply*. Her choice between two pure strategies is thereby extended to any choice of $q \in [0,1]$. In effect, this means the players now each have an infinite strategy set $[0,1]$ with payoffs computed as expected payoffs. This new game is called the *mixed extension* of the original 2×2 game.

When does the strategy pair $(p,q) \in [0,1] \times [0,1]$ define an equilibrium (which will be called a *mixed equilibrium*)? These strategies have to be best responses to each other. Suppose player I chooses p and player II has to determine her best response. We first consider her possible pure responses. Her expected payoff for *comply* is $(1-p) \cdot 0 + p \cdot 0 = 0$, which is independent of p. Her expected payoff for *cheat* is $(1-p) \cdot 10 + p \cdot (-90) = 10 - 100p$. In effect, player II faces the decision situation in Figure 6.1 with 0.9 and 0.1 replaced by $1 - p$ and p. Clearly, she prefers *comply* when $0 > 10 - 100p$, that is, $p > 0.1$, and *cheat* when $0 < 10 - 100p$, that is, $p < 0.1$, and she is indifferent if and only if $0 = 10 - 100p$, that is, $p = 0.1$.

In the mixed extension, player II can also choose to mix by choosing q so that $0 < q < 1$. We claim that this only happens when she is *indifferent*. Namely, suppose that p is very low, for example $p = 0.01$, so that *cheat* gives an expected payoff of $0.99 \cdot 10 + 0.01 \cdot (-90) = 9$. With a mixed strategy q, player II's expected payoff is $0 \cdot (1 - q) + 9q = 9q$, which is clearly maximized for $q = 1$, which is her pure best response *cheat*. Similarly, if p is high, for example $p = 0.2$, then *cheat* gives an expected payoff of $0.8 \cdot 10 + 0.2 \cdot (-90) = -10$. With a mixed strategy q, player II's expected payoff is $0 \cdot (1 - q) - 10q = -10q$, which is maximized for $q = 0$, which is her pure best response *comply*. Only when *comply* and *cheat* give the same expected payoffs, player II's expected payoff does not depend on q and she can in fact choose *any* $q \in [0,1]$ as a best response. Figure 6.3 shows a drawing of the best-response probability q against p.

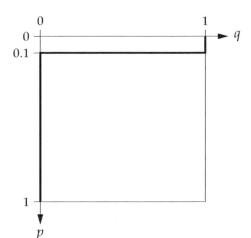

Figure 6.3 Best-response mixed-strategy probability q for *cheat* (black line) of player II in the Inspection game, against the mixed-strategy probability p for *Inspect* of player I. The mixed-strategy square is drawn so that its corners match the four cells in Figure 6.2.

In summary, the only case where player II can possibly randomize between her pure strategies is if both strategies give her the same expected payoff. As

stated and proved formally in Proposition 6.1 below, it is never optimal for a player
to assign a positive probability to a pure strategy that is inferior to other pure
strategies, for given mixed strategies of the other players.

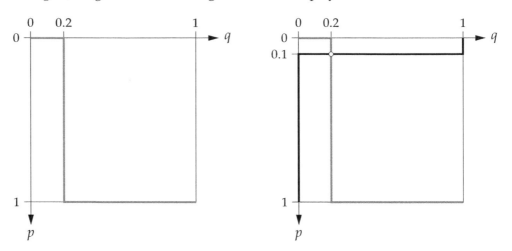

Figure 6.4 Left: Best-response mixed-strategy probability p (gray line) of player I
against q. Right: The intersection of the two best-response lines determines the
unique mixed equilibrium where $p = 0.1$ and $q = 0.2$.

Similarly, consider a mixed strategy of player II where she cheats with probability q. The expected payoffs for player I's pure strategies are $-10q$ for *Don't Inspect*
and $(-1)(1-q)-6q = -1-5q$ for *Inspect*. If player I plays *Inspect* with probability p,
then his expected payoff is therefore $(-10q)(1-p)+(-1-5q)p = -10q+p(5q-1)$.
This is a linear function of p, and has its maximum at $p = 0$ if $5q - 1 < 0$, that is,
$q < 0.2$, and at $p = 1$ if $5q - 1 > 0$, that is, $q > 0.2$, and only if $q = 0.2$ does it not
depend on p so that $p \in [0, 1]$ can then be chosen arbitrarily. This is shown in the
left diagram in Figure 6.4.

The right diagram in Figure 6.4 shows the intersection of the two best-response
lines with the unique mixed equilibrium of the game at $p = 0.1$ and $q = 0.2$
when the two mixed strategies are mutual best responses. It is similar to the
best-response diagram in Figure 3.10 for the Cournot duopoly, except for a quarter
clock-wise turn so that the vertical and horizontal parameters p and q match the
rows and columns in the 2×2 game in Figure 6.2.

In the unique mixed equilibrium with $p = 0.1$ and $q = 0.2$ of the Inspection
game, the expected payoffs are

$$-10q = -1(1-q) - \quad 6q = -2 \quad \text{for player I,}$$
$$0 = 10(1-p) - 90p = \quad 0 \quad \text{for player II.}$$

The equilibrium probabilities q and p are (independently) determined by these
equations. It is noteworthy that they depend on the *opponent's payoffs* and not on

the player's own payoffs (which can therefore be changed as long as the circular preference structure in Figure 6.2 remains intact). For example, one would expect that a more severe penalty than -90 for being caught lowers the probability of cheating in equilibrium. In fact, it does not. What does change is the probability p of inspection, which is reduced until the inspectee is indifferent.

This dependence on opponent payoffs is the most counter-intuitive property of mixed equilibrium. Moreover, mixing itself seems paradoxical when the player is indifferent in equilibrium. If player II, for example, can equally comply or cheat, why should she randomize? In particular, she could comply and get payoff zero for certain, which is simpler and safer. The answer is that precisely because there is no incentive to choose one strategy over the other, a player can mix, and only in that case can there be an equilibrium. If player II would comply for certain, then the only optimal choice of player I is not to inspect, but then complying is not optimal, so this is not an equilibrium.

6.3 Bimatrix Games

In the following, we discuss mixed equilibria for general games in strategic form. Most ideas can be explained, with much simpler notation, for two-player games. Then the strategies of player I and player II are the rows and columns of a table with two payoffs in each cell, so that the game is defined by two payoff matrices. Some basic facts about matrix manipulation are in the background material.

Background material: Matrix multiplication

If A is an $m \times k$ matrix and B is an $k \times n$ matrix, then their matrix product $A \cdot B$ (or AB for short) is the $m \times n$ matrix C with entries

$$c_{ij} = \sum_{s=1}^{k} a_{is} b_{sj} \qquad (1 \le i \le m, \quad 1 \le j \le n). \qquad (6.1)$$

When writing down a matrix product AB it is assumed to be defined, that is, A has as many columns as B has rows. Matrix multiplication is associative, that is, $(AB)C = A(BC)$, written as ABC, for matrices A, B, C.

When dealing with matrices, we use the following notation throughout. All matrices considered here are real matrices (with entries in $\mathbb{R}$). The set of $m \times n$ matrices is $\mathbb{R}^{m \times n}$. The transpose of an $m \times n$ matrix A is written $A^\top$, which is the $n \times m$ matrix with entries a_{ji} if a_{ij} are the entries of A, for $1 \le i \le m$, $1 \le j \le n$.

All vectors are column vectors, so if $x \in \mathbb{R}^n$ then x is an $n \times 1$ matrix and $x^\top$ is the corresponding row vector in $\mathbb{R}^{1 \times n}$, and row vectors are easily identified by this transposition sign. The components of a vector $x \in \mathbb{R}^n$ are $x_1, \ldots, x_n$. We write the *scalar product* of two vectors x, y in $\mathbb{R}^n$ as the matrix product $x^\top y$. In pictures (with a square for each matrix entry, here for $n = 4$):

$$x^\top y \;:\quad \boxed{} \cdot \boxed{} = \boxed{} \tag{6.2}$$

We use matrix multiplication where possible. In particular, scalars are treated like 1×1 matrices. A column vector x is multiplied with a scalar λ from the *right*, and a row vector $x^\top$ is multiplied with a scalar λ from the *left*,

$$x\lambda \;:\quad \boxed{} \cdot \boxed{} = \boxed{} , \qquad \lambda x^\top \;:\quad \boxed{} \cdot \boxed{} = \boxed{} \tag{6.3}$$

For more details see the background material on visualizing matrix multiplication.

> $\Rightarrow$ Exercise 6.1 compares $x^\top y$ and $xy^\top$.

Furthermore, we use the special vectors $\mathbf{0}$ and $\mathbf{1}$ with all components equal to 0 and 1, respectively, where their dimension depends on the context. Inequalities like $x \geq \mathbf{0}$ hold for all components; note that the resulting order $\leq$ between vectors is only a partial order.

Recall that a game in strategic form is specified by a finite set of "pure" strategies for each player, and a payoff for each player for each strategy profile, which is a tuple of strategies, one for each player. The game is played by each player independently and simultaneously choosing one strategy, whereupon the players receive their respective payoffs.

In this chapter we focus on two-player games. Unless specified otherwise, we assume that player I has m strategies and player II has n strategies. The m strategies of the row player I are denoted by $i = 1, \ldots, m$, and the n strategies of the column player II are denoted by $j = 1, \ldots, n$.

An $m \times n$ two-player game in strategic form is also called a *bimatrix game* (A, B). Here, A and B are the two $m \times n$ payoff matrices. The m rows are the pure strategies i of player I and the n columns are the pure strategies j of player II. For a row i and column j, the matrix entry a_{ij} of A is the payoff to player I, and the matrix entry b_{ij} of B is the payoff to player II. The staggered payoffs in a table such as Figure 3.1 for the Prisoner's Dilemma show A as the matrix of lower-left payoffs in each cell, and B as the matrix of upper-right payoffs in each cell, here

$$A = \begin{pmatrix} 2 & 0 \\ 3 & 1 \end{pmatrix}, \qquad B = \begin{pmatrix} 2 & 3 \\ 0 & 1 \end{pmatrix}.$$

This bimatrix game is *symmetric* because $B = A^\top$.

Background material: Visualizing matrix multiplication

In a matrix product $C = AB$, the entry c_{ij} of C in row i and column j is the scalar product of the ith row of A with the jth column of B, according to (6.2) and (6.1). There are two useful ways of visualizing the product AB. Let $A \in \mathbb{R}^{m \times k}$ and $x \in \mathbb{R}^k$. Then $Ax \in \mathbb{R}^m$. Let $A = [A_1 \cdots A_k]$, that is, $A_1, \ldots, A_k$ are the columns of A. Then $Ax = A_1 x_1 + \cdots + A_k x_k$, that is, Ax is the linear combination of the columns of A with the components $x_1, \ldots, x_k$ of x as coefficients. In pictures for $m = 3$, $k = 4$:

$$\tag{6.4}$$

Let $B \in \mathbb{R}^{k \times n}$ and $B = [B_1 \cdots B_n]$. Then the jth column of AB is the linear combination AB_j of the columns of A with the components of B_j as coefficients. We can visualize the columns of AB as follows (for $n = 2$):

The second view of AB is the same but using rows. Let $y \in \mathbb{R}^k$, so that $y^\top B$ is a row vector in $\mathbb{R}^{1 \times n}$, given as the linear combination $y_1 b_1^\top + \cdots + y_k b_k^\top$ of the rows $b_1^\top, \ldots, b_k^\top$ of B (which we can write as $B^\top = [b_1 \cdots b_k]$),

$$\tag{6.5}$$

Let $A^\top = [a_1 \cdots a_m]$. Then the ith row of AB is the linear combination $a_i^\top B$ of the rows of B with the components of the ith row $a_i^\top$ of A as coefficients:

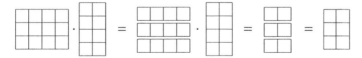

It is useful to acquire some routine in these manipulations. Most common is multiplication of a matrix with a column vector from the right as in (6.4) or with a row vector from the left as in (6.5).

A *mixed strategy* is a randomized strategy of a player. It is defined as a lottery (probability distribution) on the set of pure strategies of that player. This is played as an "active randomization", where the player selects each pure strategy according to its probability. When the other player considers a best response to such a mixed strategy, she is assumed to know the probabilities but not the outcome of the random choice, and bases her decision on the resulting expected payoffs.

A pure strategy is considered as a special mixed strategy, which chooses the pure strategy with probability one (in the same way as a deterministic outcome is considered a "random" outcome that occurs with probability one).

A mixed strategy is determined by the probabilities that it assigns to the player's pure strategies. For player I, a mixed strategy x is therefore identified with the m-tuple of probabilities $x_1, x_2, \ldots, x_m$ that it assigns to the pure strategies $1, 2, \ldots, m$ of player I, which we write as the (column) vector $x \in \mathbb{R}^m$. Similarly, a mixed strategy y of player II is an n-tuple of probabilities $y_1, \ldots, y_n$ for playing the pure strategies $j = 1, \ldots, n$, written as $y \in \mathbb{R}^n$.

The *sets of mixed strategies* of player I and player II are denoted by X and Y. The probabilities x_i in a mixed strategy x are nonnegative and sum to 1, where we can write the conditions $x_i \geq 0$ for $1 \leq i \leq m$ and $x_1 + \cdots + x_m = 1$ as $x \geq \mathbf{0}$ and $\mathbf{1}^\top x = 1$, where in this context $\mathbf{1} \in \mathbb{R}^m$ to match the dimension of x; when we similarly write $y_1 + \cdots + y_n = 1$ as $\mathbf{1}^\top y = 1$ then $\mathbf{1} \in \mathbb{R}^n$. That is,

$$X = \{x \in \mathbb{R}^m \mid x \geq \mathbf{0},\ \mathbf{1}^\top x = 1\}, \qquad Y = \{y \in \mathbb{R}^n \mid y \geq \mathbf{0},\ \mathbf{1}^\top y = 1\}. \tag{6.6}$$

For $x \in X$ and $y \in Y$, the expected payoff to player I is $x^\top A y$, and the expected payoff to player II is $x^\top B y$. It is best to think of $x^\top A y$ as $x^\top (A y)$, that is, as the scalar product of x with Ay. For each row i, the entry of Ay in row i is $(Ay)_i$, given by

$$(Ay)_i = \sum_{j=1}^{n} a_{ij}\, y_j \quad \text{for } 1 \leq i \leq m. \tag{6.7}$$

That is, $(Ay)_i$ is the *expected payoff to player I when he plays row i*. Therefore, $x^\top A y$ is the expected payoff to player I when the players use x and y, because

$$x^\top (Ay) = \sum_{i=1}^{m} x_i (Ay)_i = \sum_{i=1}^{m} x_i \sum_{j=1}^{n} a_{ij}\, y_j = \sum_{i=1}^{m} \sum_{j=1}^{n} x_i\, a_{ij}\, y_j. \tag{6.8}$$

Because the players choose their pure strategies i and j independently, the probability that they choose the pure strategy pair (i, j) is the product $x_i\, y_j$ of these probabilities, which is the coefficient of the payoff a_{ij} in (6.8).

Analogously, $x^\top B y$ is the expected payoff to player II when the players use the mixed strategies x and y. Here, it is best to read this as $(x^\top B)y$. The row vector $x^\top B$ is the vector of expected payoffs $(x^\top B)_j$ to player II for her columns j. Multiplied with the column probabilities y_j, the sum over all columns j is the expected payoff to player II, which in analogy to (6.8) is given by

$$(x^\top B)y = \sum_{j=1}^{n} (x^\top B)_j\, y_j = \sum_{j=1}^{n} \left(\sum_{i=1}^{m} x_i\, b_{ij} \right) y_j = \sum_{j=1}^{n} \sum_{i=1}^{m} x_i\, b_{ij}\, y_j.$$

6.4 The Best-Response Condition

A mixed-strategy equilibrium is a profile of mixed strategies such that no player can improve his expected payoff by unilaterally changing his own strategy. In a two-player game, an equilibrium is a pair (x, y) of mixed strategies such that x is a best response to y and vice versa. It does not seem easy to decide if x is a best response to y among all possible mixed strategies, that is, if x maximizes $x^\top Ay$ for all x in X, because X is an infinite set. However, the following proposition, known as the *best-response condition*, shows how to recognise this. This proposition is not difficult but important. We discuss it afterwards.

Proposition 6.1 (Best-response condition, two players). *Consider an $m \times n$ bimatrix game (A, B) and mixed strategies $x \in X$ and $y \in Y$. Let*

$$u = \max\{(Ay)_i \mid 1 \le i \le m\}, \qquad v = \max\{(x^\top B)_j \mid 1 \le j \le n\}. \qquad (6.9)$$

Then

$$(\forall \bar{x} \in X : x^\top Ay \ge \bar{x}^\top Ay) \quad \Leftrightarrow \quad x^\top Ay = u, \qquad (6.10)$$

that is, x is a best response to y if and only if $x^\top Ay = u$. This is equivalent to

$$x_i > 0 \implies (Ay)_i = u \quad (1 \le i \le m). \qquad (6.11)$$

Similarly, y is a best response to x if and only if $x^\top By = v$, which is equivalent to

$$y_j > 0 \implies (x^\top B)_j = v \quad (1 \le j \le n). \qquad (6.12)$$

Proof. Recall that $(Ay)_i$ is the ith component of Ay, which is the expected payoff to player I when he plays row i, as in (6.7). The maximum of these expected payoffs for the *pure* strategies of player I is u. Then

$$
\begin{aligned}
x^\top Ay &= \sum_{i=1}^{m} x_i (Ay)_i = \sum_{i=1}^{m} x_i (u - (u - (Ay)_i) \\
&= \sum_{i=1}^{m} x_i u - \sum_{i=1}^{m} x_i (u - (Ay)_i) \\
&= u - \sum_{i=1}^{m} x_i (u - (Ay)_i). \qquad (6.13)
\end{aligned}
$$

Because $x_i \ge 0$ and $u - (Ay)_i \ge 0$ for any row i, we have $\sum_{i=1}^{m} x_i(u - (Ay)_i) \ge 0$ and therefore $x^\top Ay \le u$. The expected payoff $x^\top Ay$ achieves the maximum u if and only if $\sum_{i=1}^{m} x_i(u - (Ay)_i) = 0$, that is, if $x_i > 0$ implies $(Ay)_i = u$, which shows (6.11). Condition (6.12) is shown analogously.

We repeat this argument using vector notation. First, the scalar product of two nonnegative vectors is nonnegative: for $x, w \in \mathbb{R}^m$,

$$x \ge \mathbf{0}, \quad w \ge \mathbf{0} \implies x^\top w \ge 0. \qquad (6.14)$$

Furthermore, (6.9) implies $\mathbf{1}u \geq Ay$, that is, $w = \mathbf{1}u - Ay \geq \mathbf{0}$. Hence,

$$x^\top Ay = x^\top\big(\mathbf{1}u - (\mathbf{1}u - Ay)\big) = x^\top \mathbf{1}u - x^\top(\mathbf{1}u - Ay)$$
$$= u - x^\top(\mathbf{1}u - Ay) \quad \leq u \qquad\qquad (6.15)$$

(note $x^\top \mathbf{1}u = (x^\top \mathbf{1})u = 1 \cdot u = u$, which shows the advantage of using matrix products consistently). Second, (6.14) can be extended to

$$x \geq \mathbf{0}, \quad w \geq \mathbf{0} \quad \Rightarrow \quad (x^\top w = 0 \quad \Leftrightarrow \quad x_i w_i = 0 \quad \text{for } 1 \leq i \leq m)$$

so that $x^\top Ay = u$ in (6.15) if and only if $x_i = 0$ or $u = (Ay)_i$ for $1 \leq i \leq m$, which says the same as (6.11). $\qquad\qquad\qquad\qquad\qquad\qquad\qquad\qquad\qquad\qquad\qquad\square$

The mixed strategy x is a best response to y if x gives maximum expected payoff $x^\top Ay$ to player I among all *mixed* strategies in X, as in the left statement in (6.10). In contrast, the payoff u in (6.9) is the maximum expected payoff $(Ay)_i$ among all *pure* strategies i of player I. According to (6.10), these payoffs are the same, so a player cannot *improve* his payoff by mixing. This is intuitive because mixing creates a *weighted average* $x^\top(Ay)$ of the expected payoffs $(Ay)_i$ with the probabilities x_i as weights as in (6.8).

Furthermore, (6.11) states that x is a best response to y if and only if x only chooses *pure best responses with positive probability* x_i. The condition whether a pure strategy is a best response is easy to check, because one only has to compute the m expected payoffs $(Ay)_i$ for $i = 1, \ldots, m$. For example, if player I has three pure strategies ($m = 3$), and the expected payoffs in (6.7) are $(Ay)_1 = 4$, $(Ay)_2 = 4$, and $(Ay)_3 = 3$, then only the first two strategies are pure best responses. If these expected payoffs are 3, 5, and 3, then only the second strategy is a best response. Clearly, at least one pure best response exists, because the numbers $(Ay)_i$ in (6.11) take their maximum u for at least one i.

The best-response condition is the central tool for finding mixed-strategy equilibria, which we describe here for an example; a related approach is described in Section 6.6. Consider the following 2×3 game (A, B), where we already marked the best-response payoffs and give the two payoff matrices (as determined by the table):

$$A = \begin{pmatrix} 0 & 4 & 0 \\ 2 & 2 & 1 \end{pmatrix}, \quad B = \begin{pmatrix} 3 & 0 & 2 \\ 1 & 6 & 4 \end{pmatrix}. \qquad (6.16)$$

This game has no equilibrium in pure strategies. For any pure strategy, its unique best response is a pure strategy of the other player, which does not result in an equilibrium, so that both players have to play mixed strategies that are not pure. We try to find a mixed equilibrium (x, y), were $x = (x_1, x_2)^\top$ and $y = (y_1, y_2, y_3)^\top$. (When writing down specific vectors, we will from now on omit the transposition sign for brevity and just write $x = (x_1, x_2)$, for example.)

Because player I has only two pure strategies, both have to be played with positive probability, that is, $x_1 > 0$ and $x_2 > 0$. According to the best-response condition (6.11), player I has to be indifferent between his two rows, that is,

$$(Ay)_1 = (Ay)_2 = u, \qquad 4y_2 = 2y_1 + 2y_2 + y_3 = u, \qquad 2y_2 = 2y_1 + y_3. \quad (6.17)$$

We have already seen that this cannot be achieved when y is a pure strategy, so we now consider any pair of strategies that are played with positive probability under y. Suppose first that these are the first two columns, that is, $y_1 > 0$ and $y_2 > 0$ and $y_3 = 0$. Then the unique solution to (6.17) (and $y_1 + y_2 = 1$) is $y_1 = y_2 = \frac{1}{2}$ with $u = 2$. In turn, by (6.12) player II has to be indifferent between her first two columns, that is,

$$(x^\top B)_1 = (x^\top B)_2 = v, \qquad 3x_1 + x_2 = 6x_2, \qquad 3x_1 = 5x_2 \quad (6.18)$$

which has the unique solution $x_1 = \frac{5}{8}$, $x_2 = \frac{3}{8}$ and $(x^\top B)_1 = (x^\top B)_2 = \frac{18}{8} = \frac{9}{4}$. However, we also have to make sure that $\frac{9}{4}$ is the best-response payoff to player II, a consideration that is not necessary in 2×2 games where it suffices that both players are indifferent between their two strategies. Here we have $(x^\top B)_3 = \frac{22}{8} = \frac{11}{4} > \frac{9}{4}$. That is, even though player II is indifferent between her first two columns, which is necessary because $y_1 > 0$ and $y_2 > 0$, these are not pure best responses against $x = (\frac{5}{8}, \frac{3}{8})$ as required by (6.12). So there is no mixed equilibrium where player II plays her first two columns.

Next, suppose player II chooses her first and third column, with $y_1 > 0$, $y_2 = 0$ and $y_3 > 0$. However, then the right equation in (6.17) states $0 = 2y_1 + y_3$, which is not possible. Intuitively, when player II plays only the first and third column in (6.16), then the bottom row D is the unique best response and player I cannot be made indifferent.

This leaves the final possibility that $y_1 = 0$ and $y_2 > 0$ and $y_3 > 0$ where (6.17) becomes $2y_2 = y_3$ and therefore $y_2 = \frac{1}{3}$, $y_3 = \frac{2}{3}$. Similar to (6.18), we now need

$$(x^\top B)_2 = (x^\top B)_3 = v, \qquad 6x_2 = 2x_1 + 4x_2, \qquad 2x_1 = 2x_2 \quad (6.19)$$

which has the unique solution $x_1 = x_2 = \frac{1}{2}$ with $(x^\top B)_2 = (x^\top B)_3 = v = 3$ where v is indeed the maximum of expected column payoffs because $(x^\top B)_1 = 2 < 3$. We have found the mixed equilibrium $x = (\frac{1}{2}, \frac{1}{2})$, $y = (0, \frac{1}{3}, \frac{2}{3})$. There is no equilibrium where player II chooses *all three* columns with positive probability because this implies the column indifferences (6.18) and (6.19), which have contradictory solutions x. The mixed equilibrium is therefore unique.

⇒ Exercise 6.2 asks you to find all mixed equilibria of a 3×2 game, which can be done similarly and illustrates the best-response condition. It also motivates methods for finding mixed equilibria that we develop in Section 6.6.

The best-response condition applies also to a game with more than two players, which we explain in the remainder of this section. In Definition 3.1 we defined a commonly used notation and the concepts of best response and equilibrium for an N-player game. The following definition defines its mixed extension. For the finite strategy set S_i of player i, we denote the set of mixed strategies over S_i by $\Delta(S_i)$, in agreement with Definition 5.1.

Definition 6.2 (Mixed extension). Consider an N-player game in strategic form with finite strategy set S_i for player $i = 1, \ldots, N$ and payoff $u_i(s)$ for each strategy profile $s = (s_1, \ldots, s_N)$ in $S = S_1 \times \cdots \times S_N$. Let $X_i = \Delta(S_i)$ be the set of mixed strategies σ_i of player i, where $\sigma_i(s_i)$ is the probability that σ_i selects the pure strategy $s_i \in S_i$. (If $\sigma_i(s_i) = 1$ for some s_i then σ_i is considered the same as the pure strategy s_i.) Then the *mixed extension* of this game is the N-player game in strategic form with the (infinite) strategy sets $X_1, \ldots, X_N$ for the N players, where for a mixed-strategy profile $\sigma = (\sigma_1, \ldots, \sigma_N)$ player i receives the *expected payoff*

$$U_i(\sigma) = \sum_{s \in S} u_i(s) \prod_{k=1}^{N} \sigma_k(s_k).$$

An equilibrium (or *Nash* equilibrium) σ of the mixed extension is called a *mixed equilibrium* of the original game. □

The following proposition extends the best-response condition to an N-player game. As in the case of two players, it allows us to verify that a mixed-strategy profile is an equilibrium by the condition that a player's mixed strategy only selects pure best responses with positive probability.

Proposition 6.3 (Best-response condition, N players). *Consider the mixed extension of an N-player game with the notation in Definition 6.2. Let $\sigma_i \in X_i$ be a mixed strategy of player i, let σ_{-i} be a partial profile of mixed strategies of the other $N - 1$ players, and let*

$$v_i = \max\{U_i(s_i, \sigma_{-i}) \mid s_i \in S_i\}$$

be the best possible expected payoff against σ_{-i} for a pure strategy s_i of player i. Then σ_i is a best response to σ_{-i} if and only if

$$\sigma_i(s_i) > 0 \implies U_i(s_i, \sigma_{-i}) = v_i \qquad (s_i \in S_i),$$

that is, if σ_i only selects pure best responses s_i against σ_i with positive probability.

Proof. The proof is the same as for Proposition 6.1 with the pure strategy s_i (where i denotes a player) instead of row i (where i denotes a pure strategy) and probability $\sigma_i(s_i)$ instead of x_i in (6.11). The crucial condition is that the expected payoff to player i is *linear* in his mixed-strategy probabilities $\sigma_i(s_i)$, because for a mixed-strategy profile σ,

$$
\begin{aligned}
U_i(\sigma) &= U_i(\sigma_i, \sigma_{-i}) = \sum_{s \in S} u_i(s) \prod_{k=1}^{N} \sigma_k(s_k) \\
&= \sum_{s_i \in S_i} \sum_{s_{-i} \in S_{-i}} u_i(s_i, s_{-i}) \prod_{k=1}^{N} \sigma_k(s_k) \\
&= \sum_{s_i \in S_i} \left(\sum_{s_{-i} \in S_{-i}} u_i(s_i, s_{-i}) \prod_{k=1,\, k \neq i}^{N} \sigma_k(s_k) \right) \sigma_i(s_i) \\
&= \sum_{s_i \in S_i} U_i(s_i, \sigma_{-i})\, \sigma_i(s_i).
\end{aligned} \tag{6.20}
$$

That is, the expected payoff $U_i(\sigma) = U_i(\sigma_i, \sigma_{-i})$ is exactly the expectation, with probabilities $\sigma_i(s_i)$, of the expected payoffs $U_i(s_i, \sigma_{-i})$ for the "rows" $s_i \in S_i$, by considering player i as the row player. In particular, if $N = 2$ and i is player 1, then σ_{-i} is the mixed strategy y of player 2, and $U_i(s_i, \sigma_{-i})$ is the expected payoff $(Ay)_{s_i}$ for row s_i. The proof in (6.13) with σ_i instead of x and v_i instead of u applies as before. □

> ⇒ Exercise 6.3 is about a game with three players.

We continue to focus on two-player games, partly because of the simpler notation. More importantly, for two players the best-response condition compares expected payoffs (such as $(Ay)_i$ for the rows i of the row player, say) that depend *linearly* on the mixed-strategy probabilities y_j of a single other player. In contrast, the corresponding expression $U_i(s_i, \sigma_{-i})$ in (6.20) for N players with $N > 2$ is nonlinear in the other players' mixed-strategy probabilities, which makes it much harder to find a mixed equilibrium in a general N-player game.

6.5 Existence of Mixed Equilibria

In this section, we give the original proof of John Nash (1951) that shows that any finite game has a mixed equilibrium. This proof uses Brouwer's fixed-point theorem (Theorem 6.4 below) which is about continuous functions on compact convex sets. A subset of $\mathbb{R}^d$ is *compact* if it is closed and bounded.

A subset C of $\mathbb{R}^d$ is *convex* if it contains with any two points x and y the line segment that connects x and y. As described in Figure 5.4 and (5.45), this is the case if and only for all $x, y \in C$ and $\lambda \in [0, 1]$ also $x(1 - \lambda) + y\lambda \in C$, where we now multiply the column vectors x and y from the right with the scalars $(1 - \lambda)$ and λ so that they are matrix products as in (6.3). For $\lambda \in [0, 1]$, the term $x(1 - \lambda) + y\lambda$ is called the *convex combination* of x and y (which is any point on the line segment that connects x and y). In general, a convex combination of points $z_1, z_2, \ldots, z_n$ in $\mathbb{R}^d$ is a linear combination $z_1\lambda_1 + z_2\lambda_2 + \cdots + z_n\lambda_n$ where the coefficients $\lambda_1, \ldots, \lambda_n$ are nonnegative and sum to one.

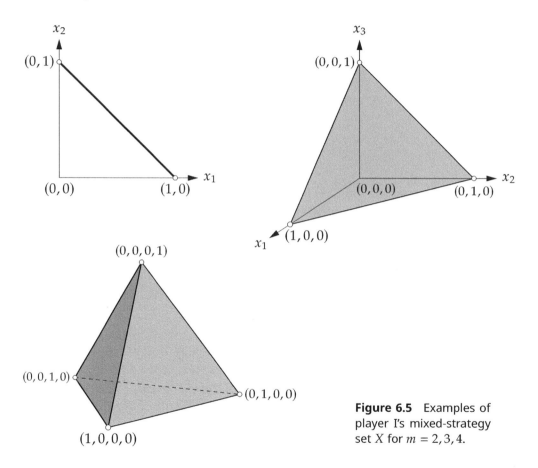

Figure 6.5 Examples of player I's mixed-strategy set X for $m = 2, 3, 4$.

The players' mixed-strategy sets are special convex sets. For an $m \times n$ game, the mixed-strategy sets of players I and II are X and Y as in (6.6). The ith *unit vector* e_i in $\mathbb{R}^m$, for $1 \le i \le m$, is the vector $(0, \ldots, 0, 1, 0, \ldots, 0)^\top$ where the ith component is one and every other component is zero. This vector belongs to X and represents the pure strategy i of player I. For any mixed strategy $x = (x_1, \ldots, x_m)^\top$ in X, clearly $x = e_1x_1 + e_2x_2 + \cdots + e_nx_m$, that is, x is the convex combination of the unit vectors $e_1, \ldots, e_m$ with the mixed-strategy probabilities $x_1, \ldots, x_m$ as coefficients (because they are nonnegative and sum to one). Similarly, any mixed strategy $y \in Y$ is the convex combination of the unit vectors in $\mathbb{R}^n$, which represent the pure strategies of player II.

Examples of X are shown in Figure 6.5 for $m = 2, 3, 4$ where X is a line segment, triangle, and tetrahedron, respectively. In general, the convex hull of the m unit vectors in $\mathbb{R}^m$ is called the *unit simplex*. It is easy to see that X and Y, and their Cartesian product $X \times Y$, are convex sets (see (6.23) below). In order to prove the existence of a mixed equilibrium, we will use the following theorem for a suitable continuous function $f : C \to C$ on the set $C = X \times Y$ of mixed strategy profiles.

Theorem 6.4 (Brouwer's fixed-point theorem). *Let C be a nonempty subset of $\mathbb{R}^d$ that is convex and compact, and let f be a continuous function from C to C. Then f has at least one* fixed point, *that is, a point z in C so that $f(z) = z$.*

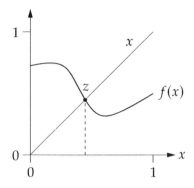

Figure 6.6 Proof of Brouwer's fixed-point theorem for $C = [0, 1]$.

Figure 6.6 illustrates Theorem 6.4 for the interval $C = [0, 1]$. If the continuous function $f : C \to C$ fulfills $f(0) = 0$, then 0 is a fixed point of f, and similarly if $f(1) = 1$ then 1 is a fixed point of f. Hence, we can assume that $f(0) > 0$ and $f(1) < 1$. Then the graph of the function f crosses the line of the identity function $x \mapsto x$ at least once for some point z where $f(z) = z$, which is the desired fixed point. More formally, the continuous function $x \mapsto x - f(x)$ is negative for $x = 0$ and positive for $x = 1$, and therefore zero for an intermediate value z with $0 < z < 1$ by the intermediate value theorem, that is, $z - f(z) = 0$ or $f(z) = z$.

The following simple examples show that the assumptions in Theorem 6.4 are necessary. Consider the function $f : \mathbb{R} \to \mathbb{R}$ defined by $f(x) = x + 1$. This is clearly a continuous function, and the set $\mathbb{R}$ of all real numbers is closed and convex, but f has no fixed point. Some assumption of the fixed-point theorem must be violated, and in this case it is compactness, because the set $\mathbb{R}$ is not bounded. Consider another function $f : [0, 1] \to [0, 1]$, given by $f(x) = x^2$. The fixed points of this function are 0 and 1. If we consider the function $f(x) = x^2$ as a function on the open interval $(0, 1)$, then this function has no longer any fixed points. In this case, the missing condition is that the function is not defined on a closed set, which is therefore not compact. Another function is $f(x) = 1 - x$ for $x \in \{0, 1\}$, where the domain of this function has just two elements, so this is a compact set. This function has no fixed points, which is possible because its domain is not convex. Finally, the function on $[0, 1]$ defined by $f(x) = 1$ for $0 \le x \le \frac{1}{2}$ and $f(x) = 0$ for

$\frac{1}{2} < x \leq 1$ has no fixed point, which is possible in this case because the function is not continuous. These examples demonstrate why the assumptions of Theorem 6.4 are necessary.

Theorem 6.4 is a major theorem in topology and much harder to prove for general dimension d. We will give a full proof in Chapter 7.

The following theorem states the existence of a mixed equilibrium. We prove it using Brouwer's fixed-point theorem along the lines of the original proof by Nash (1951), for notational simplicity for the case of two players.

Theorem 6.5 (Nash, 1951). *Every finite game has at least one mixed equilibrium.*

Proof. We will give the proof for an $m \times n$ two-player game (A, B), to simplify notation. It extends in the same manner to any finite number of players. The set C that is used in the present context is the product of the sets of mixed strategies of the players. Let X and Y be the sets of mixed strategies of player I and player II as in (6.6), and let $C = X \times Y$, which is clearly convex and compact.

Then the function $f : C \to C$ that we are going to construct maps a pair of mixed strategies (x, y) to another pair $f(x, y) = (\bar{x}, \bar{y})$. Intuitively, a mixed-strategy probability x_i of player I (and similarly y_j of player II) is changed to $\bar{x}_i$ such that it will increase if the pure strategy i does better than the current expected payoff $x^\top Ay$. If the pure strategy i does worse, then x_i will decrease unless x_i is already zero. In equilibrium, x is a best response to y and $x^\top Ay = u$ for the pure best-response payoff u according to (6.10), and all sub-optimal pure strategies have probability zero. This means that the mixed strategy x does not change, $\bar{x} = x$. Similarly, $\bar{y} = y$ if y is a best response to x. Therefore, the equilibrium property is equivalent to the fixed-point property $(x, y) = (\bar{x}, \bar{y}) = f(x, y)$.

In order to define f as described, consider the following functions $\chi : X \times Y \to \mathbb{R}^m$ and $\psi : X \times Y \to \mathbb{R}^n$. For each pure strategy i of player I, let $\chi_i(x, y)$ be the ith component of $\chi(x, y)$, and for each pure strategy j of player II, let $\psi_j(x, y)$ be the jth component of $\psi(x, y)$. The functions χ and ψ are defined by

$$\begin{aligned} \chi_i(x, y) &= \max\{0, (Ay)_i - x^\top Ay\} & (1 \leq i \leq m), \\ \psi_j(x, y) &= \max\{0, (x^\top B)_j - x^\top By\} & (1 \leq j \leq n). \end{aligned} \tag{6.21}$$

Recall that $(Ay)_i$ is the expected payoff to player I against y when he uses the pure strategy i, and that $(x^\top B)_j$ is the expected payoff to player II against x when she uses the pure strategy j, and that $x^\top Ay$ and $x^\top By$ are the expected payoffs to player I and player II. Hence, the difference $(Ay)_i - x^\top Ay$ is positive if the pure strategy i gives more than the average $x^\top Ay$ against y, zero if it gives the same payoff, and negative if it gives less. The term $\chi_i(x, y)$ is this difference, except that it is replaced by zero if the difference is negative. The term $\psi_j(x, y)$ is defined analogously. Thus, $\chi(x, y)$ is a nonnegative vector in $\mathbb{R}^m$, and $\psi(x, y)$ is a nonnegative vector in $\mathbb{R}^n$. The functions χ and ψ are continuous.

The pair of vectors (x, y) is now changed by replacing x by $x + \chi(x, y)$ in order to get $\bar{x}$, and y by $y + \psi(x, y)$ to get $\bar{y}$. Both sums are nonnegative. However, in general these new vectors are no longer mixed strategies because their components do not sum to one. For that purpose, they are "re-normalized" by the following continuous functions r and s:

$$r(x) = x\frac{1}{1^\top x}, \qquad s(y) = y\frac{1}{1^\top y}$$

for nonzero nonnegative vectors $x \in \mathbb{R}^m$ and $y \in \mathbb{R}^n$. Then $r(x) \in X$ because $1^\top r(x) = 1^\top x\frac{1}{1^\top x} = 1$. Analogously, $s(y) \in Y$. The function $f: C \to C$ is now defined by

$$f(x, y) = \big(r(x + \chi(x, y)), \; s(y + \psi(x, y))\big).$$

We claim that

$$
\begin{aligned}
x = r(x + \chi(x, y)) \quad &\Leftrightarrow \quad x \text{ is a best response to } y, \\
y = s(y + \psi(x, y)) \quad &\Leftrightarrow \quad y \text{ is a best response to } x,
\end{aligned}
\tag{6.22}
$$

which proves that (x, y) is a fixed point of f if and only if (x, y) is an equilibrium of the game (A, B). By symmetry, it suffices to prove the first equivalence in (6.22), where "$\Leftarrow$" is easy: If x is a best response to y, then the best-response condition (6.10) implies $\chi(x, y) = \mathbf{0} \in \mathbb{R}^m$ in (6.21) and therefore $x = r(x + \chi(x, y))$.

Conversely, let $x = r(x + \chi(x, y))$. In theory, this could hold if $\chi(x, y) \neq \mathbf{0}$ (so that x is not a best response to y) due to the re-normalization with r, for example if $x = (\frac{1}{3}, \frac{1}{3}, \frac{1}{3})$ and $\chi(x, y) = (\frac{1}{5}, \frac{1}{5}, \frac{1}{5})$. However, this is not possible. Let

$$
\begin{aligned}
u &= \max\{(Ay)_i \mid 1 \le i \le m\} & &= (Ay)_\ell, \\
t &= \min\{(Ay)_i \mid 1 \le i \le m, \; x_i > 0\} &= (Ay)_k \quad &(\text{with } x_k > 0).
\end{aligned}
$$

Shown analogously to (6.13), $t \le x^\top Ay$, and therefore $\chi_k(x, y) = 0$. If x is not a best response to y, then $x^\top Ay < u$, thus $\chi_\ell(x, y) > 0$ and $1^\top(x + \chi(x, y)) > 1$, that is, $\frac{1}{1^\top(x + \chi(x, y))} < 1$. Because $\chi_k(x, y) = 0$, this means that the re-normalized vector $\bar{x} = r(x + \chi(x, y))$ *shrinks* in its positive kth component, $\bar{x}_k < x_k$, which contradicts the assumption $\bar{x} = x$. This proves "$\Rightarrow$" in (6.22).

As claimed, the fixed points (x, y) are therefore exactly the equilibria of the game, and an equilibrium exists by Brouwer's fixed-point Theorem 6.4. $\qquad\square$

We give the proof of Brouwer's fixed-point theorem in Chapter 7.

6.6 Finding Mixed Equilibria in Small Games

How can we find all equilibria of a two-player game in strategic form? It is easy to determine the pure-strategy equilibria, because these are the cells in the payoff table where both payoffs are marked with a box as best-response payoffs. We now describe how to find mixed-strategy equilibria, using the best-response condition Proposition 6.1.

We first consider 2×2 games. Our first example above, the Inspection game in Figure 6.2, has no pure-strategy equilibrium. There we have determined the mixed-strategy probabilities of a player so as to make the other player indifferent between her pure strategies, because only then will she mix between these strategies. This is a consequence of Proposition 6.1: Only pure strategies that get maximum, and hence equal, expected payoff can be played with positive probability in equilibrium.

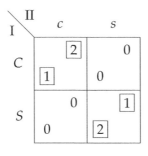 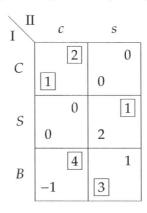

Figure 6.7 Battle of the Sexes game (left) and a 3×2 game (right), which is the same game with an extra strategy B for player I.

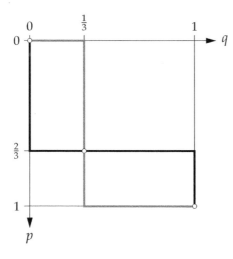

Figure 6.8 Best-response mixed strategies $(1 - p, p)$ (gray) and $(1 - q, q)$ (black) for the 2×2 Battle of the Sexes game on the left in Figure 6.7. The game has the two pure equilibria $((1, 0), (1, 0))$ and $((0, 1), (0, 1))$ and the mixed equilibrium $((\frac{1}{3}, \frac{2}{3}), (\frac{2}{3}, \frac{1}{3}))$.

A mixed-strategy equilibrium can also exist in 2×2 games that have pure-strategy equilibria. As an example, consider the Battle of the Sexes game on the left in Figure 6.7. The best-response payoffs show that it has the pure-strategy equilibria (C, c) and (S, s). Suppose now that player I plays the mixed strategy

$x = (1 - p, p)$, and player II plays the mixed strategy $y = (1 - q, q)$. Similar to Figure 6.4 for the Inspection game, Figure 6.8 shows the best responses of the player against the mixed strategy of the other player. For example, if $p = \frac{1}{2}$ then player II's expected payoffs are 1 for column c and $\frac{1}{2}$ for column s and her best response is c, which is the mixed strategy $(1, 0)$ where $q = 0$. In general, the payoffs for c and s are $2(1 - p)$ and p, so player II weakly prefers c to s if $2(1 - p) \geq p$ or equivalently $\frac{2}{3} \geq p$, that is, $p \in [0, \frac{2}{3}]$. For $p = \frac{2}{3}$ player II is indifferent and any $q \in [0, 1]$ defines a best response $(1 - q, q)$. For $p > \frac{2}{3}$ player II strictly prefers s to c which defines the unique best response $(0, 1)$. Similarly, the best-response mixed strategy $(1 - p, p)$ of player I against the mixed strategy $(1 - q, q)$ is $p = 0$ for $q \in [0, \frac{1}{3})$, any $p \in [0, 1]$ for $q = \frac{1}{3}$, and $p = 1$ for $q \in (\frac{1}{3}, 1]$ (an example to check this quickly is $q = \frac{1}{2}$, where strategy S gives player I a higher expected payoff). Apart from the pure-strategy equilibria, this gives the mixed equilibrium $((\frac{1}{3}, \frac{2}{3}), (\frac{2}{3}, \frac{1}{3}))$.

In summary, the rule to find a mixed-strategy equilibrium in a 2×2 game is to make the other player indifferent, because only in that case the other player can mix. Then the expected payoffs for the two opponent strategies have to be equal, and the resulting equation determines the player's own probability.

The mixed-strategy probabilities in the Battle of the Sexes game can be seen relatively quickly by looking at the payoff matrices: For example, player II must give twice as much weight to the left column c as to the right column s, because player I's top row C only gets payoff 1 when player II plays c, whereas his bottom row S gets payoff 2 when she plays s, and the payoffs for the strategy pairs (C, s) and (S, c) are zero. Hence, the two rows C and S have the same expected payoff only when player II mixes c and s with probabilities $\frac{2}{3}$ and $\frac{1}{3}$, respectively.

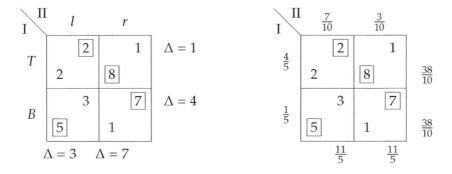

Figure 6.9 The "difference trick" for finding mixed equilibrium probabilities in a 2×2 game. The left figure shows the game, and the difference in payoffs to the other player for each pure strategy. As shown on the right, these differences are assigned to the respective *other* own strategy and are re-normalized to become probabilities. The fractions $\frac{11}{5}$ and $\frac{38}{10}$ are the resulting equal expected payoffs to player II and player I, respectively.

We now explain a quick method, the "difference trick", for finding the probabilities in a mixed equilibrium of a general 2×2 game. Consider the game on the left in Figure 6.9. When player I plays T, then l is a best response, and when player I plays B, then r is a best response. Consequently, there must be a way to mix T and B so that player II is indifferent between l and r. We consider now the difference Δ in payoffs to the other player for both rows: When player I plays T, then the difference between the two payoffs to player II is $\Delta = 2 - 1 = 1$, and when player I plays B, then that difference, in absolute value, is $\Delta = 7 - 3 = 4$, as shown on the side of the game. (Note that these differences, when considered as differences between the payoffs for l versus r, have opposite sign because l is preferred to r against T, and the other way around against B; otherwise, player II would always prefer the same strategy and could not be made indifferent.) Now the payoff difference Δ is assigned as a probability weight to the respective *other* strategy of the row player, meaning T is given weight 4 (which is the Δ computed for B), and B is given weight 1 (which is the Δ for T). The probabilities for T and B are then chosen proportional to these weights, so one has to divide each weight by 5 (which is the sum of the weights) in order to obtain probabilities. This is shown on the right in Figure 6.9. The expected payoffs to player II are also shown there, at the bottom, and are for l given by $(4 \cdot 2 + 1 \cdot 3)/5 = \frac{11}{5}$, and for r by $(4 \cdot 1 + 1 \cdot 7)/5 = \frac{11}{5}$, so they are indeed equal as claimed. Similarly, the two payoff differences for player I in columns l and r are 3 and 7, respectively, so l and r should be played with probabilities that are proportional to 7 and 3, respectively. With the resulting probabilities $\frac{7}{10}$ and $\frac{3}{10}$, the two rows T and B get the same expected payoff $\frac{38}{10}$.

> $\Rightarrow$ Exercise 6.4 asks you to prove that this "difference trick" always works.

Next, we consider games where one player has two and the other more than two strategies. The 3×2 game on the right in Figure 6.7 is like the Battle of the Sexes game, except that player I has an additional strategy B. Essentially, such a game can be analyzed by considering the 2×2 games obtained by restricting both players to two strategies only, and checking if the resulting equilibria carry over to the whole game. First, the pure-strategy equilibrium (C, c) of the original Battle of the Sexes game is also an equilibrium of the larger game, but (S, s) is not because S is no longer a best response to s because B gives a larger payoff to player I. Second, consider the mixed-strategy equilibrium of the smaller game where player I chooses C and S with probabilities $\frac{1}{3}$ and $\frac{2}{3}$ (and B with probability zero), so that player II is indifferent between c and s. Hence, the mixed strategy $(\frac{2}{3}, \frac{1}{3})$ of player II is still a best response, against which C and S both give the same expected payoff $\frac{2}{3}$. However, this is not enough to guarantee an equilibrium, because C and S have to be best responses, that is, their payoff must be at least as large as that for the additional strategy B. That payoff is $(-1) \cdot \frac{2}{3} + 3 \cdot \frac{1}{3} = \frac{1}{3}$, so it is indeed not larger than the payoff for C and S. In other words, the mixed equilibrium of the smaller game is also a mixed equilibrium of the larger game, given by the pair of mixed strategies $((\frac{1}{3}, \frac{2}{3}, 0), (\frac{2}{3}, \frac{1}{3}))$.

Other "restricted" 2×2 games are obtained by letting player I play only C and B, or only S and B. In the first case, the difference trick gives the mixed strategy $(\frac{3}{5}, \frac{2}{5})$ of player II (for playing c, s) so that player I is indifferent between C and B, where both strategies receive expected payoff $\frac{3}{5}$. However, the expected payoff to the third strategy S is then $\frac{4}{5}$, which is higher, so that C and B are not best responses, which means we cannot have an equilibrium where only C and B are played with positive probability by player I. Another reason why no equilibrium strategy of player II mixes C and B is that against both C and B, player II's best response is always c, so that player I could not make player II indifferent by playing only C and B.

Finally, consider S and B as pure strategies that player I mixes in a possible equilibrium. This requires, via the difference trick, the mixed strategy $(0, \frac{3}{4}, \frac{1}{4})$ of player I (as a probability vector for playing his three pure strategies C, S, B) so that player II is indifferent between c and s with payoff 1 for both strategies. Then if player II uses the mixed strategy $(\frac{1}{2}, \frac{1}{2})$ (in order to make player I indifferent between S and B), the expected payoffs for the three rows C, S, B are $\frac{1}{2}$, 1, and 1, respectively, so that indeed player I only uses best responses with positive probability, which gives a third equilibrium of the game.

Why is there no mixed equilibrium of the 3×2 game in Figure 6.7 where player I mixes between his three pure strategies? The reason is that player II has only a single probability at her disposal (which determines the complementary probability for the other pure strategy), which does not give enough freedom to satisfy two equations, namely indifference between C and S as well as between S and B (the third indifference between C and B would then hold automatically). We have already computed the probabilities $(\frac{2}{3}, \frac{1}{3})$ for c, s that are needed to give the same expected payoff for C and S, against which the payoff for row B is different, namely $\frac{1}{3}$. This alone suffices to show that it is not possible to make player I indifferent between all three pure strategies, so there is no equilibrium where player I mixes between all of them. In certain games, which are called "degenerate" and which are treated in Section 6.8, it is indeed the case that player II, say, mixes only between two pure strategies, but where "by accident" three strategies of player I have the same optimal payoff. Equilibria of degenerate games are more complicated to determine, as discussed in Section 6.8.

> ⇒ Attempt Exercise 6.2 if you have not done this earlier. Compare your solution with the approach for the 3×2 game in Figure 6.7.

6.7   The Upper-Envelope Method

The topic of this section is the *upper-envelope diagram* for finding mixed equilibria, in particular in games where one player has only two strategies.

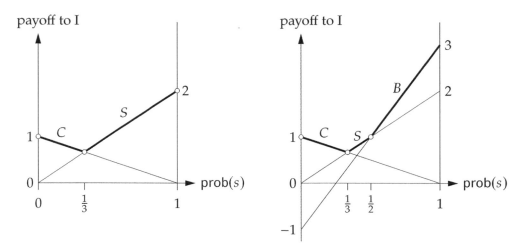

Figure 6.10   Upper envelope of expected payoffs to player I against the mixed strategy of player II for the two games in Figure 6.7. The white circles indicate mixed strategies of player II that are part of an equilibrium.

The graphs in Figure 6.10 show the upper-envelope diagrams for the two games in Figure 6.7. This diagram is a plot of the expected payoff of one player against the mixed strategy of the other player. We use this typically when that mixed strategy is determined by a single probability, which applies when the player has only two pure strategies. In this example, player II has in both games only the two strategies c and s, and her mixed strategy is $(1 - q, q)$. The horizontal axis represents the probability $q = \text{prob}(s)$ for playing the second pure strategy s of player II. In the vertical direction, we plot the resulting expected payoffs to the other player, here player I, for each of his pure strategies.

Consider first the left game, where player I has only the two strategies C and S. For C, his expected payoff against $(1 - q, q)$ is $1 - q$, and for S it is $2q$. The *upper envelope* is the maximum of the resulting two lines, shown in bold.

The right game has an additional strategy B for player I, with payoff -1 against the left column c where $q = 0$, and payoff 3 against the right column s where $q = 1$. For $q \in [0, 1]$, the resulting expected payoff is $-(1 - q) + 3q$, that is, $4q - 1$. This expected payoff becomes a new part of the upper envelope for $q \geq \frac{1}{2}$, because then the expected payoffs for rows S and B are equal to 1 when $q = \frac{1}{2}$, where the two lines cross, and for $q > \frac{1}{2}$ the expected payoff for row B is strictly better than for the other rows.

The lines that describe the expected payoffs are particularly easy to obtain graphically, according to the following *goalpost method* for constructing the upper envelope:

- Identify a player who has only two pure strategies (here player II).

- Plot the probability of the *second* strategy of the player along a horizontal axis from 0 to 1, here $q = \text{prob}(s)$. This defines the mixed strategy $(1 - q, q)$ of player II. In that way, the left end 0 of the interval $[0, 1]$ for q corresponds to the left strategy of player II, and the right end 1 to her right strategy in the payoff table.

- Erect a vertical line (a "goalpost") at each endpoint of the interval $[0, 1]$.

- For each pure strategy of the other player (here player I), do the following: At the left goalpost, mark the payoff to player I against the left strategy of player II as a *height* on the goalpost, and similarly at the right goalpost for the right strategy of player II; then connect these two heights by a straight line. For row C, the left and right heights are 1 and 0; for row S, they are 0 and 2; and for row B (right game only), they are -1 and 3. These are exactly the entries in the rows of the payoff matrix to player I.

- The maximum of the plotted lines is the upper envelope.

The fact that this construction works, in particular that one obtains straight lines for the expected payoffs, is due to the algebraic description of line segments explained in Figure 5.4 above. Namely, consider some row with payoffs a and b to player I, like $a = -1$ and $b = 3$ for row B. Then his expected payoff against the mixed strategy $(1 - q, q)$ is $(1 - q)a + qb$. This expected payoff is the vertical coordinate ("height") in the goalpost diagram, whereas the horizontal coordinate is just q. The two endpoints on the goalposts have coordinates $(0, a)$ and $(1, b)$, and the line segment that is drawn to connect them consists of the points $(1 - q)(0, a) + q(1, b)$ for $q \in [0, 1]$, which are the points $(q, (1 - q)a + qb)$, as claimed.

In the left diagram in Figure 6.10, the two lines of expected payoffs for rows C and S obviously intersect exactly where player I is indifferent between these two rows, which happens when $q = \frac{1}{3}$. However, the diagram not only gives information about this indifference, but about player I's preference in general: Any pure strategy that is a *best response* of player I is obviously the *topmost* line segment of all the lines describing the expected payoffs. Here, row C is a best response whenever $0 \leq q \leq \frac{1}{3}$, and row S is a best response whenever $\frac{1}{3} \leq q \leq 1$, with indifference between C and S exactly when $q = \frac{1}{3}$. The upper envelope of expected payoffs is the *maximum* of the lines that describe the expected payoffs for the different pure strategies, and it is indicated by the bold line segments in the diagram. Marking this upper envelope in bold is therefore the final step of the goalpost method.

The upper envelope is of particular use when a player has more than two pure strategies, because mixed equilibria of a 2×2 are anyhow quickly determined with the "difference trick". Consider the upper-envelope diagram on the right of Figure 6.10, for the 3×2 game in Figure 6.7. The extra strategy B of player I gives the line connecting height -1 on the left goalpost to height 3 on the right goalpost, which shows that B is a best response for all q so that $\frac{1}{2} \leq q \leq 1$, with indifference between S and B when $q = \frac{1}{2}$. The upper envelope consists of three line segments for the three pure strategies C, S, and B of player I. Moreover, there are only two points where player I is indifferent between any two such strategies, namely between C and S when $q = \frac{1}{3}$, and between S and B when $q = \frac{1}{2}$. These two points, indicated by small circles on the upper envelope, give the only two mixed-strategy probabilities of player II where player I can mix between two pure strategies (the small circle for $q = 0$ corresponds to the pure equilibrium strategy c). There is a third indifference between C and B, for $q = \frac{2}{5}$, where the two lines for C and B intersect, but this intersection point is below the line for S and hence not on the upper envelope. Consequently, this intersection point does not have to be investigated as a possible mixed-strategy equilibrium strategy where player I mixes between the two rows C and B because they are not best responses.

Given these candidates of mixtures for player I, we can quickly find if and how the two respective pure strategies can be mixed in order to make player II indifferent using the difference trick, as described in the previous section.

Hence, the upper envelope restricts the pairs of strategies that need to be tested as possible strategies that are mixed in equilibrium. Exercise 6.5, for example, describes a 2×5 game. Here it is player I who has two strategies, and the upper envelope describes the expected payoffs to player II against player I's mixed strategy $(1-p, p)$. Because player II has five pure strategies, the upper envelope may consist of up to five line segments; there may be fewer line segments if some pure strategy of player II is never a best response. Consequently, there are up to four intersection points where these line segments join up on the upper envelope, where player II is indifferent between her corresponding two strategies and can mix between them. Without the upper envelope, there would not be four but ten (which are the different ways to pick two out of five) pure strategy pairs to be checked as possible pure strategies that player II mixes in equilibrium, which would be much more laborious to investigate.

> $\Rightarrow$ Exercise 6.2 can also be solved with the upper-envelope method. This method is of particular help for larger games as in Exercise 6.5.

In principle, the upper-envelope diagram is also defined for the expected payoffs against a mixed strategy that mixes between more than two pure strategies. This diagram is harder to draw and visualize in this case because it requires higher dimensions. An example is shown in Figure 9.2 in Chapter 9, where we develop the upper-envelope method for finding equilibria further.

> ⇒ The simple Exercise 6.6 helps understanding dominated strategies with the upper-envelope diagram.

6.8 Degenerate Games

In this section, we treat games that have to be analyzed more carefully in order to find all their equilibria. We completely analyze an example, and then give the definition of a *degenerate game* that applies to 2×2 games like in this example. We then discuss larger games, and give the general Definition 6.7 for when an $m \times n$ game is degenerate.

Figure 6.11 is the strategic form, with best response payoffs, of the Threat game in Figure 4.3. It is similar to the Battle of the Sexes game in that it has two pure equilibria, here (T, l) and (B, r). As in the Battle of the Sexes game, we should expect an additional mixed-strategy equilibrium, which indeed exists. Assume this equilibrium is given by the mixed strategy $(1 - p, p)$ for player I and $(1 - q, q)$ for player II. The difference trick gives $q = \frac{1}{2}$ so that player I is indifferent between T and B and can mix, and for player I the difference trick gives $p = 0$ so that player II is indifferent between l and r.

In more detail, if player I plays T and B with probabilities $1 - p$ and p, then the expected payoff to player II is $3(1 - p)$ for her strategy l, and $3(1 - p) + 2p$ for r, so that she is indifferent between l and r only when $p = 0$. Indeed, because r weakly dominates l, and player II's payoff for r is larger when player I plays B, any positive probability for B would make r the unique best response of player II (against which B is the unique best response of player I). Hence, the only equilibrium where B is played with positive probability is the pure-strategy equilibrium (B, r).

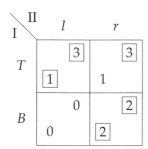

Figure 6.11 Strategic form of the Threat game, which is a degenerate game.

Consequently, a mixed-strategy equilibrium of the game is $\left((1, 0), (\frac{1}{2}, \frac{1}{2})\right)$. However, in this equilibrium only player II uses a properly mixed strategy and player I plays the pure strategy T. The best-response condition requires that player II is indifferent between her pure strategies l and r because both are played with positive probability, and this implies that T is played with probability one, as just described. For player I, however, the best-response condition requires that T has maximal expected payoff, which does *not* have to be the same as the expected payoff for B because B does not have positive probability! That is to say, player I's indifference between T and B, which we have used to determine player II's mixed strategy $(1-q, q)$ as $q = \frac{1}{2}$, is too strong a requirement. All we need is that T is a best response to this mixed strategy, so the expected payoff 1 to player I when he plays T has to be *at least as large* as his expected payoff $2q$ when he plays B, that is, $1 \geq 2q$. In other words, the equation $1 = 2q$ has to be replaced by an *inequality*.

The inequality $1 \geq 2q$ is equivalent to $0 \leq q \leq \frac{1}{2}$. With these possible values for q, we obtain an infinite set of equilibria $((1,0), (1-q, q))$. The two extreme cases for q, namely $q = 0$ and $q = \frac{1}{2}$, give the pure-strategy equilibrium (T, l) and the mixed equilibrium $((1,0), (\frac{1}{2}, \frac{1}{2}))$ that we found earlier.

Note: An equilibrium where at least one player uses a mixed strategy that is not a pure strategy is always called a mixed equilibrium. Also, when we want to find "all mixed equilibria" of a game, we usually include the pure-strategy equilibria among them (which are at any rate easy to find).

The described set of equilibria (where player II uses a mixed-strategy probability q from some interval) has an intuition in terms of the possible "threat" in this game that player II plays l (with an undesirable outcome after player I has chosen B, in the game tree in Figure 4.3(a)). First, r weakly dominates l, so if there is any positive probability that player I ignores the threat and plays B, then the threat of playing l at all is not sustainable as a best response, so any equilibrium where player II does play l requires that player I's probability p for playing B is zero. Second, if player I plays T, then player II is indifferent between l and r and she can in principle play anything, but in order to maintain the threat, T must be a best response, which requires that the probability $1-q$ for l is sufficiently high. Player I is indifferent between T and B (and therefore T still optimal) when $1 - q = \frac{1}{2}$. The threat works whenever $1 - q \geq \frac{1}{2}$. In other words, "threats" can work even when the threatened action with the undesirable outcome is uncertain, as long as its probability is sufficiently high.

The complications in this game arise because the game in Figure 6.11 is "degenerate" according to the following definition.

Definition 6.6. A 2×2 game is called *degenerate* if some player has a pure strategy with two pure best responses of the other player. $\qquad\square$

A degenerate game is complicated to analyze because one player can use a pure strategy (like T in the threat game) so that the other player can mix between her two best responses, but the mixed-strategy probabilities do not have to fulfill the equation that the other player is indifferent between his strategies. Instead of that equation, it suffices to fulfill the *inequality* that the first player's pure strategy is a best response. Note that the mentioned equation, as well as the inequality, is a consequence of the best-response condition, Proposition 6.1.

⇒ Exercise 6.7 is about a degenerate game.

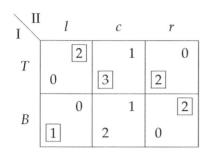 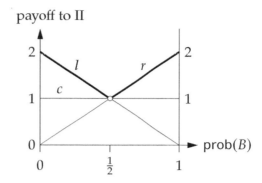

Figure 6.12 Degenerate 2×3 game and upper envelope of expected payoffs to player II.

Degeneracy occurs for larger games as well, which requires a more general definition. Consider the 2×3 game in Figure 6.12. Each pure strategy has a unique best response, so the condition in Definition 6.6 does not hold. This game does not have a pure-strategy equilibrium, so both players have to use mixed strategies. We apply the upper-envelope method, plotting the payoff to player II as a function of the probability that player I plays B, shown on the right in Figure 6.12. Here, the lines of expected payoffs cross in a single point, and all three pure strategies of player II are best responses when player I uses the mixed strategy $(\frac{1}{2}, \frac{1}{2})$.

How should we analyze this game? Clearly, we only have an equilibrium if both players mix, which requires the mixed strategy $(\frac{1}{2}, \frac{1}{2})$ for player I. Then player I has to be made indifferent, and in order to achieve this, player II can mix between any of her three best responses l, c, and r. We now have the same problem as earlier in the Threat game, namely too much freedom: Three probabilities for player II, call them y_l, y_c, and y_r, have to be chosen so that $y_l + y_c + y_r = 1$, and so that player I is indifferent between his pure strategies T and B, which gives the equation $3y_c + 2y_r = y_l + 2y_c$. These are three unknowns subject to two equations, with an underdetermined solution.

We give a method for finding out all possible probabilities, which can be generalized to the situation that three probabilities are subject to only two linear equations, like here. First, sort the coefficients of the probabilities in the equation $3y_c + 2y_r = y_l + 2y_c$, which describes the indifference of the expected payoffs to the other player, so that each probability appears only *once* and so that its coefficient is *positive*, which here gives the equation $y_c + 2y_r = y_l$. Of the three probabilities, normally two appear on one side of the equation and one on the other side. The "extreme solutions" of this equation are obtained by setting either of the probabilities on the side of the equation with two probabilities to zero, here either $y_c = 0$ or $y_r = 0$. In the former case, this gives the solution $(\frac{2}{3}, 0, \frac{1}{3})$ for (y_l, y_c, y_r), in the latter case the solution $(\frac{1}{2}, \frac{1}{2}, 0)$. This implies that in any equilibrium $0 \le y_c \le \frac{1}{2}$ (and similarly $0 \le y_r \le \frac{1}{3}$). By chosing $y_c \in [0, \frac{1}{2}]$ as the free parameter, the remaining probabilities are given by $y_l = y_c + 2y_r = y_c + 2(1 - y_l - y_c) = 2 - 2y_l - y_c$ or $y_l = \frac{2}{3} - \frac{1}{3}y_c$, and $y_r = 1 - y_l - y_c = 1 - (\frac{2}{3} - \frac{1}{3}y_c) - y_c = \frac{1}{3} - \frac{2}{3}y_c$.

Another way to find the extreme solutions to the underdetermined mixed-strategy probabilities of player II is to simply ignore one of the three best responses and apply the difference trick: Assume that player II uses only his best responses l and c. Then the difference trick, to make player I indifferent, gives probabilities $(\frac{1}{2}, \frac{1}{2}, 0)$ for l, c, r. If player II uses only his best responses l and r, the difference trick gives the mixed strategy $(\frac{2}{3}, 0, \frac{1}{3})$. These are exactly the two mixed strategies just described. If player II uses only his best responses c and r, then the difference trick does not work because player I's best response to both c and r is always T.

The following is a definition of degeneracy for general $m \times n$ games.

Definition 6.7 (Degenerate game). A two-player game is called *degenerate* if some player has a mixed strategy that assigns positive probability to exactly k pure strategies so that the other player has more than k pure best responses to that mixed strategy. □

In a degenerate game, a mixed strategy with "too many" best responses creates the difficulty that we have described with the above examples. Consider, in contrast, the "normal" situation that the game is nondegenerate.

Proposition 6.8. *In any equilibrium of a two-player game that is not degenerate, both players use mixed strategies that mix the same number of pure strategies.*

> ⇒ You are asked to give the easy and instructive proof of Proposition 6.8 in Exercise 6.8.

In other words, in a nondegenerate game we have pure-strategy equilibria (both players use exactly one pure strategy), or mixed-strategy equilibria where both players mix between exactly two strategies, or equilibria where both player mix between exactly three strategies, and so on. In general, in a nondegenerate game both players mix between exactly k pure strategies. Each of these strategies has a probability, and these probabilities are subject to k equations in order to make the other player indifferent between his or her pure strategies: One of these equations states that the probabilities sum to one, and the other $k - 1$ equations state the equality between the first and second, second and third, etc., up to the

$(k-1)$st and kth strategy of the other player. These equations have unique solutions. (It can be shown that non-unique solutions can only occur in a degenerate game.) These solutions have to be checked for the equilibrium property: The resulting probabilities have to be nonnegative, and the unused pure strategies of the other player must not have a higher payoff. Finding mixed-strategy probabilities by equating the expected payoffs to the other player no longer gives unique solutions in degenerate games, as we have demonstrated.

Should we or should we not care about degenerate games? In a certain sense, degenerate games can be ignored when considering games that are given in strategic form, and where each cell of the payoff table arises from a different circumstance of the interactive situation that is modeled by the game. In that case, it is "unlikely" that two payoffs are identical, which is necessary to get two pure best responses to a pure strategy, which is the only case where a 2×2 game can be degenerate. A small change of the payoff will result in a unique preference of the other player. Similarly, it is unlikely that more than two lines that define the upper envelope of expected payoffs cross in one point, or that more than three planes of expected payoffs against a mixed strategy that mixes three pure strategies (like in the two planes in Figure 9.2) cross in one point, and so on. In other words, "generic" (or "almost all") games in strategic form are not degenerate. On the other hand, degeneracy is very likely when looking at a game tree, because a payoff of the game tree will occur repeatedly in the strategic form, as demonstrated by the Threat game.

We conclude with a useful proposition that can be used when looking for all equilibria of a degenerate game. In the Threat game in Figure 6.11, there are two equilibria $((1,0),(1,0))$ and $((1,0),(\frac{1}{2},\frac{1}{2}))$ that are obtained, essentially, by "ignoring" the degeneracy. These two equilibria have the same strategy $(1,0)$ of player I. Similarly, the game in Figure 6.12 has two mixed equilibria $((\frac{1}{2},\frac{1}{2}),(\frac{1}{2},\frac{1}{2},0))$ and $((\frac{1}{2},\frac{1}{2}),(\frac{2}{3},0,\frac{1}{3}))$, which again share the same strategy of player I.

The following proposition states that in a two-player game, two mixed equilibria that *have the same strategy of one player* can be combined with convex combinations to obtain further equilibria, as it is the case in the above examples. In the proposition player II's strategy y is fixed, but it holds similarly when player I's strategy is fixed. Fixing strategies is important here, because the set of equilibria is in general not convex.

Proposition 6.9. *Consider a bimatrix game (A, B) with mixed-strategy sets X and Y as in (6.6). Let (x, y) and $(\hat{x}, y)$ be equilibria of the game, where $x, \hat{x} \in X$ and $y \in Y$. Then $(x(1-p) + \hat{x}p), y)$ is also an equilibrium of the game, for any $p \in [0, 1]$.*

Proof. Let $\bar{x} = x(1-p) + \hat{x}p$. Clearly, $\bar{x} \in X$ because $\bar{x}_i \geq 0$ for all pure strategies i of player I, and

$$\mathbf{1}^\top \bar{x} = \mathbf{1}^\top (x(1-p) + \hat{x}p) = \mathbf{1}^\top x(1-p) + \mathbf{1}^\top \hat{x}p = (1-p) + p = 1. \qquad (6.23)$$

The pair $(\bar{x}, y)$ is an equilibrium if $\bar{x}$ is a best response to y and vice versa. For any mixed strategy $\tilde{x}$ of player I, we have $x^\top Ay \geq \tilde{x}^\top Ay$ and $\hat{x}^\top Ay \geq \tilde{x}^\top Ay$, and consequently

$$\bar{x}^\top Ay = (1-p)x^\top Ay + p\hat{x}^\top Ay \geq (1-p)\tilde{x}^\top Ay + p\tilde{x}^\top Ay = \tilde{x}^\top Ay\,,$$

which shows that $\bar{x}$ is a best response to y. In other words, these inequalities hold because they are preserved under convex combinations. Similarly, y is a best response to x and $\hat{x}$, that is, for any $\tilde{y} \in Y$ we have $x^\top By \geq x^\top B\tilde{y}$ and $\hat{x}^\top By \geq \hat{x}^\top B\tilde{y}$. Again, taking convex combinations shows $\bar{x}^\top By \geq \bar{x}^\top B\tilde{y}$. Hence, $(\bar{x}, y)$ is an equilibrium, as claimed.   □

We come back to the example of Figure 6.12 to illustrate Proposition 6.9: This game has two equilibria (x, y) and $(x, \hat{y})$ where $x = (\frac{1}{2}, \frac{1}{2})$, $y = (\frac{1}{2}, \frac{1}{2}, 0)$, and $\hat{y} = (\frac{2}{3}, 0, \frac{1}{3})$, so this is the situation of Proposition 6.9, except that the two equilibria have the same mixed strategy of player I rather than of player II. Consequently, $(x, \bar{y})$ is an equilibrium for any convex combination $\bar{y}$ of y and $\hat{y}$. The equilibria (x, y) and $(x, \hat{y})$ are "extreme" in the sense that y and $\hat{y}$ have as many zero probabilities as possible, which means that y and $\hat{y}$ are the endpoints of a line segment of equilibrium strategies $\bar{y} = y(1-p) + \hat{y}p$ (that is, $(x, \bar{y})$ is an equilibrium).

> ⇒ Express the set of equilibria for the Threat game in Figure 6.11 with the help of Proposition 6.9.

6.9   Further Reading

In Section 6.2, the Inspection game is used to motivate mixed strategies. Inspection games are an applied area of game theory; see, for example, Avenhaus and Canty (1996).

The seminal paper by Nash (1951) defined "equilibrium points" in finite games and proved their existence. As mentioned, the best-response condition is stated as "trivial" by Nash (1951, p. 287), but it is central for finding mixed equilibria. Nash's "equilibrium point" is now universally called "Nash equilibrium", which is the equilibrium concept considered throughout this book, except for the generalization of correlated equilibrium in Chapter 12. For textbook treatments of mixed-strategy equilibria see Osborne (2004, Chapter 4) or Maschler, Solan, and Zamir (2013, Chapter 5).

The use of the upper envelope seems to be folklore, and is at any rate straightforward. For Definition 6.7 see van Damme (1987, p. 52), and for a detailed comparison of conditions of degeneracy for two-player games (which are all equivalent to this definition) see von Stengel (2021).

6.10   Exercises for Chapter 6

Exercise 6.1. Let $x \in \mathbb{R}^m$ and $y \in \mathbb{R}^n$ and remember that all vectors are column vectors. If $m = n$, what is the difference between $x^\top y$ and $xy^\top$? What if $m \neq n$?

Exercise 6.2. Find all mixed equilibria (which always includes any pure equilibria) of this 3×2 game.

	l		r	
T		1		0
	0		6	
M		0		2
	2		5	
B		3		4
	3		3	

Exercise 6.3. Consider the three-player game tree in Exercise 4.5. Recall that some or all of the strategies in a mixed equilibrium may be pure strategies.

(a) Is the following statement true or false? It is easiest to argue directly with the game tree.

For each of the players I, II, or III, the game has a mixed equilibrium in which that player plays a mixed strategy that is not a pure strategy.

(b) Is the following statement true or false? Explain.

In every mixed equilibrium of the game, at least one player I, II, or III plays a pure strategy.

Exercise 6.4. In this 2×2 game, A, B, C, D are the payoffs to player I, which are real numbers, no two of which are equal. Similarly, a, b, c, d are the payoffs to player II, which are real numbers, also no two of which are equal.

	left		*right*	
Top		a		b
	A		B	
Bottom		c		d
	C		D	

(a) Under which conditions does this game have a mixed equilibrium which is not a pure-strategy equilibrium?

[*Hint: Consider the possible patterns of best responses and resulting possible dominance relations, and express this by comparing the payoffs, as, for example, $A > C$.*]

(b) Under which conditions in (a) is this the only equilibrium of the game?

(c) Consider one of the situations in (b) and compute the probabilities $1 - p$ and p for playing *Top* and *Bottom*, respectively, and $1 - q$ and q for playing *left* and *right*, respectively, that hold in equilibrium.

(d) For the solution in (c), give a simple formula for the quotients $(1-p)/p$ and $(1-q)/q$ in terms of the payoff parameters. Try to write this formula such that the denominator and numerator are both positive.

(e) Show that the mixed equilibrium strategies do not change if player I's payoffs A and C are replaced by $A+E$ and $C+E$, respectively, for some constant E, and similarly if player II's payoffs a and b are replaced by $a+e$ and $b+e$, respectively, for some constant e.

Exercise 6.5. Consider the following 2×5 game:

II \\ I	a	b	c	d	e
T	2 / 0	4 / 2	3 / 1	5 / 0	0 / 3
B	7 / 1	4 / 0	5 / 4	0 / 1	8 / 0

(a) Draw the expected payoffs to player II for all her strategies a,b,c,d,e, in terms of the probability p, say, that player I plays strategy B. Indicate the best responses of player II, depending on that probability p.

(b) Using the diagram in (a), find all mixed (including pure) equilibria of the game.

Exercise 6.6. Consider the Quality game in Figure 3.2(b).

(a) Use the "goalpost" method twice, to draw the upper envelope of best-response payoffs to player I against the mixed strategy of player II, and vice versa.

(b) Explain which of these diagrams shows that a pure strategy dominates another pure strategy.

(c) Explain how that diagram shows that if a pure strategy s dominates a strategy t, then t is never a best response, even against a *mixed* strategy of the other player.

Exercise 6.7. Find all equilibria of this degenerate game, which is a variant of the Inspection game in Figure 6.2. The difference is that player II, the inspectee, derives no gain from acting illegally (playing r) even if she is not inspected (player I choosing T).

II \\ I	l	r
T	0 / 0	0 / -10
B	0 / -1	-90 / -6

Exercise 6.8. Prove Proposition 6.8.

7

Brouwer's Fixed-Point Theorem

As we showed in Section 6.5, Nash (1951) proved the existence of a mixed equilibrium in a finite game with the help of Brouwer's fixed-point Theorem 6.4. Fixed-point theorems are powerful tools for proving the existence of many equilibrium concepts in economics. Brouwer's theorem is the first and most important of these.

This chapter is about proving Brouwer's fixed-point theorem. We present a standard elementary proof (the most complex proof in this book), without the machinery of algebraic topology, along the following route:

- The compact convex domain of the considered continuous function is first assumed to be the *unit simplex* Δ. In a simplex, every point is the unique convex combination of the vertices of the simplex, which for Δ are the unit vectors.

- A finite number of "grid points" of Δ is considered, which are the vertices in a so-called simplicial subdivision or *triangulation* of Δ. Although such triangulations have a clear intuition in low dimensions, we carefully prove all required properties about them.

- Suitable *labels* identify "completely labeled sub-simplices" of the triangulation as *approximate fixed points*. *Sperner's lemma* (Theorem 7.10) states that a completely labeled sub-simplex exists.

- The *KKM lemma* (Theorem 7.12) of Knaster, Kuratowski, and Mazurkiewicz (1929) is proved via a limit process from Sperner's lemma and implies the existence of a fixed point of a continuous function on Δ.

- The nontrivial part of this limit process is to find finer and finer triangulations, for which we present (at the end of the chapter) the "Freudenthal" triangulation.

- A general compact convex set C is "embedded" into the unit simplex in order to prove the fixed-point theorem for a continuous function on C.

The main step is Sperner's lemma. In an ingenious induction proof, Sperner (1928) counted the completely labeled sub-simplices in order to show that their number is odd, and therefore not zero. At its core, Sperner's argument is due to a

> **Background material:** Undirected graphs
>
> A *graph* G is given by a finite set of *nodes* and a set of *edges*. Every edge is an unordered pair $\{u, v\}$ of nodes u and v, which are called the *endpoints* of the edge, where $u \neq v$. The *degree* of a node u is the number of its *neighbors* in G, that is, the number of nodes v so that $\{u, v\}$ is an edge. A *path* in G of *length* $m \geq 0$ is a sequence of nodes $u_0, u_1, \ldots, u_m$ so that $\{u_k, u_{k+1}\}$ is an edge for $0 \leq k < m$. The *end nodes* of the path are u_0 and u_m. If $u_0 = u_m$ then the path is called a *cycle*.

simple observation from graph theory (see the background material on graphs), namely that the number of odd-degree nodes in a graph is even. This *parity argument* holds because the sum of the degrees of all nodes is twice the number of edges (and hence an even number), because each edge contributes to the degree of both its endpoints.

In addition to the parity-counting by Sperner (1928), we give a clean *constructive* proof of Sperner's lemma. This alternative step in the proof combines the graphs across different dimensions in such a way that there is a single path from a vertex of the simplex to a completely labeled sub-simplex (see Figure 7.10). The dimension of the sub-simplices that define the path may go up and down multiple times along the path.

Section 7.2 introduces the concept of a label of a point in the unit simplex, and illustrates its use for the fixed-point theorem when the simplex is an interval. Simplices and triangulations are introduced in Section 7.3. Section 7.4 proves Sperner's lemma. The KKM lemma is presented in Section 7.5.

In Section 7.6, we prove Brouwer's fixed-point theorem on a general compact convex set C by first "embedding" C with an injection into the unit simplex Δ so that we can replace C by its image D, which is now a subset of Δ. Secondly, we extend the given continuous function f on D to a function on Δ whose fixed points are exactly the fixed points of f. For points x in Δ but not in D, the new function simply chooses the point $p(x)$ in D that is *closest* to x. We carefully prove that this "projection" function p is continuous, in fact "contractive" (see Theorem 7.18), and emphasize the underlying geometric intuition. Even though these considerations are simple, they produce, in passing, important results from convex analysis such as the theorem of the "separating hyperplane" (Corollary 7.16 and Figure 7.13).

Section 7.7 treats in detail the nontrivial problem of finding finer and finer triangulations of a simplex in arbitrary dimension, which has been postponed in the proof. We present the elegant triangulation due to Freudenthal (1942), which has two components: the subdivision of the unit cube into small "cubelets", which works well in any dimension d, and the subdivision of the cube and each cubelet into $d!$ simplices obtained from the possible "monotone paths" on the cube.

7.1 Prerequisites and Learning Outcomes

This chapter continues from Section 6.5, where we demonstrated Brouwer's fixed-point theorem on an interval and the necessity of its assumptions. The main geometric tool is convexity, with line segments defined in Figure 5.4 and convex combinations defined in Section 6.5. We assume basics of linear algebra when explaining affine combinations, for example in Lemma 7.2.

We also assume familiarity with continuity of functions in $\mathbb{R}^d$, with the notions of closed and compact sets, and that a continuous real-valued function on a compact domain assumes its minimum and maximum. We will recall as background material the Euclidean norm and distance, and the Bolzano–Weierstrass theorem on the existence of converging subsequences.

After studying this chapter, you should be able to:

- understand affine and convex combinations, simplices, and triangulations;

- define the dimension of a convex set, and the faces of a simplex;

- state the main theorems of this chapter: Sperner's lemma, the KKM lemma, Brouwer's fixed-point theorem both on the unit simplex and on a general compact convex set, and explain the difference between these domains;

- understand the significance of graphs with maximum node degree two in the proof of Sperner's lemma, and their connection to Lemma 7.6;

- explain the role of the projection onto a closed convex set and why it is contractive and therefore continuous (Theorem 7.18);

- explain the Freudenthal triangulation.

7.2 Labels

In this chapter we give a proof of Brouwer's fixed-point Theorem 6.4 for a continuous function f on a compact and convex set C. The main effort will be to prove the theorem when C is the *unit simplex* Δ in $\mathbb{R}^d$ defined by

$$\Delta = \{x \in \mathbb{R}^d \mid x \geq 0,\ \mathbf{1}^\top x = 1\}. \tag{7.1}$$

The mixed strategy sets X and Y in (6.6) are unit simplices, as shown in Figure 6.5. (However, $X \times Y$ is not a unit simplex.)

Recall that the ith *unit vector* e_i in $\mathbb{R}^d$ has its ith component equal to one and all other components equal to zero. The unit vectors are the vertices (extreme points or "corners") of the unit simplex Δ.

Let $f : \Delta \to \Delta$ and $x \in \Delta$, and write $f(x) = (f_1(x), \ldots, f_d(x))^\top$. For a point in $\mathbb{R}^d$, a *label* is an element of $\{1, \ldots, d\}$, and depending on f we assign a set of labels

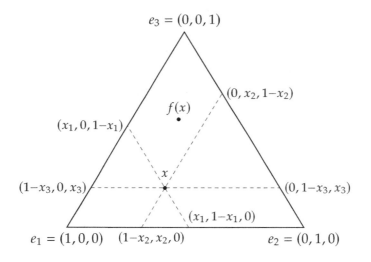

Figure 7.1 Illustration of (7.2). The dashed lines show points (z_1, z_2, z_3) where $z_i = x_i$. Here x has labels 1 and 2 because $x_1 \geq f_1(x)$ and $x_2 \geq f_2(x)$.

to each $x \in \Delta$, according to

$$x \text{ has } label\ i \quad \Leftrightarrow \quad x_i \geq f_i(x) \quad (1 \leq i \leq d). \tag{7.2}$$

Any label i of $x = (x_1, \ldots, x_d)$ identifies a component x_i of x that in $f(x)$ is not increased (increased means being moved closer to the vertex e_i of Δ); see Figure 7.1.

A point $x \in \Delta$ is *completely labeled* if x has every label $i \in \{1, \ldots, d\}$. This condition characterizes the fixed points of f.

Lemma 7.1. *Let $f : \Delta \to \Delta$ and $x \in \Delta$. Then $f(x) = x$ if and only if x is completely labeled.*

Proof. If $f(x) = x$ then $f_i(x) = x_i$ for all $i = 1, \ldots, d$ and x is therefore completely labeled. Conversely, suppose x is completely labeled, $x_i \geq f_i(x)$ for all i. If $x_i > f_i(x)$ for some i then $\sum_{k=1}^{d} x_k > \sum_{k=1}^{d} f_k(x) = 1$, which contradicts $x \in \Delta$. Hence, $x_i = f_i(x)$ for all i, that is, $f(x) = x$, as claimed. $\square$

Our strategy for proving Brouwer's fixed-point theorem is as follows:

- Discretize Δ with finitely many points on some kind of "grid".

- For simplicity, assign to each gridpoint a *single* label rather than a set of labels as in (7.2), so that in (7.2), "$\Leftrightarrow$" is replaced by "$\Rightarrow$".

- Show the existence of a *set of* "neighboring gridpoints" that together have all labels, which will locate an "approximate fixed point".

- By choosing a finer and finer grid, a suitable limit of such approximate fixed points defines a fixed point of the considered function.

We explain this strategy for the example in Figure 6.6 where f is defined on $[0, 1]$. The interval $[0, 1]$ corresponds to Δ in (7.1) for $d = 2$, which is the line

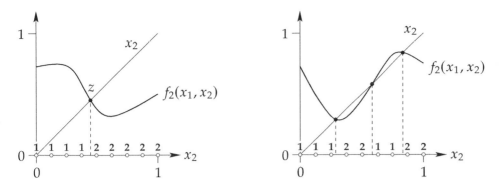

Figure 7.2 Two examples of a labeled subdivision of Δ for $d = 2$. A *completely labeled* sub-segment contains a fixed point of f.

segment that connects the two points $(1, 0)$ and $(0, 1)$ in $\mathbb{R}^2$, because the elements (x_1, x_2) of Δ are uniquely given by $(1 - x_2, x_2)$ for $x_2 \in [0, 1]$. Similarly, we can identify a function $f : \Delta \to \Delta$ with its second component $f_2(x_1, x_2)$ because $f_1(x_1, x_2) = 1 - f_2(x_1, x_2)$. Then (x_1, x_2) has label 1 if $x_1 \geq f_1(x_1, x_2)$ or equivalently $x_2 \leq f_2(1 - x_2, x_2)$, and label 2 if $x_2 \geq f_2(1 - x_2, x_2)$. Furthermore, we can assume

$$(1, 0) \text{ has label 1,} \qquad (0, 1) \text{ has label 2.} \qquad (7.3)$$

The left diagram in Figure 7.2 shows the example from Figure 6.6 with the line segment Δ shown as the interval $[0, 1]$ that contains x_2. A number of "gridpoints" in Δ are shown as white dots. Each is given a label according to the sign of $x_2 - f_2(1 - x_2, x_2)$. These points *subdivide* Δ into sub-segments, each of which has a labeled endpoint. A sub-segment that has two endpoints with *both* labels 1 and 2 is called *completely labeled*. Such a sub-segment contains a fixed point z of f because the sign of $x_2 - f_2(1 - x_2, x_2)$ changes when x_2 is in that sub-interval. In Figure 7.2, condition (7.3) holds automatically because $(1, 0)$ has only label 1 and $(0, 1)$ has only label 2. In general, (7.3) implies that there is at least one completely labeled sub-segment because when changing x_2 from 0 to 1, the label of a gridpoint $(1 - x_2, x_2)$ changes at some point from 1 to 2.

In the right diagram in Figure 7.2, the function f has three fixed points, which are again identified by the completely labeled sub-segments. In both diagrams, the fixed points would be located more accurately with a "finer" grid.

7.3 Simplices and Triangulations

In Figure 7.2, the unit simplex Δ in $\mathbb{R}^d$ is a line segment which is subdivided into smaller sub-segments. The generalization of this, for general d, is a subdivision of Δ into smaller sub-simplices, called a *simplicial subdivision* or *triangulation*, which will take the role of the indicated "grid".

The basic operation from geometry that we need is drawing a line through two points x and y in $\mathbb{R}^d$, as shown in Figure 5.4, and iterating that process (where a third point not on that line together with any point on the line creates a plane, and so on). The points on the line through x and y are the special linear combinations $x(1-\lambda)+y\lambda$ for $\lambda \in \mathbb{R}$ called *affine* combinations. In general, an *affine combination* of points $v^1,\ldots,v^n$ in $\mathbb{R}^d$ is of the form $\sum_{i=1}^n v^i\lambda_i$ where $\lambda_1,\ldots,\lambda_n$ are reals with $\sum_{i=1}^n \lambda_i = 1$. It is called a *convex combination* if $\lambda_i \geq 0$ for $1 \leq i \leq n$. A set of points is *convex* if it is closed under forming convex combinations. The *convex hull* $\operatorname{conv} V$ of a set of points V is the (with respect to set inclusion) smallest convex set C so that $V \subseteq C$.

Given points are *affinely independent* if none of these points is an affine combination of the others. This has an easy equivalent characterization in terms of linear independence in one higher dimension.

Lemma 7.2. *The vectors $v^1,\ldots,v^n$ in $\mathbb{R}^d$ are affinely independent if and only if the vectors*

$$\begin{pmatrix}1\\v^1\end{pmatrix},\ldots,\begin{pmatrix}1\\v^n\end{pmatrix} \tag{7.4}$$

in $\mathbb{R}^{1+d}$ are linearly independent.

Proof. Suppose $v^1,\ldots,v^n$ in $\mathbb{R}^d$ are not affinely independent. That is, some vector, say v^1, is an affine combination of the others, with reals $\lambda_2,\ldots,\lambda_n$ so that

$$\begin{aligned}1 &= \lambda_2+\cdots+\lambda_n\\ v^1 &= v^2\lambda_2+\cdots+v^n\lambda_n\ .\end{aligned}$$

This is equivalent to

$$\begin{pmatrix}1\\v^1\end{pmatrix}(-1)+\begin{pmatrix}1\\v^2\end{pmatrix}\lambda_2+\cdots+\begin{pmatrix}1\\v^n\end{pmatrix}\lambda_n=\begin{pmatrix}0\\\mathbf{0}\end{pmatrix} \tag{7.5}$$

which shows that the vectors in (7.4) are not linearly independent. Conversely, if they are not linearly independent, with

$$\begin{pmatrix}1\\v^1\end{pmatrix}\lambda_1+\begin{pmatrix}1\\v^2\end{pmatrix}\lambda_2+\cdots+\begin{pmatrix}1\\v^n\end{pmatrix}\lambda_n=\begin{pmatrix}0\\\mathbf{0}\end{pmatrix} \tag{7.6}$$

where $\lambda_1,\ldots,\lambda_n$ are not all zero, say $\lambda_1 \neq 0$, then multiplication of (7.6) with $-1/\lambda_1$ gives (7.5) and therefore the affine dependence of $v^1,\ldots,v^n$. (The additional first row with entries 1 in (7.4) is needed for linear independence, for example if some v^i is the zero vector, which has no special role in affine combinations.) $\qquad\square$

A convex set has *dimension* n if and only if it has $n+1$, but no more, affinely independent points. A *polytope* is the convex hull of a finite set of points in some

Euclidean space $\mathbb{R}^d$. If these points are affinely independent, then the polytope is called a *simplex*.

The unit simplex Δ in (7.1) is the convex hull of the unit vectors $e_1, \ldots, e_d$ in $\mathbb{R}^d$. They are affinely independent, so the unit simplex has dimension $d - 1$, and for that reason is sometimes denoted by Δ_{d-1}. For example (see Figure 7.1), Δ_2 is the convex hull of the three unit vectors in $\mathbb{R}^3$ and a triangle, which has dimension two (even though it is a subset of $\mathbb{R}^3$, because all $(x_1, x_2, x_3)^\top$ in Δ_2 fulfill the equation $x_1 + x_2 + x_3 = 1$).

In agreement with the definition of dimension, the empty set has dimension -1. A set of dimension zero is any singleton. A simplex of dimension one is the convex hull of two distinct points and therefore a line segment. A simplex of dimension two is a triangle. A simplex of dimension three is a tetrahedron. Every simplex of a given dimension (in any Euclidean space of possibly much larger dimension) is in an affine (and hence continuous) bijection with the unit simplex of the same dimension, as follows.

Lemma 7.3. *Let S be a simplex of dimension $d - 1$ given as the convex hull of the affinely independent points $v^1, \ldots, v^d$ in some $\mathbb{R}^k$. Then there is a linear map $\mathbb{R}^d \to \mathbb{R}^k$ that restricted to Δ in (7.1) is a bijection $\Delta \to S$. Its inverse $J : S \to \Delta$ is the restriction of an affine map $\mathbb{R}^k \to \mathbb{R}^d$.*

Proof. If $x \in S$, then there are nonnegative reals $\lambda_1, \ldots, \lambda_d$ such that

$$\begin{pmatrix} 1 \\ x \end{pmatrix} = \begin{pmatrix} 1 \\ v^1 \end{pmatrix} \lambda_1 + \cdots + \begin{pmatrix} 1 \\ v^d \end{pmatrix} \lambda_d . \tag{7.7}$$

By affine independence and Lemma 7.2, the vector $\lambda = (\lambda_1, \ldots, \lambda_d)^\top$ of coefficients in (7.7) is unique, and different coefficient vectors define different points $x \in S$. These coefficient vectors λ belong to the unit simplex Δ, which is therefore in a bijection $\Delta \to S$ via the linear map $\lambda \mapsto x$ in (7.7).

The inverse of this bijection is seen to be the restriction of an affine map $\mathbb{R}^k \to \mathbb{R}^d$ as follows. We first construct a linear map $\mathbb{R}^{1+k} \to \mathbb{R}^d$, which is uniquely defined by its (arbitrary) images on a basis of $\mathbb{R}^{1+k}$. For this purpose, extend $\begin{pmatrix} 1 \\ v^1 \end{pmatrix}, \ldots, \begin{pmatrix} 1 \\ v^d \end{pmatrix}$ to such a basis, map $\begin{pmatrix} 1 \\ v^i \end{pmatrix}$ to the unit vector e_i in $\mathbb{R}^d$ for $1 \le i \le d$, and each remaining basis vector to $\mathbf{0}$. Because this is a linear map, it can be written as $\begin{pmatrix} \alpha \\ x \end{pmatrix} \mapsto \lambda = A \begin{pmatrix} \alpha \\ x \end{pmatrix}$ for $\alpha \in \mathbb{R}$ and $x \in \mathbb{R}^k$ for some $d \times (1 + k)$ matrix A. By (7.7), $\lambda \in \Delta$ if $\alpha = 1$ and $x \in S$. For $\alpha = 1$ we obtain an affine map $\mathbb{R}^k \to \mathbb{R}^d$, $x \mapsto \lambda = A \begin{pmatrix} 1 \\ x \end{pmatrix}$. The restriction of this map to S is the desired affine bijection

$J : S \rightarrow \Delta$. (If $x \notin S$ then $A \begin{pmatrix} 1 \\ x \end{pmatrix}$ depends on the choice of the additional basis vectors, so A itself is not unique, but this is irrelevant here; the restriction $J : S \rightarrow \Delta$ is unique.)   □

In (7.7), the unique coefficients $\lambda_1, \ldots, \lambda_d$ are called the *barycentric coordinates* of x in S (which depend on the chosen order of the points $v^1, \ldots, v^d$ that define the simplex S). The reason for the name "barycentric" is that x is the center of weight (barycenter) of the weights λ_i placed at the points v^i for $1 \leq i \leq d$.

Consider a simplex S of dimension $d - 1$ given as the convex hull of the d affinely independent points $v^1, \ldots, v^d$. Then any subset of these points (including the empty set) is also affinely independent, and its convex hull is also a simplex. Each such simplex is a called a *face* of S. Any singleton face $\{v^i\}$, or v^i itself, is called a *vertex* of S. A face of dimension 1 is called an *edge*. A face of dimension $d - 2$ of S is called a *facet* of S. A vertex v of a general polytope P is an *extreme point* of P, that is, $v \in P$ and v is not a convex combination of other points of P. Exercise 7.2 shows that each vertex of a simplex is indeed such an extreme point.

Definition 7.4. Let P be a polytope. A *triangulation* or *simplicial subdivision* of P is a finite set $\Sigma = \{S_1, \ldots, S_N\}$ of simplices (called *sub-simplices*) S_i for $i = 1, \ldots, N$ so that (a) $P = \bigcup_{i=1}^{N} S_i$, (b) no sub-simplex is a subset of another, and (c) the intersection of any two sub-simplices is a face of both. A *vertex* in Σ is any vertex of any sub-simplex in Σ. A *sub-facet* in Σ is any facet of any sub-simplex in Σ. □

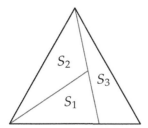 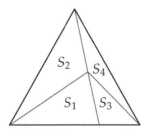 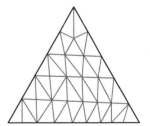

Figure 7.3 Left: $\{S_1, S_2, S_3\}$ is not a triangulation of the triangle because $S_1 \cap S_3$ is not a face of S_3. Center and right: triangulations of the triangle.

In Definition 7.4(c), the intersection of two sub-simplices may be the empty set, which is a face of any polytope. Figure 7.3 illustrates the concept of simplicial subdivisions of a triangle. The left picture fulfills conditions (a) and (b) of Definition 7.4 but not (c), which does hold in the middle picture. The right picture shows that the sub-simplices in a triangulation may vary in shape and size.

The following lemmas state some, rather intuitive, properties of triangulations.

Lemma 7.5. *Let S be a simplex with vertices $v^1, \ldots, v^d$ and let Σ be a triangulation of S. Then all sub-simplices S_i in Σ have the same dimension $d - 1$ as S.*

Proof. The vertices of any sub-simplex belong to S, which has dimension $d - 1$, so no sub-simplex can have higher dimension. Suppose one sub-simplex, call it S_1, has lower dimension, with vertices $w^1, \ldots, w^k$ and $k < d$. Let $x = \sum_{i=1}^{k} w^i \frac{1}{k}$ be the barycenter of S_1. At least one vertex v^i of S is not an affine combination of $w^1, \ldots, w^k$. For $\varepsilon \in \mathbb{R}$, let

$$z(\varepsilon) = x(1 - \varepsilon) + v^i \varepsilon. \tag{7.8}$$

Geometrically, the points $z(\varepsilon)$ lie on the line through x and v^i, see Figure 7.4.

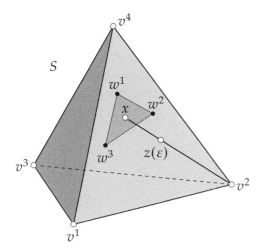

Figure 7.4 Illustration of (7.8) for $d = 4$, $k = 3$, and $i = 2$ in (7.9).

Then for $\varepsilon \neq 0$ we have $v^i = z(\varepsilon)\frac{1}{\varepsilon} + x\left(1 - \frac{1}{\varepsilon}\right)$, which shows that $z(\varepsilon)$ is not an affine combination of $w^1, \ldots, w^k$ because otherwise v^i would be. That is,

$$z(\varepsilon) \notin S_1 \qquad (\varepsilon \neq 0). \tag{7.9}$$

Consider a decreasing sequence $\varepsilon_1, \varepsilon_2, \ldots$ in $(0, 1]$ that tends to 0, for example $\varepsilon_n = \frac{1}{n}$, where $z(\varepsilon_n) \in S$ by (7.8) because S is convex. Because S is the union of the sub-simplices in Σ, there is for each n some sub-simplex that contains $z(\varepsilon_n)$. Because Σ is finite, one sub-simplex, call it S_2, contains infinitely many of the points $z(\varepsilon_n)$ (and in fact all points in between in that sequence because they are on a line segment and S_2 is convex). Because $\varepsilon_n \to 0$ as $n \to \infty$, the points $z(\varepsilon_n)$ converge to x, which implies $x \in S_2$ because S_2 is closed. Hence, $x \in S_1 \cap S_2$. By condition (c) in Definition 7.4, the intersection $S_1 \cap S_2$ is a face of both S_1 and S_2, which includes all vertices $w^1, \ldots, w^k$ of S_1 because they are used with positive weight $\frac{1}{k}$ in the representation $x = \sum_{i=1}^{k} w^i \frac{1}{k}$. Hence, $S_1 = S_1 \cap S_2$, that is, $S_1 \subseteq S_2$. Moreover, $z(\varepsilon) \in S_2 - S_1$ for some $\varepsilon > 0$, which implies $S_1 \neq S_2$. This contradicts Definition 7.4(b). Hence, all sub-simplices have the same dimension $d - 1$ as S, as claimed. $\qquad\square$

Lemma 7.6. *Let S be a simplex, let Σ be a triangulation of S, and let F be a sub-facet in Σ. Then exactly one of the following two cases occurs:*

(a) *F is a subset of a facet of S, and F is the facet of exactly one sub-simplex in Σ.*

(b) *F is not a subset of a facet of S, and F is the facet of exactly two sub-simplices in Σ.*

Proof. Let S have dimension $d - 1$ and vertices $v^1, \ldots, v^d$. By Lemma 7.5, every sub-simplex also has dimension $d - 1$. Let F be a facet of some sub-simplex T in $Σ$ and let $w^1, \ldots, w^{d-1}$ be the vertices of F. Let $x = \sum_{i=1}^{d-1} w^i \frac{1}{d-1}$ be the barycenter of F, and let w^d be the remaining vertex of T. Consider the line through x and w^d given by the points

$$y(\varepsilon) = x(1 - \varepsilon) + w^d \varepsilon$$

for $\varepsilon \in \mathbb{R}$, which for $\varepsilon \in [0,1]$ belong to T. Suppose F is the facet of another sub-simplex T', which has another dth vertex w_0, different from w_d. The d vertices $w^0, w^1, \ldots, w^{d-1}$ of T' are affinely independent. Hence, there are unique $\lambda_0, \lambda_1, \ldots, \lambda_{d-1}$ with $w^d = \sum_{i=0}^{d-1} w^i \lambda_i$, where $\lambda_0 \neq 0$ because $w^d \notin F$. We claim that $\lambda_0 < 0$, that is, w_0 and w_d are on opposite sides of the facet F. Namely,

$$y(\varepsilon) = x(1 - \varepsilon) + \sum_{i=0}^{d-1} w^i \lambda_i \varepsilon$$
$$= w^0 \lambda_0 \varepsilon + \sum_{i=1}^{d-1} w^i \left(\frac{1}{d-1}(1 - \varepsilon) + \lambda_i \varepsilon \right)$$
$$= w^0 \lambda_0 \varepsilon + \sum_{i=1}^{d-1} w^i \left(\frac{1}{d-1} + \left(\lambda_i - \frac{1}{d-1} \right) \varepsilon \right). \tag{7.10}$$

If $\lambda_0 > 0$, this would imply $y(\varepsilon) \in (T \cap T') - F$ for sufficiently small positive ε, which is not possible because T and T' intersect in a common facet which can only be F. Hence, $\lambda_0 < 0$ as claimed.

The two cases (a) and (b) are shown as follows. If F is a subset of a facet of S, then $y(\varepsilon) \notin S$ for all $\varepsilon < 0$: Otherwise, for $\varepsilon < 0$ so that $y(\varepsilon)$ and $y(-\varepsilon)$ are in S, we have $x = y(\varepsilon)\frac{1}{2} + y(-\varepsilon)\frac{1}{2} = \sum_{i=1} v^i \mu_i$ where $\mu_i > 0$ for all $i = 1, \ldots, d$, that is, x is in the interior of S instead of on a facet of S. This implies that there is no second sub-simplex T' with facet F as considered above, because T' (and hence S) would contain such $y(\varepsilon)$ with $\varepsilon < 0$ and $\lambda_0 < 0$ by (7.10). So the sub-simplex T is unique as stated in (a).

On the other hand, if F is not a subset of a facet of S, then x is in the interior of S, and there are $y(\varepsilon)$ in S for $\varepsilon < 0$ and therefore there is a second sub-simplex T' with facet F. By the same reasoning as applied to T, there is no other sub-simplex with a vertex on the same side of F as the vertex w^0 of T'. This shows case (b), and obviously exactly one of the two cases (a) or (b) applies. □

The following lemma will be used in an inductive proof in the next section.

Lemma 7.7. *Let $n \geq 2$, let S be a simplex with vertices $v^1, \ldots, v^n$, let F be the facet of S with vertices $v^1, \ldots, v^{n-1}$, let Σ be a triangulation of S, and let Σ' be the set of sub-facets in Σ that are subsets of F. Then Σ' is a triangulation of F.*

Proof. Let $\Sigma_1 = \{T \cap F \mid T \in \Sigma\}$. Then $F = \bigcup_{T' \in \Sigma_1} T'$ because every x in F (as an element of S) belongs to some sub-simplex T in Σ. For every $T \in \Sigma$, we claim that $T \cap F$ is a face of T. Namely, every vertex v of T is a unique convex combination $v = \sum_{i=1}^n v^i \lambda_i$ of the vertices v^i of S, where $v \in F$ if and only if $\lambda_n = 0$. Then $T \cap F$ is the convex hull of these vertices v that belong to F. Therefore, $T \cap F$ is a face of T as claimed, and is a simplex of dimension at most $n - 2$.

Let Σ_2 be the set of those simplices in Σ_1 that have dimension $n - 2$. Because all simplices in Σ have dimension $n - 1$ by Lemma 7.5, $\Sigma_2 = \Sigma'$.

We show that every $x \in F$ belongs to some simplex T' in Σ', that is, in Σ_2. Suppose there is some $x \in F$ where that is not the case. Then $x \in S_1$ for some simplex $S_1 \in \Sigma_1$ where S_1 has dimension less than $n - 2$. Let S_1 be the simplex in Σ_1 of largest dimension (by assumption at most $n - 3$) that contains x. By the same argument as used in the proof of Lemma 7.5 with $d = n - 1$, there is some $S_2 \in \Sigma_1$ which is a superset of S_1 and has at least one further vertex and higher dimension than S_1, a contradiction. This shows that $F = \bigcup_{T' \in \Sigma'} T'$ as claimed.

For no two sub-facets in Σ' is one a subset of the other because they have the same dimension but different sets of vertices. Moreover, the intersection of two such sub-facets is determined by the intersection of their vertices, and hence a face of both these sub-facets. This shows that Σ' is a triangulation of F. $\square$

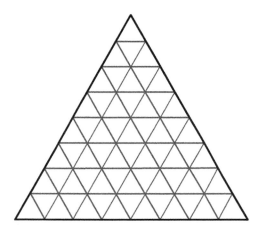

Figure 7.5 Triangulation of a regular triangle into small similar triangles.

The triangulation of a simplex S takes the role of the "grid" mentioned earlier, with its vertices as gridpoints. The goal is to be able to create for any $\varepsilon > 0$ an "ε-fine grid" where the vertices in every sub-simplex are no more than ε apart. Figure 7.5 shows a good way to do this when S is a triangle. However, triangulating a higher-dimensional simplex to achieve this is not as easy. For example, consider a regular tetrahedron where each of its facets, which are triangles, is subdivided into four regular triangles obtained by joining the midpoints of each edge of the tetrahedron. Then "chopping off" the small tetrahedron at each corner leaves a central polytope that is not a tetrahedron but an octahedron. This octahedron has to be subdivided further into tetrahedra in some new way. In addition, it is not clear how such a construction would extend to higher dimensions. For the moment, we assume that in any dimension we can find a sufficiently "fine" triangulation of a simplex. A specific construction that achieves this is described in Section 7.7.

7.4   Sperner's Lemma

In this section, we prove Sperner's lemma, which is often called a combinatorial version of Brouwer's fixed-point theorem, and which is a crucial step in its proof. Sperner's lemma applies to a simplex S with d vertices and triangulation Σ of S, where each vertex in Σ is given a label in $\{1, \ldots, d\}$, which fulfills the following condition.

Definition 7.8 (Sperner labeling). Let S be a simplex with vertices $v^1, \ldots, v^d$, let Σ be a triangulation of S, and let V be the set of vertices in Σ. Then a function $V \to \{1, \ldots, d\}$ that assigns to each x in V a *label* is said to fulfill the *Sperner condition*, and is called a *Sperner labeling*, if whenever $x = \sum_{k=1}^{d} v^k \lambda_k$ and x has label i, then $\lambda_i > 0$. $\qquad\square$

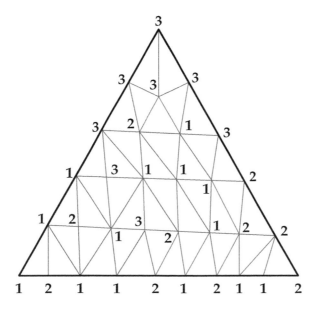

Figure 7.6 A triangulation of the triangle S with a Sperner labeling. The vertices v^1, v^2, v^3 of S are uniquely identified by their respective labels $1, 2, 3$.

Note that in Definition 7.8, every point x in the simplex S has the unique barycentric coordinates $\lambda_1, \ldots, \lambda_d$ as in (7.7). The Sperner condition implies that each vertex v^i of S has label i (and therefore every vertex has a different label). In general, the Sperner condition states that if x is on a face F of S (and no smaller face), then x is only allowed to have one of the labels of the vertices of F. For example, if x is on the edge that connects v^1 and v^2, then x can only have labels 1 or 2. For $d = 2$ the Sperner condition is stated in (7.3). For $d = 3$ the simplex S is a triangle, with an example of a triangulation with a Sperner labeling shown in Figure 7.6.

For a function $f : \Delta \to \Delta$ on the unit simplex Δ in (7.1) and $x \in \Delta$, one of the labels of x defined in (7.2) can always be chosen to fulfill the Sperner condition.

Lemma 7.9. *Consider the $(d-1)$-dimensional unit simplex Δ in (7.1) and a function $f : \Delta \to \Delta$. Then every point $x \in \Delta$ can be given a label i so that $x_i \geq f_i(x)$ and the Sperner condition holds.*

Proof. Because Δ is the convex hull of the unit vectors $e_1, \ldots, e_d$, the barycentric coordinates of $x \in \Delta$ are just $x_1, \ldots, x_d$. Let $I = \{i \mid x_i > 0\}$. If $f_i(x) > x_i$ for all $i \in I$, then

$$\sum_{i=1}^{d} f_i(x) \geq \sum_{i \in I} f_i(x) > \sum_{i \in I} x_i = \sum_{i=1}^{d} x_i = 1,$$

which contradicts $f(x) \in \Delta$. So there is at least one $i \in I$ with $x_i \geq f_i(x)$, that is, $x_i > 0$ and x has label i, as required. $\qquad\square$

Sperner's lemma asserts that a triangulation of a simplex with a Sperner labeling has at least one completely labeled sub-simplex. A suitable induction proof shows the stronger statement that the number of completely labeled sub-simplices is *odd*. Figure 7.7 shows this for the subdivided triangle considered earlier.

Theorem 7.10 (Sperner's lemma). *Let S be a simplex with vertices $v^1, \ldots, v^d$, let Σ be a triangulation of S, and consider a Sperner labeling of the vertices in V. Then Σ has an odd number of sub-simplices T that are completely labeled (that is, every label in $\{1, \ldots, d\}$ is the label of some vertex of T).*

Proof. In the statement of the theorem, the vertices $v^1, \ldots, v^d$ of S can be elements of some space $\mathbb{R}^k$ for any $k \geq d - 1$. Consider the following statement:

(i)   Let $1 \leq n \leq d$, let F_n be the simplex with vertices $v^1, \ldots, v^n$, and let $\Sigma_n = \{F \mid F$ is a face of dimension $n - 1$ of some $T \in \Sigma$ and $F \subseteq F_n\}$. Then Σ_n is a triangulation of F_n.

By assumption, (i) is true for $n = d$. If (i) holds for some $n > 1$, then Lemma 7.7 (with $d = n$) implies that Σ_{n-1} is a triangulation of F_{n-1}. Hence, by induction, (i) is true for $n = d, d - 1, \ldots, 1$, as claimed. We now prove by induction:

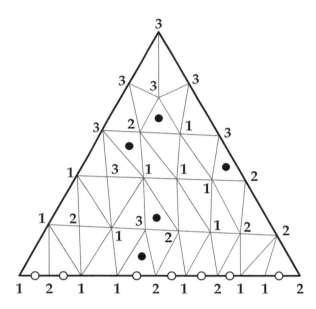

Figure 7.7 Illustration of Sperner's lemma (Theorem 7.10). The black (full) dots mark completely labeled sub-simplices, and the white (hollow) dots mark completely labeled sub-facets, as used in the induction proof of Theorem 7.10. Both white and black dots are odd in number.

(ii) For $n = 1, 2, \ldots, d$, let Σ_n be the triangulation of F_n as defined in (i). Call $T \in \Sigma_n$ *completely labeled* if the vertices of T have all labels $1, \ldots, n$. Then Σ_n has an odd number of completely labeled simplices T.

Statement (ii) is trivial for $n = 1$ where $F_1 = \{v_1\}$. To prove the induction step, assume (ii) holds for some $n \geq 1$ where $n < d$, and we want to prove (ii) for $n + 1$. For any face F of a simplex T in Σ, let

$$L(F) = \{\, i \mid i \text{ is a label of some vertex of } F \,\}.$$

Consider the following sets:

$$
\begin{aligned}
C &= \{F \in \Sigma_n \mid L(F) = \{1, \ldots, n\}\}, \\
C' &= \{F \mid F \text{ is a facet of some } T \in \Sigma_{n+1},\ L(F) = \{1, \ldots, n\},\ F \notin \Sigma_n\}, \\
D &= \{T \in \Sigma_{n+1} \mid L(T) = \{1, \ldots, n+1\}\}, \\
A &= \{T \in \Sigma_{n+1} \mid L(T) = \{1, \ldots, n\}\}.
\end{aligned}
\tag{7.11}
$$

The case $n = 2$ is shown in Figures 7.7 and 7.8. The set $F_3 = F_{n+1}$ is the triangle considered earlier, with a triangulation Σ_3 of F_3 into triangles; the vertices in Σ_3 may have labels 1, 2, and 3. The completely labeled elements of Σ_3 define the set D and are marked with black dots. The set $F_2 = F_n$ is the bottom facet of the triangle F_3, with a triangulation Σ_2 of F_2 into line segments; the vertices in Σ_2 may have labels 1 and 2. The completely labeled elements of Σ_2 define the set C and are marked with white dots. By inductive assumption, $|C|$ is odd.

The set C' is the set of sub-facets F whose n vertices have all labels $1, \ldots, n$ and which are not in Σ_n; in Figure 7.8 they are also marked with white dots. (We

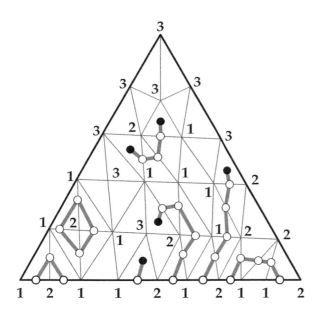

Figure 7.8 The sets defined in (7.11) for $n = 2$. The elements of C (white dots) are in Σ_2 with labels $1, 2$, and the elements of D (black dots) are sub-simplices in Σ_3 with labels $1, 2, 3$. They are connected by paths with further nodes in C' (white dots, sub-facets inside the triangle with labels $1, 2$), which define the graph G_3. Hence, $|C \cup D|$ is even, and thus $|D|$ is odd because $|C|$ is odd.

use the terminology when $n + 1 = d$ and call the elements of Σ_{n+1} "sub-simplices" and any facet of a sub-simplex a "sub-facet", which includes the elements of Σ_n.) The set A is the set of *almost completely* labeled sub-simplices where for any $T \in A$ the $n + 1$ vertices of T have all labels $1, \ldots, n$ but not label $n + 1$.

We now define a new *graph* G_{n+1} with node set $C \cup C' \cup D$, where any two nodes are joined by an edge if they are subsets of the same sub-simplex, which belongs to A or D. Figure 7.8 shows the nodes of the graph G_3 as white or black dots, and its edges as thick gray lines. The following cases can occur for such an edge of G_{n+1}:

Case A. One endpoint of the edge is a sub-facet $F \in C$, so that F is a subset of the facet F_n of F_{n+1}. Hence, by Lemma 7.6(a), F is the face of exactly one sub-simplex T in $D \cup A$. Then:

(a) If $T \in D$, then T is completely labeled, and has exactly one facet with labels $1, \ldots, n$, which is F. The edge of the graph is $\{F, T\}$, and both F and T have degree one. The edge itself defines a path of length one in the graph. An example of such a single-edge path is shown in Figure 7.8 at the bottom in the fourth triangle from the left.

(b) If $T \in A$, then T has $n + 1$ vertices that together have labels $1, \ldots, n$, so one of these labels occurs twice. Hence, T has exactly two sub-facets that have labels $1, \ldots, n$, one of which is F and the second, say F', belongs to C'. The edge of the graph is $\{F, F'\}$, with F' as the only neighbor of F, so F again has degree one. In Figure 7.8, these are all the other cases where $F \in \Sigma_2$ (that is, F is a white dot at the bottom of the triangle), where T (which represents in effect the edge of the graph) has either labels $1, 2, 1$ or $1, 2, 2$.

Case B. One endpoint of the edge of the graph is a sub-facet $F \in C'$. Then $F \notin \Sigma_n$ by definition of C', and F is therefore not a subset of *any* facet of F_{n+1} because of the Sperner condition: the vertices of F have the labels $1, \ldots, n$, and every facet of F_{n+1} other than F_n is missing one of these labels. Hence, by Lemma 7.6(b), such a sub-facet F in C' is the facet of exactly two sub-simplices in $D \cup A$. In the graph, F has degree two. For each of the two sub-simplices $T \in D \cup A$ that have F as a facet, either (a) or (b) applies as before: (a) if $T \in D$, then F is the unique neighbor of T, the edge of the graph is $\{F, T\}$, and T has degree one in the graph; (b) if $T \in A$, then A has exactly one other facet F' with labels $1, \ldots, n$, where $F' \in C \cup C'$, and the edge of the graph is $\{F, F'\}$.

Case C. One endpoint of the edge of the graph is a completely labeled sub-simplex $T \in D$. This case has been covered before: T has a unique facet F with labels $1, \ldots, n$, which is the unique neighbor of T, so that T has degree one in the graph, and $F \in C \cup C'$; the edge of the graph is $\{T, F\}$.

In summary, the graph G_{n+1} has nodes in C and D which have degree one, and nodes in C' that have degree two, and no other nodes. Clearly, such a graph is a set of *paths* and *cycles* (as can be seen in Figure 7.8; note the cycle of length four on the left). The *end nodes* of the paths are exactly the degree-one nodes, which are the elements of $C \cup D$ (the cycles have no end nodes). Because every path has two end nodes, the number $|C| + |D|$ of these end nodes is even. (Alternatively, as stated at the beginning of this chapter, every graph has an even number of odd-degree nodes, which are here the nodes of degree one.) Because $|C|$ is odd, this implies that $|D|$ is also odd, which completes the induction proof for (ii). For $n = d$, (ii) states the theorem.

The fact that the number of end nodes of paths is even is called a *parity argument*. The original proof in Sperner (1928), and very similarly in Knaster, Kuratowski, and Mazurkiewicz (1929), uses also a parity argument without the graphs G_n, namely the following equation:

$$|D| + 2|A| = |C| + 2|C'|. \tag{7.12}$$

On the left-hand side in (7.12), each completely labeled sub-facet in $C \cup C'$ is counted as either the unique facet of a sub-simplex T in D, or as one of the two facets of a sub-simplex T in A; that is, we look at all the sub-simplices in $D \cup A$ and count for each one its (one or two) sub-facets that have the labels $1, \ldots, n$. On the right-hand side of (7.12), the sub-facet is either an element of C and then the facet of exactly one sub-simplex in $D \cup A$, or an element of C', and then it is counted twice because it belongs to two sub-simplices in $D \cup A$; here we have used Lemma 7.6. Clearly, (7.12) implies $|D| = |C| + 2|C'| - 2|A|$, and therefore that $|D|$ is odd because $|C|$ is odd, which again completes the induction step for (ii).

We use the graphs G_n for $n = 2, \ldots, d$ because the picture in Figure 7.8 shows the parity argument immediately via the end nodes of paths, compared to the more abstract equation (7.12) (although the reason is the same). Moreover, the graphs lead to a *constructive* proof of Theorem 7.10, as we show next.   □

We have completed the proof of Sperner's lemma in two ways, where the second way based on (7.12) seems much shorter. In the remainder of this section on Sperner's lemma, we combine the graphs $G_2, \ldots, G_d$ to show that they can actually be used to *find* a completely labeled sub-simplex in the given triangulation.

At first sight, the use of these graphs does not seem constructive, and only proves the induction step of claim (ii) in the proof. For example, consider the triangulation Σ_2 of the bottom edge of the triangle in Figure 7.8. The first completely labeled line segment on the left starts a path that ends at another such line segment, not at a completely labeled triangle. It is easy to continue along the edge (which is just a sequence of line segments) until finding a path that ends at a black dot inside the triangle (there are three such paths in the figure). However, the triangle may just be the next face F_3 in a larger simplex, for example a tetrahedron, and some other black dots (for example near the top right) may be needed to find a completely labeled sub-simplex in the next triangulation Σ_4.

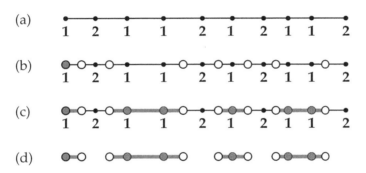

Figure 7.9 Triangulation of F_2, which is the bottom edge of the triangle in Figures 7.6–7.8, with line segments as sub-simplices, and the vertices in Σ_2, marked by small dots, as sub-facets. The Sperner labeling is (a), analogous to Figure 7.6. The completely labeled sub-simplices of Σ_2 are marked with white dots in (b), analogous to the black dots in Figure 7.7; the completely labeled sub-facet is the leftmost vertex marked in gray, analogous to any white dot in Figure 7.7. The graph G_2 is shown in (c), analogous to Figure 7.8, and the graph alone in (d), which in this dimension is only a set of paths.

A more systematic approach is needed, for which we consider the *first* induction step from $n = 1$ to $n = 2$ of claim (ii) in the proof, shown in Figure 7.9. The facet F_1 consists only of a single vertex v^1, marked as a gray node on the left. This defines the graph G_1, which has no edges, but one node (and hence an odd number of nodes) given by v^1, which is also the single sub-simplex of Σ_1 and completely labeled; the graph G_1 is the only graph which does not consist of paths and cycles. In Σ_2, the sub-simplices are line segments, and their sub-facets are single vertices.

The set C' in (7.12) consists of those vertices that have label 1, marked in gray in Figure 7.9, in addition to the leftmost gray node as the unique element of C. This is one of the end nodes of the paths in G_2. All other end nodes are the completely labeled sub-simplices in Σ_2, marked as before with white dots. (There are no cycles because F_2 is just a line segment.)

We now state our constructive version of Sperner's lemma, obtained by just taking the union of the considered graphs, which concatenates their paths.

Proposition 7.11. *Consider a triangulation Σ of a simplex S with vertices $v^1, \ldots, v^d$ and a Sperner labeling as in Theorem 7.10, $d \geq 2$. Let $G_1, \ldots, G_d$ be the graphs defined in the proof of the theorem, and define the graph G by*

$$G = G_1 \cup \cdots \cup G_d \tag{7.13}$$

where the union is taken separately for the sets of nodes and edges, respectively. Then every node in G has degree one or two, so that G consists of paths and cycles. The end nodes of the paths are given by v^1 and the completely labeled sub-simplices of Σ.

Proof. Let $1 \leq n < d$. According to the construction in (7.12), the common nodes of G_n and G_{n+1} are the sub-simplices in Σ_n with n vertices that have all labels $1, \ldots, n$. For $n > 1$, these nodes have degree one in both G_n and G_{n+1}, and therefore degree two in $G_n \cup G_{n+1}$ and in G. Hence, the degree-one nodes in G are given by v^1 (which is the only node in G_1, where it has degree zero, and which has degree one in G_2), and the completely labeled sub-simplices in G_d, as claimed. □

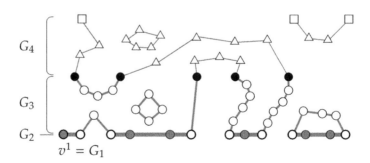

Figure 7.10 Example of G in (7.13) for $d = 4$. The graph G_2 is from Figure 7.9. The graph G_3 from Figure 7.8 is drawn as a "layer" between the completely labeled elements of Σ_2 and Σ_3, and the graph G_4 as another such layer. The triangles indicate sub-facets in Σ_4 with labels $1, 2, 3$ (apart from the black dots, which also belong to G_3), and the squares are sub-simplices in Σ_4 with all labels $1, 2, 3, 4$.

An example of G is shown in Figure 7.10, for a triangulated tetrahedron which has F_3 in Figure 7.8 as its bottom facet. The concatenated paths from $G_2, \ldots, G_d$ may result in new cycles, as on the bottom right. The path that starts at v^1 ends at a completely labeled sub-simplex in Σ, which is therefore found by following the path. The example shows that the path may move up and down through the different dimensions, such as visiting a triangle in G_4 and coming back to a single vertex in G_2. The construction of G in (7.13) is very clean because it does not need any further addition to the triangulation Σ and its labeled vertices.

7.5 The Knaster–Kuratowski–Mazurkiewicz Lemma

In this section, we use Sperner's lemma to prove a theorem due to Knaster, Kuratowski, and Mazurkiewicz (1929) known as the "KKM lemma" after the initials of the authors. The KKM lemma considers a simplex S with vertices $v^1, \ldots, v^d$ and closed sets $A_1, \ldots, A_d$. The assumption, which is closely related to the Sperner condition, is that for any set of vertices, the face of S with these vertices v^i is covered by the corresponding closed sets A_i, as illustrated in Figure 7.11. The conclusion is that then these sets have a common point. In some contexts, such closed sets can be defined more naturally than a function for which one tries to find a fixed point, so that the KKM lemma is a useful tool in its own right. At any rate, it leads almost immediately to a proof of Brouwer's fixed-point theorem on the unit simplex (Theorem 7.13). As background material, we recall the concept of Euclidean norm and distance, and the classic theorem of Bolzano–Weierstrass on the existence of converging subsequences, which is used in the proof of the KKM lemma.

Background material: Euclidean norm and distance

For $a = (a_1, \ldots, a_d)^\top \in \mathbb{R}^d$, the *Euclidean norm* $\|a\|$ of a is defined as

$$\|a\| = \sqrt{a^\top a} = \sqrt{a_1^2 + \cdots + a_d^2}.$$

The Euclidean *distance* between two points $a, b \in \mathbb{R}^d$ is defined as $\|a - b\|$.

Theorem 7.12 (The KKM lemma). *Let S be a simplex with vertices $v^1, \ldots, v^d$, and let $A_1, \ldots, A_d$ be closed sets such that for all $I \subseteq \{1, \ldots, d\}$ the face $F_I = \operatorname{conv}\{v^i \mid i \in I\}$ of S is a subset of $\bigcup_{i \in I} A_i$. Then $\bigcap_{i=1}^d A_i \neq \emptyset$.*

Proof. Let $\varepsilon > 0$, and consider a triangulation Σ_ε of S so that for any sub-simplex T in Σ_ε and any $x, y \in T$ we have $\|x - y\| \leq \varepsilon$; we will show in Section 7.7 that such an ε-*fine* triangulation exists for any $\varepsilon > 0$. For any vertex x in Σ_ε, let $I \subseteq \{1, \ldots, d\}$ be inclusion-minimal so that $x \in F_I$. By assumption, $F_I \subseteq \bigcup_{i \in I} A_i$ and therefore $x \in A_i$ for some $i \in I$, where we choose i as a label of x. By construction, this is a Sperner labeling. According to Sperner's lemma (Theorem 7.10), there is a completely labeled sub-simplex in Σ_ε with vertices $w_\varepsilon^1, \ldots, w_\varepsilon^d$ so that w_ε^i has label i and therefore belongs to A_i for $1 \leq i \leq d$.

Background material: Every bounded sequence has a converging subsequence

The theorem of Bolzano–Weierstrass states: Let $x^{(1)}, x^{(2)}, \ldots$ be a bounded sequence of points in $\mathbb{R}^d$, that is, there is some M so that $\|x^{(n)}\| \leq M$ for all n. Then this sequence has a converging subsequence.

This important theorem has a nice proof that we recall here. We first consider a bounded sequence of *real* numbers $y^{(1)}, y^{(2)}, \ldots$ where $|y^{(n)}| \leq M$ for all n. We show that this sequence has a *monotonic* subsequence $y^{(k_1)}, y^{(k_2)}, \ldots$, that is, either $y^{(k_1)} \geq y^{(k_2)} \geq \cdots$ or $y^{(k_1)} \leq y^{(k_2)} \leq \cdots$ holds. Imagine the real numbers $y^{(1)}, y^{(2)}, \ldots$ as the heights of "hotels at the seaside". From the top of each hotel with height $y^{(k)}$ you can look beyond the subsequent hotels with heights $y^{(k+1)}, y^{(k+2)}, \ldots$ as long as they have *lower* height, and see the sea at infinity if these are all lower. In other words, a hotel has "seaview" if it belongs to the set $K = \{\, k \mid y^{(k)} > y^{(j)} \text{ for all } j > k \,\}$ (presumably, these are very expensive hotels). If K is infinite, then we take the elements of K in ascending order as the superscripts $k_1, k_2, k_3, \ldots$ that give our subsequence $y^{(k_1)}, y^{(k_2)}, y^{(k_3)}, \ldots$, which is clearly decreasing. If, however, K is finite with maximal element m (take $m = 0$ if K is empty), then for each $k > m$ we have $k \notin K$ and hence there exists some $j > k$ with $y^{(k)} \leq y^{(j)}$. Starting with $y^{(k_1)}$ for $k_1 = m + 1$ we let $k_2 > k_1$ with $y^{(k_1)} \leq y^{(k_2)}$. Then find another $k_3 > k_2$ so that $y^{(k_2)} \leq y^{(k_3)}$, and so on, which gives a weakly increasing subsequence $y^{(k_1)} \leq y^{(k_2)} \leq y^{(k_3)} \leq \cdots$. In either case, the original sequence has a monotonic subsequence.

This monotonic subsequence $y^{(k_1)}, y^{(k_2)}, \ldots$ is bounded because the original sequence is bounded. It is easy to show that the subsequence converges to the infimum of its members if it is decreasing, or to the supremum if it is weakly increasing. This shows the Bolzano–Weierstrass theorem for a sequence of real numbers in $\mathbb{R}^1$.

To show the theorem for a sequence of points $x^{(n)} = (x_1^{(n)}, \ldots, x_d^{(n)})$ in $\mathbb{R}^d$ for $n = 1, 2, 3, \ldots$, note that the components $x_i^{(n)}$ for $i = 1, \ldots, d$ of these points are bounded because $|x_i^{(n)}| \leq \|x^{(n)}\| \leq M$. First, find a converging subsequence of the first components $x_1^{(1)}, x_1^{(2)}, x_1^{(3)}, \ldots$ with limit z_1, say. For ease of notation, replace the original sequence of points with this subsequence, which now converges in its first component. Second, find a subsequence so that the second components $x_2^{(1)}, x_2^{(2)}, x_2^{(3)}, \ldots$ have limit z_2, so that now both the first and second components converge. By continuing in this manner, we obtain a subsequence with limit $(z_1, z_2, \ldots, z_d)$, which completes the proof. $\square$

Consider a sequence of positive reals $\varepsilon(1), \varepsilon(2), \ldots$ that tends to 0 (for example, $\varepsilon(n) = \frac{1}{n}$), an $\varepsilon(n)$-fine triangulation $\Sigma_{\varepsilon(n)}$ of S, and the corresponding sequence of points $w^1_{\varepsilon(1)}, w^1_{\varepsilon(2)}, \ldots$ in the simplex S that have label 1 considered earlier. Because

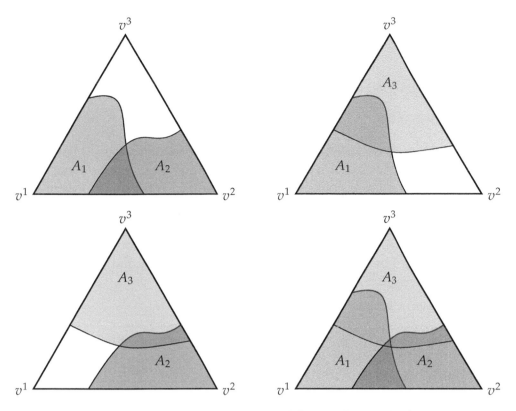

Figure 7.11   Illustration of the assumptions of the KKM lemma for $d = 3$, for the sets $I = \{1, 2\}, \{1, 3\}, \{2, 3\}$, and $\{1, 2, 3\}$.

S is bounded and closed, this sequence of points has a converging subsequence with limit z in S. For all $i = 1, \ldots, d$, we have $\|w^1_{\varepsilon(n)} - w^i_{\varepsilon(n)}\| \leq \varepsilon(n)$, so that the points $w^i_{\varepsilon(n)}$ in that subsequence with label i also converge to z as $n \to \infty$, where $w^i_{\varepsilon(n)} \in A_i$. Because A_i is closed, $z \in A_i$. Therefore, the intersection $\bigcap_{i=1}^{d} A_i$ contains z and is not empty, as claimed. $\square$

The following theorem is an immediate consequence of the KKM lemma, applied to the unit simplex Δ. The closed sets A_i are those points x in Δ that have label i according to our original definition (7.2).

Theorem 7.13 (Brouwer's fixed-theorem on the unit simplex). *A continuous function $f : \Delta \to \Delta$ on the $(d - 1)$-dimensional unit simplex Δ in (7.1) has a fixed point.*

Proof. For $i = 1, \ldots, d$, let $A_i = \{ x \in \Delta \mid x_i \geq f_i(x) \}$. Then A_i is a closed set because f and hence f_i is continuous. Let $x \in \Delta$ and $I = \{ i \mid x_i > 0 \}$. Because the vertices of Δ are the unit vectors, the face F_I in Theorem 7.12 is the smallest facet of Δ that contains x. As shown in Lemma 7.9, there is some i so that $x_i > 0$ and

$x \in A_i$. Because x is arbitrary, this shows $F_I \subseteq \bigcup_{i \in I} A_i$ for any $I \subseteq \{1, \ldots, d\}$. By Theorem 7.12, there is some $z \in \bigcap_{i=1}^{d} A_i$, where $f(z) = z$ by Lemma 7.1.   □

7.6 Brouwer's Fixed-Point Theorem on a General Compact Convex Set

Theorem 7.13 states that a continuous function on the unit simplex Δ has a fixed point. In this section, we prove this for a continuous function f on a compact convex set C that is a subset of some $\mathbb{R}^k$, which is the general form of Brouwer's fixed-point theorem. The first step is to *embed* C into Δ with a bijective linear map. Assuming that C is a subset of Δ, the second step is to construct a continuous function $p : \Delta \to C$ so that $p(x) = x$ for $x \in C$. Then the continuous function $h : \Delta \to \Delta$ defined by $h(x) = f(p(x))$ for $x \in \Delta$ has a fixed point, which is easily seen to be a fixed point of the given function $f : C \to C$.

The following lemma is about the embedding of C into Δ.

Lemma 7.14. *Let $C \subset \mathbb{R}^k$ and let C be a compact and convex set of dimension $d - 1$. Then there is an injective affine map $J : C \to \Delta$ from C to the $(d-1)$-dimensional unit simplex Δ in (7.1).*

Proof. Because C has dimension $d - 1$, it has d, but no more, affinely independent points $w^1, \ldots, w^d$. If $d = 1$ then C is just a point, so we can assume $d > 1$. By Lemma 7.2, each point $x \in C$ is the affine combination $x = \sum_{i=1}^{d} w^i \lambda_i$ where this equation and the equation $\sum_{i=1}^{d} \lambda_i = 1$ uniquely determine the coefficients $\lambda_1, \ldots, \lambda_d$. Hence, these coefficients are defined by a linear map from $\mathbb{R}^k$ (which contains C) to $\mathbb{R}^d$ and therefore depend continuously on x. Their minimum $\min_{1 \le i \le d} \lambda_i$ is a continuous function of x and therefore assumes its minimum M on the compact set C, that is,

$$x = \sum_{i=1}^{d} w^i \lambda_i, \quad \sum_{i=1}^{d} \lambda_i = 1 \quad \Rightarrow \quad \lambda_i \ge M \quad (x \in C, \ 1 \le i \le d). \quad (7.14)$$

For $x = w^2$ we have $\lambda_1 = 0$, so $M \le 0$. Let $w = \sum_{i=1}^{d} w^i \frac{1}{d}$ be the barycenter of the points $w^1, \ldots, w^d$. Our aim is to "stretch" the points w^i to new points v^i (typically outside of C) "away from w" so that every $x \in C$ is a convex and not just affine combination of $v^1, \ldots, v^d$. For some yet to be determined $\alpha > 0$, let

$$v^i = w + (w^i - w)\alpha \quad (1 \le i \le d)$$

and therefore

$$w^i = w + (v^i - w)\tfrac{1}{\alpha} = w \left(1 - \tfrac{1}{\alpha}\right) + v^i \tfrac{1}{\alpha}.$$

Then w is also the barycenter of the points $v^1, \ldots, v^d$ because

$$\sum_{i=1}^{d} v^i \tfrac{1}{d} = \sum_{i=1}^{d} (w + (w^i - w)\alpha)\tfrac{1}{d} = w.$$

Hence, for $x \in C$ we have

$$x = \sum_{i=1}^{d} w^i \lambda_i = \sum_{i=1}^{d} \left(w \left(1 - \tfrac{1}{\alpha}\right) + v^i \tfrac{1}{\alpha}\right) \lambda_i = \sum_{i=1}^{d} v_i \left(\tfrac{1}{d}\left(1 - \tfrac{1}{\alpha}\right) + \lambda_i \tfrac{1}{\alpha}\right) \tag{7.15}$$

where we want to achieve that for $1 \leq i \leq d$,

$$\tfrac{1}{d}\left(1 - \tfrac{1}{\alpha}\right) + \lambda_i \tfrac{1}{\alpha} \geq 0 \tag{7.16}$$

which holds if $\lambda_i \geq \tfrac{1-\alpha}{d}$ and therefore if $M \geq \tfrac{1-\alpha}{d}$ because of (7.14). For this purpose, it suffices to choose $\alpha = 1 - Md$ (note that $M \leq 0$). In addition, $\sum_{i=1}^{d} \left(\tfrac{1}{d}\left(1 - \tfrac{1}{\alpha}\right) + \lambda_i \tfrac{1}{\alpha}\right) = 1$, so (7.16) and (7.15) imply that C is a subset of the simplex T with vertices $v^1, \ldots, v^d$. By Lemma 7.3, there is an affine bijection $T \rightarrow \Delta$, which restricted to C gives the desired affine injection $J : C \rightarrow \Delta$.   □

Let C be a compact convex set and let $D = J(C)$ be the image of C under the affine injection J in Lemma 7.14. The set D is a closed convex subset of Δ. Next, we construct a continuous function $p : \Delta \rightarrow D$ that restricted to D is the identity, that is, $p(x) = x$ for $x \in D$. The following function p is easiest to use.

Let D be any nonempty closed convex subset of $\mathbb{R}^d$. The *projection* onto D is the function $p : \mathbb{R}^d \rightarrow D$ defined for $x \in \mathbb{R}^d$ by

$$p(x) = \arg \min_{z \in D} \|x - z\|. \tag{7.17}$$

Condition (7.17) means that $p(x)$ is the point z in D that is closest in Euclidean distance to x, where clearly $p(x) = x$ if $x \in D$. Because D is nonempty and closed and the distance is bounded from below by zero, such a closest point in D exists. We will shortly see that this point is also unique, so that $p(x)$ is well defined. We will then show that p is a continuous function. We first need some properties of the Euclidean norm.

General vectors $a, b \in \mathbb{R}^d$ are called *orthogonal* if $a^\top b = 0$. (This includes the case that a or b is the zero vector, which is always covered in the following observations, typically in some trivial way.) Then the triangle with corners $a, \mathbf{0}, b$ has a right angle at $\mathbf{0}$, and the theorem of Pythagoras applies,

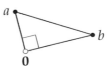

$$a^\top b = 0 \qquad \Leftrightarrow \qquad \|a - b\|^2 = \|a\|^2 + \|b\|^2,$$

which holds because for any a, b we have $b^\top a = a^\top b \in \mathbb{R}$ and

$$\|a - b\|^2 = (a - b)^\top (a - b) = a^\top a - a^\top b - b^\top a + b^\top b = \|a\|^2 - 2a^\top b + \|b\|^2. \tag{7.18}$$

Hence,

$$a^\top b = 0 \quad \Rightarrow \quad \|a - b\| \geq \|a\|, \quad \|a - b\| \geq \|b\|. \tag{7.19}$$

The following lemma is concerned with the distance of a point x to the points on a *ray* (or halfline) with endpoint c through some other point b, given by the set $\{c + (b - c)\lambda \mid \lambda \geq 0\}$. As shown in Figure 7.12, this depends on the sign of $(x - c)^\top(b - c)$, which indicates the *angle* between the vectors $x - c$ and $b - c$; the angle is acute if $(x - c)^\top(b - c) > 0$, obtuse if $(x - c)^\top(b - c) < 0$, and a right angle if $(x - c)^\top(b - c) = 0$. Exactly in the latter two cases the closest point to x on the ray is its endpoint c.

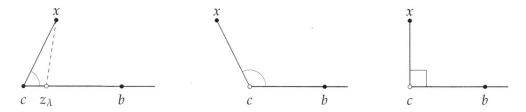

Figure 7.12   Illustration of Lemma 7.15 (a) (left) and (b) (center and right).

Lemma 7.15. *Let* $x, b, c \in \mathbb{R}^d$ *with* $b \neq c$, *and* $z_\lambda = c + (b - c)\lambda$ *for* $\lambda \in \mathbb{R}$. *Then*

(a) $(x - c)^\top(b - c) > 0 \quad \Leftrightarrow \quad \|x - z_\lambda\| < \|x - c\| \quad$ *for all sufficiently small positive* λ,

(b) $(x - c)^\top(b - c) \leq 0 \quad \Leftrightarrow \quad \|x - z_\lambda\| > \|x - c\| \quad$ *for all* $\lambda > 0$.

Proof. By (7.18),

$$\begin{aligned}
\|x - z_\lambda\|^2 &= (x - z_\lambda)^\top(x - z_\lambda) \\
&= (x - c - (b - c)\lambda)^\top(x - c - (b - c)\lambda) \\
&= \|x - c\|^2 - 2(x - c)^\top(b - c)\lambda + \|b - c\|^2\lambda^2 \\
&= \|x - c\|^2 - \left(2(x - c)^\top(b - c) - \|b - c\|^2\lambda\right)\lambda.
\end{aligned}$$

Hence, if $(x - c)^\top(b - c) > 0$, then $\|x - z_\lambda\| < \|x - c\|$ for sufficiently small positive λ, and if $(x - c)^\top(b - c) \leq 0$, then $\|x - z_\lambda\| > \|x - c\|$ for all $\lambda > 0$. This shows "$\Rightarrow$" in (a) and (b). To show the converse implication "$\Leftarrow$" in (a), $\|x - z_\lambda\| < \|x - c\|$ for some $\lambda > 0$ and $(x - c)^\top(b - c) \leq 0$ would contradict "$\Rightarrow$" in (b). Similarly, "$\Leftarrow$" in (b) holds because $\|x - z_\lambda\| > \|x - c\|$ for some sufficiently small $\lambda > 0$ and $(x - c)^\top(b - c) > 0$ contradicts "$\Rightarrow$" in (a).   □

Lemma 7.15 implies that for a nonempty closed convex set D, the point c in D that is closest to x is unique, so that $p(x) = c$ in (7.17) is well defined: Namely, if b is any other point in D and $(x - c)^\top(b - c) > 0$, then by Lemma 7.15(a) there is a point z_λ that is closer to x than c, where for $0 < \lambda \leq 1$ this point belongs to D because D is convex, so c is not the closest point to x in D, a contradiction; and if $(x - c)^\top(b - c) \leq 0$, then Lemma 7.15(b) with $\lambda = 1$ (and thus $z_\lambda = b$) shows that b has greater distance $\|x - b\|$ from x than c. Moreover, we have shown the following.

Corollary 7.16. *Let D be a nonempty closed convex subset of $\mathbb{R}^d$ and $x \in \mathbb{R}^d$ and let $c = p(x)$ be the projection of x onto D. Then $(x - c)^\top(b - c) \leq 0$ for all $b \in D$.*

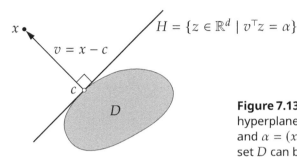

Figure 7.13 For $x \notin D$ in Corollary 7.16, the hyperplane H with normal vector $v = x - c$ and $\alpha = (x - c)^\top c$ separates x from D. (The set D can be unbounded.)

Corollary 7.16 is also known as the theorem of the *separating hyperplane*. A *hyperplane H in $\mathbb{R}^d$* is defined as $H = \{z \in \mathbb{R}^d \mid v^\top z = \alpha\}$ for a nonzero *normal vector v* in $\mathbb{R}^d$ and a real number α. The corresponding *halfspace* is defined as $\{z \in \mathbb{R}^d \mid v^\top z \leq \alpha\}$. Suppose that in Corollary 7.16 we have $x \notin D$ and therefore $x \neq c$ and $(x - c)^\top(x - c) = \|x - c\|^2 > 0$. With $v = x - c$ and $\alpha = (x - c)^\top c$ the corollary states that D is a subset of the halfspace defined by H, whereas x is clearly not in the halfspace and therefore "separated" from D by H. See also Figure 7.13. The separating hyperplane theorem is a very useful tool, but one has to ensure that its assumptions hold, in particular that the set D is closed.

A second easy consequence of Lemma 7.15 is that the shortest connecting line segment of a point x to a line L is orthogonal to that line.

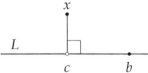

Lemma 7.17. *Let L be a line in $\mathbb{R}^d$, that is, the set of affine combinations of two points in $\mathbb{R}^d$, let $x \in \mathbb{R}^d$, and let $c = p(x)$ be the projection of x onto L. Then $(x - c)^\top(b - c) = 0$ for all $b \in L$.*

Proof. Let b be any point on L other than c. Then $L = \{c + (b - c)\lambda \mid \lambda \in \mathbb{R}\}$. If $(x - c)^\top(b - c) > 0$, then c is not the closest point to x by Lemma 7.15(a). If $(x - c)^\top(b - c) < 0$, let $b' = c + (b - c)(-1)$ so that $b' - c = -(b - c)$ and replace b by b' to observe the same, again by Lemma 7.15(a). Hence, $(x - c)^\top(b - c) = 0$ as claimed. □

Our goal is to show that the projection map p in (7.17) is continuous. This is a consequence of the following theorem.

Theorem 7.18. *Let D be a nonempty closed convex subset of $\mathbb{R}^d$, and let $p : \mathbb{R}^d \to D$ be the projection onto D defined in (7.17). Then for all $x, y \in \mathbb{R}^d$*

$$\|p(x) - p(y)\| \le \|x - y\| . \qquad (7.20)$$

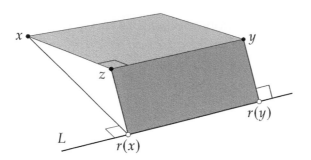

Figure 7.14 Projections $r(x)$ and $r(y)$ onto a line L.

Proof. Let $x, y \in \mathbb{R}^d$. We first show the claim when D is a line L, that is,

$$\|r(x) - r(y)\| \le \|x - y\| , \qquad (7.21)$$

where $r(x)$ and $r(y)$ denote the projections of x and y onto L. The claim is apparent from Figure 7.14, where we consider the point $z = r(x) + y - r(y)$, so that the points $z, y, r(y), r(x)$ form a rectangle where $\|z - y\| = \|r(x) - r(y)\|$, and the points x, z, y are three corners of a rectangle whose side length $\|z - y\|$ is not longer than the diagonal length $\|x - y\|$. Formally,

$$
\begin{aligned}
&(x - z)^{\top}(y - z) \\
= \; &\big(x - r(x) - y + r(y)\big)^{\top}\big(r(y) - r(x)\big) \\
= \; &\big(x - r(x)\big)^{\top}\big(r(y) - r(x)\big) - \big(y - r(y)\big)^{\top}\big(r(x) - r(y)\big) \\
= \; &0 - 0
\end{aligned}
$$

where we use Lemma 7.17 twice, with $r(x)$ as the closest point to x on L and $r(y)$ as the closest point to y on L. Then (7.19) implies $\|x - y\| = \|x - z - (y - z)\| \ge \|y - z\| = \|y - (r(x) + y - r(y))\| = \|r(y) - r(x)\| = \|r(x) - r(y)\|$, which shows (7.21).

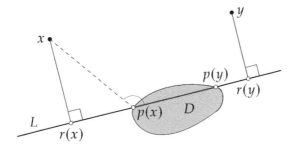

Figure 7.15 Projections $p(x)$ and $p(y)$ of x and y onto a convex set D, with the line L through $p(x)$ and $p(y)$. The angle between x, $p(x)$, and $p(y)$ cannot be acute, so $r(x)$ is to the left of (or equal to) $p(x)$. Similarly, $r(y)$ is to the right of (or equal to) $p(y)$.

Secondly, consider now a general nonempty closed convex set D, shown in Figure 7.15 for an example in dimension two. Let $p(x)$ and $p(y)$ be the projections of x and y onto D, where we can assume $p(x) \neq p(y)$ because otherwise (7.20) holds trivially. Let L be the line through $p(x)$ and $p(y)$, and let $r(x)$ and $r(y)$ be the projections of x and y onto L.

The line segment that connects $p(x)$ and $p(y)$ is a subset of D because D is convex. We show that it is a subset of the line segment that connects $r(x)$ and $r(y)$, and therefore at most as long. Let $c = p(x)$ and $b = p(y)$. By Corollary 7.16, $(x - c)^\top (b - c) \leq 0$, that is, the angle at c shown in Figure 7.15 is a right or obtuse angle. Let $r(x) = p(x) + (p(y) - p(x))\lambda = z_\lambda$. Then $\lambda \leq 0$, because if $\lambda > 0$ then Lemma 7.15(b) would imply $\|x - r(x)\| > \|x - p(x)\|$, which is a contradiction because $p(x) \in L$ and $r(x)$ is the closest point to x on L. Similarly, $r(y) = p(y) + (p(x) - p(y))\mu$ where $\mu \leq 0$. Then

$$
\begin{aligned}
\|r(x) - r(y)\| &= \|p(x) + (p(y) - p(x))\lambda - p(y) - (p(x) - p(y))\mu\| \\
&= \|(p(x) - p(y))(1 - \lambda - \mu)\| \\
&= \|(p(x) - p(y))\| \, (1 - \lambda - \mu) \\
&\geq \|(p(x) - p(y))\|
\end{aligned}
$$

because $\lambda \leq 0$ and $\mu \leq 0$, which with (7.21) implies (7.20). $\qquad\square$

Theorem 7.18 states that the projection p onto D is "contractive", and therefore indeed continuous:

Corollary 7.19. *Let D be a nonempty closed convex subset of $\mathbb{R}^d$. Then the projection map p of $\mathbb{R}^d$ onto D in (7.17) is continuous.*

Proof. Continuity of $p : \mathbb{R}^d \to D$ means that for all $x, y \in \mathbb{R}^d$

$$
\forall \varepsilon > 0 \ \exists \delta > 0 \ : \ \|x - y\| < \delta \ \Rightarrow \ \|p(x) - p(y)\| < \varepsilon.
$$

By Theorem 7.18, it suffices to take $\delta = \varepsilon$. $\qquad\square$

The convexity of the set D in Corollary 7.19 is essential. For example, if D is just a set of two distinct points a and b, then the projection p onto D is not well defined for points that are equidistant to a and b, and at any rate "jumps" from a to b near these points. Moreover, the function $D \to D$ that exchanges a and b is trivially continuous but has no fixed points, which it would have if p was continuous, according to the proof of the following theorem. This theorem is Brouwer's fixed-point theorem in full generality.

Theorem 7.20. *Let C be a nonempty compact and convex subset of $\mathbb{R}^k$, and let $f : C \to C$ be continuous. Then f has a fixed point $z \in C$, that is, $f(z) = z$.*

Proof. Suppose C has dimension $d - 1$. By Lemma 7.14, there is an affine injective map $J : C \to \Delta$ with image $D = J(C) = \{ J(z) \mid z \in C \} \subseteq \Delta$, and its inverse function $J^{-1} : D \to C$ is a linear bijection. The set D is compact, and convex because J preserves convex combinations. Define the function $g : D \to D$ by

$$g\left(J(z)\right) = J\left(f(z)\right), \tag{7.22}$$

which is unambiguous because J is injective, and which defines g on all of D because $D = J(C)$. Then for all $z \in C$

$$f(z) = z \quad \Leftrightarrow \quad g\left(J(z)\right) = J(z) \tag{7.23}$$

where "$\Rightarrow$" is immediate from (7.22), and "$\Leftarrow$" holds because $g\left(J(z)\right) = J(z)$ implies

$$f(z) = J^{-1}\left(J\left(f(z)\right)\right) = J^{-1}\left(g\left(J(z)\right)\right) = J^{-1}\left(J(z)\right) = z .$$

Condition (7.23) states that the fixed points of f are in one-to-one correspondence with the fixed points of g via the bijection J.

Let $p : \Delta \to D$ be the projection map onto D as defined in (7.17), which is continuous by Corollary 7.19 and fulfills $p(x) = x$ for all $x \in D$. Define the function

$$h : \Delta \to \Delta, \qquad h(x) = g\left(p(x)\right) \quad (x \in \Delta),$$

which is a continuous function on the unit simplex Δ because p and g are continuous. By Theorem 7.13, h has a fixed point $x \in \Delta$, that is, $h(x) = g\left(p(x)\right) = x$, where $h(x) \in D$ because g takes only values in D, and therefore $p(x) = x$, which implies $g(x) = x$. That is, the fixed points of h are exactly the fixed points of g. By (7.23), the desired fixed point of f is given by $z = J^{-1}(x)$. $\qquad\square$

7.7   The Freudenthal Triangulation

This section provides the missing piece in the proof of Brouwer's fixed-point theorem, namely an arbitrarily fine triangulation of a simplex, which is used to prove the KKM lemma (Theorem 7.12). As mentioned at the end of Section 7.3, this is only obvious when the simplex has dimension one or two. In the following we describe an elegant triangulation, due to Freudenthal (1942), that works in any dimension. Its construction rests on two main ideas:

- The d-dimensional *unit cube*

$$C_d = [0, 1]^d = \{x \in \mathbb{R}^d \mid 0 \le x_i \le 1, \ 1 \le i \le d\}$$

  can be easily subdivided into small "cubelets".

- The cube C_d has a canonical triangulation into $d!$ simplices that can in the same way be applied to the cubelets, which gives then a triangulation of one of the simplices in the triangulation of C_d.

We continue to use d has the dimension of the space $\mathbb{R}^d$ of the points that we consider. The unit simplex Δ in (7.1) is a subset of $\mathbb{R}^d$ but has dimension $d - 1$. For the Freudenthal triangulation, we use a less symmetric simplex T_d than Δ, which does not matter because any two simplices of a given dimension are in an affine bijection via their barycentric coordinates by Lemma 7.3. The simplex T_d has $d + 1$ vertices in $\mathbb{R}^d$ and has therefore dimension d, which by Lemma 7.2 is maximal. (A polytope with this property is also called *full-dimensional*.)

The simplex T_d is the convex hull of certain vertices of the unit cube C_d. For example, if $d = 3$, then T_d is the convex hull of the vectors

$$\begin{pmatrix} 0 \\ 0 \\ 0 \end{pmatrix}, \begin{pmatrix} 1 \\ 0 \\ 0 \end{pmatrix}, \begin{pmatrix} 1 \\ 1 \\ 0 \end{pmatrix}, \begin{pmatrix} 1 \\ 1 \\ 1 \end{pmatrix}. \tag{7.24}$$

The first of these vectors is $\mathbf{0}$ in $\mathbb{R}^d$, and the next vector is always obtained by changing one component from 0 to 1, where the last vector is the all-one vector $\mathbf{1}$ in $\mathbb{R}^d$. We then generalize T_d in two ways:

(a) The different orders in which the components are changed from 0 to 1 generate $d!$ different simplices, which together define a triangulation of the unit cube.

(b) In a scaled-up version of T_d which we call $T_d(k)$, each entry 1 in (7.24) is replaced by a positive integer k. We then consider all points with integer coordinates in $T_d(k)$, and shifted unit cubes with these integer points as vertices. The subdivision of each such cube into simplices as described in (a), as far as they are subsets of $T_d(k)$, then creates the desired triangulation of $T_d(k)$.

The construction in (b) exploits the fact that a cube with edge length k is easily subdivided into cubes of side length 1, which in turn are then subdivided into sub-simplices as in (a). Relative to the scaled-up simplex $T_d(k)$, the distances of any two points in a sub-simplex are $\frac{1}{k}$ of the distances in $T_d(k)$, so that by shrinking $T_d(k)$ back to T_d one obtains an arbitrarily fine triangulation of T_d by choosing k large enough. (It is easier to deal with integer coordinates than with multiples of $\frac{1}{k}$.)

Figure 7.16 shows this construction for a triangle $T_d(k)$ where $d = 2$ and $k = 8$, which is essentially a "shifted" triangulation of the symmetric triangulation in Figure 7.5.

The general construction of the Freudenthal triangulation is as follows. Fix a dimension d. Let k be a positive integer. Consider the $d + 1$ vectors in $\mathbb{R}^d$

$$v^0(k) = \begin{pmatrix} 0 \\ 0 \\ 0 \\ \vdots \\ 0 \end{pmatrix}, \; v^1(k) = \begin{pmatrix} k \\ 0 \\ 0 \\ \vdots \\ 0 \end{pmatrix}, \; v^2(k) = \begin{pmatrix} k \\ k \\ 0 \\ \vdots \\ 0 \end{pmatrix}, \; \cdots \; , \; v^d(k) = \begin{pmatrix} k \\ k \\ k \\ \vdots \\ k \end{pmatrix}. \tag{7.25}$$

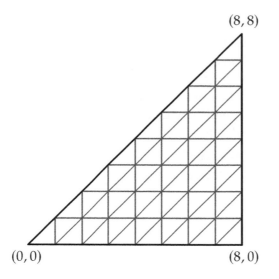

(8,8)

(0,0)   (8,0)

Figure 7.16 The Freudenthal triangulation of the two-dimensional simplex $T_2(k)$ with vertices $(0,0)$, $(k,0)$ and (k,k) for $k = 8$.

These are affinely independent because the corresponding vectors $z^i = \begin{pmatrix} 1 \\ v^i(k) \end{pmatrix}$ in $\mathbb{R}^{1+d}$ as considered in (7.4) are linearly independent because $\sum_{i=0}^{d} z^i \lambda_i = \mathbf{0}$ implies $\lambda_d = 0$, then $\lambda_{d-1} = 0$, and so on until $\lambda_0 = 0$. Their convex hull therefore defines a simplex, which we call $T_d(k)$.

Let $k = 1$. The vertices of $T_d(1)$ are defined by starting from the all-zero vector v^0 and changing one coordinate at a time from 0 to 1 until we arrive at the all-one vector $v^d = \mathbf{1} \in \mathbb{R}^d$. In order to construct $T_d(1)$ we have chosen the coordinates in their natural order $1, \ldots, d$. As a generalization, consider an *arbitrary* order $\pi(1), \ldots, \pi(d)$ of the coordinates, that is, a permutation π of $1, \ldots, d$ (a bijection on the set $\{1, \ldots, d\}$). We denote the set of all permutations of $1, \ldots, d$ by $\mathsf{Perm}(d)$. For $\pi \in \mathsf{Perm}(d)$, let $v_\pi^0 = \mathbf{0} \in \mathbb{R}^d$, and for $i = 1, \ldots, d$ define

$$v_\pi^i = v_\pi^{i-1} + e_{\pi(i)} \,. \tag{7.26}$$

That is, the ith vector v_π^i is obtained from the previous vector by adding the unit vector $e_{\pi(i)}$, which changes the coordinate at position $\pi(i)$ from 0 to 1. For example, if $(\pi(1), \ldots, \pi(d)) = (2, 3, 1, 4)$, then

$$v_\pi^0 = \begin{pmatrix} 0 \\ 0 \\ 0 \\ 0 \end{pmatrix}, \quad v_\pi^1 = \begin{pmatrix} 0 \\ 1 \\ 0 \\ 0 \end{pmatrix}, \quad v_\pi^2 = \begin{pmatrix} 0 \\ 1 \\ 1 \\ 0 \end{pmatrix}, \quad v_\pi^3 = \begin{pmatrix} 1 \\ 1 \\ 1 \\ 0 \end{pmatrix}, \quad v_\pi^4 = \begin{pmatrix} 1 \\ 1 \\ 1 \\ 1 \end{pmatrix}. \tag{7.27}$$

This defines a "monotonic path" of vertices of C_d that starts at the all-zero vector $\mathbf{0}$ and ends at the all-one vector $\mathbf{1}$. (It is easy to see that the vertices of the unit cube C_d are the 2^d vectors in $\mathbb{R}^d$ where each component is either 0 or 1.) The convex hull of these vectors defines the simplex

$$T_d^\pi = \mathsf{conv}\, \{v_\pi^0, v_\pi^1, \ldots, v_\pi^d\} \,.$$

The following lemma describes this simplex in terms of $d+1$ linear inequalities rather than as a convex hull of vertices.

Lemma 7.21. *Let $x \in \mathbb{R}^d$ and $\pi \in \text{Perm}(d)$. Then $x \in T_d^\pi$ if and only if*

$$1 \geq x_{\pi(1)} \geq x_{\pi(2)} \geq \cdots \geq x_{\pi(d)} \geq 0. \tag{7.28}$$

Proof. Suppose (7.28) holds. Define $y_0, y_1, \ldots, y_d$ by

$$\begin{aligned}
y_0 &= 1 - x_{\pi(1)} \\
y_1 &= x_{\pi(1)} - x_{\pi(2)} \\
&\vdots \\
y_{d-1} &= x_{\pi(d-1)} - x_{\pi(d)} \\
y_d &= x_{\pi(d)}
\end{aligned} \tag{7.29}$$

which implies

$$x_{\pi(j)} = \sum_{i=j}^{d} y_i \qquad (1 \leq j \leq d)$$

and by (7.29) and (7.28)

$$\sum_{i=0}^{d} y_i = 1, \qquad y_i \geq 0 \qquad (1 \leq i \leq d). \tag{7.30}$$

By (7.26), as the example (7.27) illustrates, the components of the vectors v_π^i are given by

$$(v_\pi^i)_{\pi(j)} = \begin{cases} 0 & \text{if } i < j \\ 1 & \text{if } i \geq j \end{cases} \qquad (0 \leq i \leq d, \ 1 \leq j \leq d).$$

Hence, by (7.29),

$$\sum_{i=0}^{d} (v_\pi^i)_{\pi(j)} y_i = \sum_{i=j}^{d} y_i = x_{\pi(j)} \qquad (1 \leq j \leq d)$$

and therefore

$$x = \sum_{i=0}^{d} v_\pi^i y_i. \tag{7.31}$$

That is, x is by (7.30) a convex combination of the vectors $v_\pi^0, \ldots, v_\pi^d$, which shows that $x \in T_d^\pi$. Moreover, $y_0, \ldots, y_d$ in (7.29) are the barycentric coordinates of x.

Conversely, any vertex $x = v_\pi^i$ of T_d^π for $0 \leq i \leq d$ fulfills (7.28), where the only inequality that holds strictly (if $0 < i < d$) is when the component of x changes from 1 to 0. Therefore, the inequalities (7.28) also hold for any convex combination of these vertices, that is, for any element of T_d^π. $\qquad \square$

The simplex $T_d(1)$ is T_d^π for the identity permutation π. It is easy to see that $T_d(k) = \{ xk \mid x \in T_d(1) \}$ and therefore by Lemma 7.21

$$z \in T_d(k) \qquad \Leftrightarrow \qquad k \geq z_1 \geq z_2 \geq \cdots \geq z_d \geq 0 . \tag{7.32}$$

Lemma 7.22. *The set $\{ T_d^\pi \mid \pi \in \mathrm{Perm}(d) \}$ is a triangulation of the unit cube C_d.*

Proof. Let $x = (x_1, \ldots, x_d)^\top \in C_d$, that is, $1 \geq x_i \geq 0$ for $1 \leq i \leq d$, and let $\pi \in \mathrm{Perm}(d)$. Sort the components x_i of x so that (7.28) holds (the permutation π is not unique if some components of x are equal to each other). By Lemma 7.21, $x \in T_d^\pi$. This shows $C_d = \bigcup_{\pi \in \mathrm{Perm}(d)} T_d^\pi$.

If $\pi, \pi' \in \mathrm{Perm}(d)$ and $\pi \neq \pi'$ then $T_d^\pi \not\subseteq T_d^{\pi'}$, by taking some $x \in T_d^\pi$ with strict inequalities in (7.28) where clearly $x \notin T_d^{\pi'}$. It remains to show that the intersection of any two sub-simplices T_d^π and $T_d^{\pi'}$ is a face of both. Any x in $T_d^\pi \cap T_d^{\pi'}$ fulfills (7.28) as well as

$$x_{\pi'(1)} \geq x_{\pi'(2)} \geq \cdots \geq x_{\pi'(d)} . \tag{7.33}$$

Any inequality that holds in reverse order in (7.28) and (7.33) has to hold as an equality. This includes any implied equality; for example, if for $d = 3$ we have $x_1 \geq x_2 \geq x_3$ and $x_2 \geq x_3 \geq x_1$, then this implies $x_1 = x_3$ as well as $x_2 = x_3$. The intersection $T_d^\pi \cap T_d^{\pi'}$ is then characterized by (7.28) where some inequalities are replaced by equalities. The barycentric coordinates $y_1, \ldots, y_{d-1}$ in (7.29) of x that correspond to these equalities are then set to zero, which means that the corresponding vertex is not used in the representation of x in (7.31). Hence, x is only a convex combination of the remaining vertices. These are the vertices of $T_d^\pi \cap T_d^{\pi'}$, which is therefore a common face of the two sub-simplices T_d^π and $T_d^{\pi'}$, as claimed. It may be the case that all inequalities in (7.28) are equalities, so that the common face has just the vertices $\mathbf{0}$ and $\mathbf{1}$. $\qquad\square$

There are $d!$ many permutations π of the numbers $1, \ldots, d$. Any two simplices T_d^π and $T_d^{\pi'}$ have the same shape and size because they only differ in the order of their coordinates, and they intersect in a set of dimension $d - 1$ or lower which has volume zero in $\mathbb{R}^d$. Because the unit cube C_d has volume 1, Lemma 7.22 implies that each simplex T_d^π has volume $\frac{1}{d!}$.

We now use the triangulation of the unit cube C_d in Lemma 7.22 to construct a triangulation of the simplex $T_d(k)$ in (7.25), for some integer $k \geq 1$.

For a polytope P in $\mathbb{R}^d$ such as C_d or T_d^π and an integer "grid" vector $g = (g_1, \ldots, g_d)^\top \in \mathbb{Z}^d$, let $g + P = \{ g + x \mid x \in P \}$. Let

$$\Sigma = \{ g + T_d^\pi \mid \pi \in \mathrm{Perm}(d),\ g \in \mathbb{Z}^d,\ g + T_d^\pi \subseteq T_d(k) \}. \tag{7.34}$$

In other words, we consider any simplex T_d^π of the triangulation of C_d, translate it by adding some integer vector g, and if the resulting set $g + T_d^\pi$ is a subset of $T_d(k)$ then this is a sub-simplex in Σ. By (7.32) this can only hold if $0 \leq g_i \leq k$ for

$1 \le i \le d$ because otherwise $g + \mathbf{0} \notin T_d(k)$ (where $\mathbf{0} \in T_d^\pi$), so Σ is finite. Figure 7.17 shows Σ for $d = 3$ and $k = 2$.

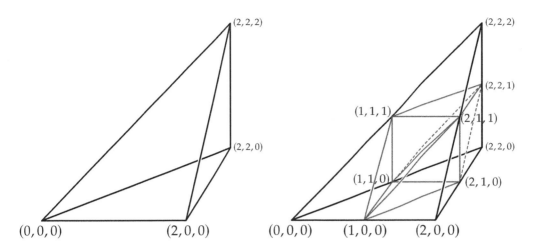

Figure 7.17 The Freudenthal triangulation Σ of the three-dimensional simplex $T_3(k)$ with vertices $(0,0,0)$, $(k,0,0)$, $(k,k,0)$ and (k,k,k) for $k = 2$.

Theorem 7.23. Σ *as defined in* (7.34) *is a triangulation of* $T_d(k)$.

Proof. By (7.34), $\bigcup_{S \in \Sigma} S \subseteq T_d(k)$. To show the converse inclusion, let $z \in T_d(k)$. We want to find $g \in \mathbb{Z}^d$ and $\pi \in \text{Perm}(d)$ so that $z \in g + T_d^\pi$, and crucially, $g + T_d^\pi \subseteq T_d(k)$ as required in (7.34). If this holds then $z = g + x$ for some $x \in T_d^\pi$, which implies

$$z_i = g_i + x_i \qquad (g_i \in \mathbb{Z},\ x_i \in [0,1],\ 1 \le i \le d). \tag{7.35}$$

In (7.35), the choice of g_i and x_i is not unique if $z_i \in \mathbb{Z}$ because then either $x_i = 0$ or $x_i = 1$ are possible. Then we normally choose $x_i = 0$, except when $z_i = k$ where we choose $g_i = k - 1$ and $x_i = 1$, to ensure that $g + T_d^\pi \subseteq T_d(k)$. With $\lfloor z_i \rfloor = \max\{\, t \in \mathbb{Z} \mid t \le z_i \,\}$, this is achieved by

$$g_i = \begin{cases} \lfloor z_i \rfloor & \text{if } z_i < k \\ k - 1 & \text{if } z_i = k, \end{cases} \qquad x_i = z_i - g_i \qquad (1 \le i \le d), \tag{7.36}$$

which clearly implies (7.35).

Similarly, the integral parts g_i of z_i fulfill

$$g_i \ge g_{i+1} \qquad (1 \le i < d) \tag{7.37}$$

because if $z_i = k$ then $z_{i+1} \le k$ and $g_i = k - 1 \ge g_{i+1}$, and if $z_i < k$ then by (7.36) we have $x_i < 1$ so that $g_i < g_{i+1}$ would imply $z_i = g_i + x_i < g_i + 1 \le g_{i+1} \le z_{i+1}$, which contradicts (7.32).

Given (7.35), we want to find $\pi \in \mathrm{Perm}(d)$ so that $x \in T_d^\pi$ and $g + T_d^\pi \subseteq T_d(k)$. The second condition is not automatic: For example, if $d = 2$ and $k = 1$ then $\Sigma = \{T_2(1)\}$ and for $z = x = (\frac{1}{2}, \frac{1}{2})$ both $T_2(1)$ and $T_2^\pi(1)$ for $(\pi(1), \pi(2)) = (2, 1)$ contain x, but $\mathbf{0} + T_2^\pi(1) \notin \Sigma$. The following is a safe choice of π. Sort the components x_i of x with a permutation π so that (7.28) holds and so that

$$1 \le h < j \le d, \quad x_{\pi(h)} = x_{\pi(j)} \quad \Rightarrow \quad \pi(h) < \pi(j). \tag{7.38}$$

This can always be achieved, if necessary by exchanging $\pi(h)$ and $\pi(j)$ if $x_{\pi(h)} = x_{\pi(j)}$, which maintains (7.28). Then $x \in T_d^\pi$ by Lemma 7.21. We want to show that $g + y \in T_d(k)$ for any $y \in T_d^\pi$, that is, according to (7.32),

$$g_i + y_i \ge g_{i+1} + y_{i+1} \quad (1 \le i < d) \tag{7.39}$$

(by (7.36), we have $k \ge g_1 + y_1$ and $g_d + y_d \ge 0$). By (7.37), condition (7.39) holds if $y_i \ge y_{i+1}$. Otherwise, let $y \in T_d^\pi$ and suppose that $y_i < y_{i+1}$ for some $1 \le i < d$. Let $i + 1 = \pi(h)$ and $i = \pi(j)$. Then $y_{\pi(h)} = y_{i+1} > y_i = y_{\pi(j)}$, which implies $h < j$ by Lemma 7.21. Then $x_{\pi(h)} \ge x_{\pi(j)}$, where $x_{\pi(h)} = x_{\pi(j)}$ would imply $i + 1 = \pi(h) < \pi(j) = i$ by (7.38) which is not possible, that is, $x_{i+1} = x_{\pi(h)} > x_{\pi(j)} = x_i$. By (7.37), either $g_i = g_{i+1}$ or $g_i > g_{i+1}$. If $g_i = g_{i+1}$ then $z_{i+1} = g_{i+1} + x_{i+1} > g_i + x_i = z_i$, which violates (7.32). Hence, $g_i + y_i \ge g_i \ge g_{i+1} + 1 \ge g_{i+1} + y_{i+1}$. This shows (7.39), and therefore, as claimed, $g + T_d^\pi \subseteq T_d(k)$. That is, the sub-simplices in Σ cover $T_d(k)$.

In order to show that Σ is a triangulation of $T_d(k)$, it remains to show that no sub-simplex is a subset of another, and that the intersection of two sub-simplices is a face of both. Both properties are shown similarly to Lemma 7.22, which implies them for two sub-simplices $g + T_d^\pi$ and $g' + T_d^{\pi'}$ if $g = g'$. If $g \ne g'$ then the two sub-simplices are subsets of two different cubes $g + C_d$ and $g' + C_d$ which have disjoint non-empty interiors, so neither sub-simplex is a subset of the other. The intersection of $g + T_d^\pi$ and $g' + T_d^{\pi'}$ is a subset of $(g + C_d) \cap (g' + C_d)$, which is non-empty only if $|g_i - g_i'| \le 1$ for $1 \le i \le d$. By Lemma 7.21, $z \in (g + T_d^\pi) \cap (g' + T_d^{\pi'})$ if and only if $z = g + x = g' + x'$ so that (7.28) holds and, analogously,

$$1 \ge x'_{\pi'(1)} \ge x'_{\pi'(2)} \ge \cdots \ge x'_{\pi'(d)} \ge 0. \tag{7.40}$$

If $g_i = g_i' + 1$, then the equation $z_i = g_i + x_i = g_i' + x_i'$ imposes the additional equations $x_i = 1$ and $x_i' = 0$ (and similarly $x_i = 0$ and $x_i' = 1$ if $g_i + 1 = g_i'$). For those subscripts i where $g_i = g_i'$ we have $x_i = x_i'$, which induce further equations in (7.28) and (7.40) if these components of x and x' appear in reverse order in these inequalities. As argued in the proof of Lemma 7.22, these equations mean that certain barycentric coordinates $y_0, y_1, \ldots, y_d$ in (7.29) are zero (which may now include y_0 if $x_{\pi(1)} = 1$ and y_d if $x_{\pi(d)} = 0$), which excludes the corresponding vertices of the two sub-simplices from their intersection. The remaining vertices (if any) are then the vertices of the intersection, which is therefore a common face of both sub-simplices. This completes the proof of the triangulation property. $\square$

The simplex $T_d(k)$ has $d + 1$ vertices $v^i(k)$ in (7.25) that are vertices of a cube in $\mathbb{R}^k$ with edge length k. The triangulation Σ in (7.34) subdivides $T_d(k)$ into simplices T_d^π (which triangulate the unit cube C_d) shifted by integer vectors $g \in \mathbb{Z}^d$. The goal is to obtain an arbitrarily fine triangulation of the unit simplex Δ, as used in the KKM lemma (Theorem 7.12). This achieved as follows: When mapping $T_d(k)$ to Δ with the affine bijection J in Lemma 7.3, we obtain a corresponding triangulation $\{ J(T) \mid T \in \Sigma \}$ of Δ. For each sub-simplex $J(g + T_d^\pi)$ of that triangulation, the maximum of the distances between its vertices only depends on π, not on g. That maximum (over all $\pi \in \mathrm{Perm}(d)$ and all vertex pairs) is proportional to $\frac{1}{k}$ because J is affine and $T_d(k) = \{ xk \mid x \in T_d(1) \}$. For any $\varepsilon > 0$, one can therefore choose k large enough so that no two points in any sub-simplex of the triangulation of Δ have distance greater than ε, which gives the desired ε-fine triangulation.

The proof of Brouwer's fixed-point Theorem 7.20 is thereby complete.

7.8   Further Reading

The fixed-point theorem is stated at the end of Brouwer (1911, Satz 4). In this and related papers Brouwer developed and used the concept of "degree" (number of inverse images of a point) of a simplicial approximation of a continuous function. For a detailed discussion see Brouwer (1976). A historical survey of the wide-ranging impact of Brouwer's fixed-point theorem and its subsequent extensions is Park (1999).

Brouwer's fixed-point theorem is a major theorem in topology. The basic notion of topology is continuity, but more primitively topology can be seen as a qualitative study of "position". The first paper published in that regard is the solution by Euler (1741) to the problem of the "bridges of Königsberg" (see also Biggs, Lloyd, and Wilson, 1976). Euler's paper marks the foundation of both graph theory and topology; it introduces the concept of a graph, and the degree of a node plays a central role. It is fitting that Sperner's lemma relates to the even number of odd-degree nodes in a graph.

Theorem 7.10 is due to Sperner (1928), and was used by Knaster, Kuratowski, and Mazurkiewicz (1929) to prove the fixed-point theorem on a simplex with the help of their Theorem 7.12. As we mentioned in the proof of Theorem 7.10, these proofs use the counting modulo two in equation (7.12). The term "parity argument" has been coined by Papadimitriou (1994).

The path-following argument is due to Cohen (1967). Cohen's proof is not given in enough detail to see that the path may also come back to several dimensions further below, as in the example in Figure 7.10. For a more precise description see Kuhn (1969). We have defined the "internal" nodes of the graph as the sub-facets in C' rather than as the elements of A in (7.11). When staying in a fixed dimension, it is more natural to consider only sub-simplices, the elements of D and A, as nodes of a graph, which are joined by an edge if they have a common sub-facet in C'. However, for the construction of G_2, in particular, the edges are more easily seen as line segments and the nodes as vertices in Σ_2. A key argument in the construction of G in (7.13) is that the end nodes of some paths in a graph G_n have different dimension, and we change to the higher dimension only at the end node.

The image of "seaview hotels" to prove the Bolzano–Weierstrass theorem is taken from Bryant (1990).

The Freudenthal simplicial subdivision is due to Freudenthal (1942) and has been re-discovered several times, for example by Kuhn (1960).

7.9   Exercises for Chapter 7

Exercise 7.1. Let $C = \{(x,y) \in \mathbb{R}^2 \mid x^2 + y^2 \le 1\}$, and consider the following functions $g, h, f : C \to C$:

$$g(x,y) = (x, \sqrt{1-x^2}), \qquad h(x,y) = (-y, x), \qquad f(x,y) = h(g(x,y)).$$

Let $G = g(C) = \{g(x,y) \mid (x,y) \in C\}$. Describe the geometric shape of C and G, and how the functions g and h act. Does each of the functions g, h, f have a fixed point in C? Justify why, or why not. Determine all fixed points, if any, of each of g, h, and f.

Exercise 7.2. For a general polytope P, a vertex of P is defined as an extreme point of P, that is, it is not a convex combination of other points of P.

(a) Show that for a simplex S given as the convex hull of the affinely independent points $v^1, \ldots, v^d$, each point v^i is indeed a vertex of S according to this definition.

(b) Show that the set of vertices of a simplex is unique.

Exercise 7.3. Give examples of

(a) two simplices whose intersection is not a simplex;

(b) a nonempty convex set that has no extreme points;

(c) a bounded convex set that has at least one extreme point but that is not equal to the convex hull of its extreme points.

Exercise 7.4. Show that the following generalization of Sperner's lemma (Theorem 7.10) is *false*: Let S be a simplex with vertices $v^1, \ldots, v^d$, let Σ be a triangulation of S, and consider a labeling of the vertices in V so that v_i has label i for $1 \le i \le d$. Then Σ has at least one completely labeled sub-simplex.

8

Zero-Sum Games

Zero-sum games are games of two players where the interests of the players are directly opposed: One player's loss is the other player's gain. Competitions between two players in sports or in parlor games can be thought of as zero-sum games. The combinatorial games studied in Chapter 1 are also zero-sum games.

In a zero-sum game, the sum of payoffs to the players in any outcome of playing the game is zero. An $m \times n$ bimatrix game (A, B) is zero-sum if $B = -A$. It is completely specified by the matrix A of payoffs to the row player, and the game is therefore often called a *matrix game*. Consequently, we can analyse matrix games just like bimatrix games. However, zero-sum games have additional strong properties that do not hold for general bimatrix games. These properties are, which are the subject of this chapter, are:

- In a zero-sum game, an equilibrium strategy is a *max-min strategy* that maximizes the worst-case expected payoff to the player, and is therefore *independent* of what the opponent does.

- Every equilibrium has the same unique payoff to the players, which is called the *value* of the game.

- If a player has several equilibrium strategies, these are *exchangeable* and can be used in any equilibrium. Moreover, the set of a player's equilibrium strategies is *convex*.

Apart from these special properties, we treat zero-sum games in a chapter of its own, because:

- Historically, the main result about zero-sum games of von Neumann (1928) precedes the equilibrium concept of Nash (1951) for games with general payoffs.

- The concept of equilibrium as "no incentive to deviate" has a stronger form in a zero-sum game, namely that a player can make his strategy *public* without a disadvantage. As the games of Matching Pennies and Rock-Paper-Scissors show (see Figure 3.12), this requires the concept of *mixed strategies*. (Zero-sum games may not always require mixed strategies, for example if they have perfect information like the games in Chapters 1 and 4. A game where one player has only two strategies is dominance solvable, see Exercise 8.2.)

• This chapter is mathematically *self-contained*, with an independent two-page proof of von Neumann's *minimax theorem*, which (see Proposition 8.3) is equivalent to the existence of an equilibrium in a zero-sum game. The minimax theorem is less powerful than Brouwer's fixed-point theorem, and slightly less powerful than the duality theorem of linear programming, but will be useful for proving the existence of "correlated equilibria" in Chapter 12.

In Section 8.2, we analyze the soccer penalty as an introductory example of a zero-sum game, and show how to find a max-min strategy graphically (Figures 8.2 and 8.3).

Section 8.3 states the important properties of zero-sum games. The central concept is that of a *max-min* strategy, which is a (typically mixed) strategy that maximizes the *minimum* expected payoff to the player for all possible responses of the opponent. This (rather paranoid) approach can be applied to any game, but in a zero-sum game means that the other player uses a best response, and is therefore justified.

The payoffs to the maximizing row player in a matrix game are costs to the column player who wants to minimize them. In terms of these costs (rather than the payoffs to the column player), a max-min strategy is called a *min-max* strategy of the minimizer. If the max-min payoff equals the min-max cost, then the max-min strategy and min-max strategy define an equilibrium, and an equilibrium is such a pair of strategies (Proposition 8.3). Because finite games have an equilibrium, the maxmin = minmax equality holds, known as the *minimax* theorem of von Neumann (1928) (Theorem 8.3). It also defines the unique equilibrium payoff (to the maximizer, and cost to the minimizer) as the *value* of the game.

In Section 8.4 we give a short proof of the minimax theorem that is independent of the existence of an equilibrium. It connects two related optimization problems that are a special case of pair of "dual linear programs". We consider them only for this special case, where we prove directly, by induction, that the two problems have the same optimal value.

In zero-sum game, an equilibrium strategy is the same as a max-min strategy (in terms of the player's own payoff), also called an *optimal* strategy, which does not depend on what the other player does, a very strong property. Further properties of optimal strategies are discussed in Section 8.5.

8.1 Prerequisites and Learning Outcomes

As prerequisites we require the definition of bimatrix games and mixed strategies from Chapter 6, and manipulating matrices and linear inequalities between vectors. The proof of the minimax theorem in Section 8.4 uses that a continuous function on a compact domain assumes its minimum and maximum.

After studying this chapter, you should be able to:

- Explain the concept of max-min strategies and their importance for zero-sum games, and apply this to given games.

- Find max-min and min-max strategies in small matrix games like in Exercise 8.1, and exploit the "maxmin = minmax" theorem to find the (max-min or min-max) strategy that is easier to find via the lower or upper envelope of expected payoffs.

- State and prove the equivalence of existence of an equilibrium and the minimax theorem in a zero-sum game (Proposition 8.3).

- State the main properties of zero-sum games: independence of "optimal strategies" from the what the opponent does, and its implications (unique value, exchangeability, and convexity of the set of equilibrium strategies).

- Explain the main idea behind the induction proof of the minimax theorem in Section 8.3.

8.2 Example: Soccer Penalty

A zero-sum game is a bimatrix game (A, B) with $B = -A$. Because it is completely specified by the matrix A of payoffs to the row player I, it is also called a *matrix game* (with payoff matrix) A. When the players choose row i and column j, respectively, player II pays the matrix entry a_{ij} to player I, which is therefore a cost that player II wants to minimize, also in expectation when player II faces a mixed strategy. Player I is called the *maximizer* and player II the *minimizer*.

Figure 8.1 shows the penalty kick in soccer (football in Britain) as a zero-sum game between striker and goalkeeper. For simplicity, assume that the striker's possible strategies are L and R, that is, to kick into the left or right corner, and that the goalkeeper's strategies are l, w, and r, where l and r mean that he jumps immediately into the left or right corner, respectively, and w that he waits and sees where the ball goes and jumps afterwards. In Figure 8.2, we give an example of resulting probabilities that the striker scores a goal, depending on the choices of the two players. Clearly, the striker (player I) tries to maximize and the goalkeeper (player II) to minimize that probability, and we only need this payoff to player I to specify the game. We call the maximizer "Max" and the minimizer "min" (where we keep our tradition of using lower case letters for the column player).

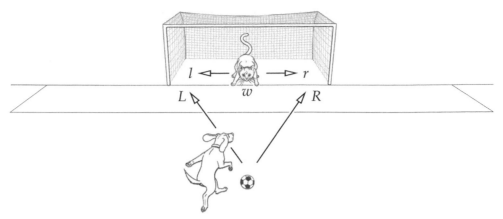

Figure 8.1 Strategies for striker and goalkeeper in the soccer Penalty game. In addition to jumping left or right, the goalkeeper can wait (w). An extra strategy for the striker of shooting down the middle is considered in Exercise 8.1.

Because the players care about the *probability* that a goal is scored, this payoff (to player I, cost to player II) automatically fulfills the condition of an expected-utility function, because an expected probability is just a probability. (This is a nice "practical" example of the construction of the values of an expected-utility function as probabilities, see Figure 5.1.) In a general zero-sum game, payoffs and costs are not necessarily defined via probabilities, but are nevertheless assumed to represent the preference of *both* players with respect to taking expectations. This is a stronger assumption than in a non-zero-sum game where the payoffs can be adjusted independently to reflect different attitudes to risk.

The top left payoff table in Figure 8.2 shows the game as a bimatrix game with two payoffs per cell, and the pure best-response payoffs marked as usual with boxes. The top right shows the more concise *matrix game* with only one number per cell. A best-response payoff for the maximizer is shown with a box, a best-response (minimal) cost for the minimizer with a circle. This represents the same information as the two payoffs per cell, but more concisely.

The best responses show that the game has no pure-strategy equilibrium. This is immediate from the sports situation: obviously, the goalkeeper's best response to a kick into the right or left corner is to jump into that corner, whereas the striker would rather kick into the opposite corner to where the goalkeeper jumps. We therefore have to find a mixed equilibrium, and active randomization is obviously advantageous for both players.

We first find the equilibria of the game as we would do this for a 2×3 game, except that we only use the payoffs to player I. It is more than appropriate to use the goalpost method for solving the Penalty game! The bottom picture in Figure 8.2 shows the expected payoffs to player I as a function of his probability, say p, of choosing R, for the reponses l, w, and r of player II. These payoffs to

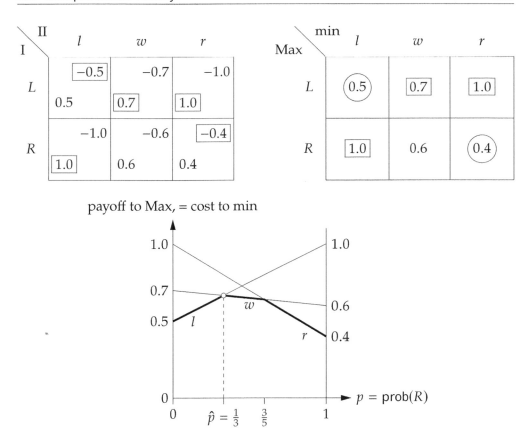

Figure 8.2 Top left: the Penalty game as a zero-sum bimatrix game with scoring probabilities as payoffs to the striker (player I), which are the negative of the payoffs to the goalkeeper (player II). Top right: the same game as a matrix game, which shows only the payoffs to the maximizer, which are costs to the minimizer. A box indicates a best-response payoff to the maximizer, a circle a best-response (minimal) cost to the minimizer. Bottom: lower envelope of best-response costs as a function of the mixed strategy $(1 - p, p)$ of Max. For $p = \hat{p} = \frac{1}{3}$ this is his *max-min* strategy.

player I are costs to player II, so the best-response costs to player II are given by the minimum of these lines, which defines the *lower envelope* shown by bold line segments. There are two intersection points on the lower envelope, of the lines for l and w when $p = \frac{1}{3}$, and of the lines for w and r when $p = \frac{3}{5}$. These intersection points are easily found with the difference trick (where it does not matter if one considers costs or payoffs, because the differences are the same in absolute value): Namely, l and w have cost difference 0.2 in row L and 0.4 in row R, so that L and R should be played in the ratio $0.4 : 0.2$, that is, for $p = \frac{1}{3}$, to get indifference between l and w. The indifference between w and r when $p = \frac{3}{5}$ is found similarly.

Hence, player II's best response to the mixed strategy $(1 - p, p)$ is l for $0 \le p \le \frac{1}{3}$, and w for $\frac{1}{3} \le p \le \frac{3}{5}$, and r for $\frac{3}{5} \le p \le 1$. Only for $p = \frac{1}{3}$ and for $p = \frac{3}{5}$ can player II mix, and only in the first case, when l and w are best responses, can player I, in turn, be made indifferent. This gives the unique mixed equilibrium of the game, $((\frac{2}{3}, \frac{1}{3}), (\frac{1}{6}, \frac{5}{6}, 0))$. The resulting expected costs for the three columns l, w, r are $\frac{2}{3}$, $\frac{2}{3}$, $\frac{4}{5}$, where indeed player II assigns positive probability only to the smallest-cost columns l and w. The expected payoffs for both rows are $\frac{2}{3}$. (This unique equilibrium payoff of the game is the *value* of the game defined in Theorem 8.4.)

So far, our analysis is not new. We now consider the lower envelope of expected payoffs to player I as a function of his own mixed strategy $x = (1 - p, p)$ from that player's own perspective: It represents the worst possible expected payoff to player I, given the possible responses of player II. It is easy to see that it suffices to look at pure responses of player II (see also Proposition 8.2 below). That is, against the mixed strategy x the minimum is found over all pure responses of the minimizer, here columns l, w, and r.

If player I wants to "secure" the maximum of this worst possible payoff, he should maximize over the lower envelope with a suitable choice of p. This maximum is easily found for $\hat{p} = \frac{1}{3}$, shown by a dashed line in Figure 8.2. The strategy $\hat{x} = (1 - \hat{p}, \hat{p})$ is called the *max-min strategy* of the maximizer, which we found here graphically as the highest point on the lower envelope. The "hat" accent $\hat{x}$ may be thought of as representing that "peak".

The max-min strategy $(1 - \hat{p}, \hat{p})$ is one of the intersection points on the lower envelope. The intersection point at $p = \frac{3}{5}$ has a lower payoff to the maximizer because the best-response line w slopes downwards from $\frac{1}{3}$ to $\frac{3}{5}$. Hence, $p = \frac{3}{5}$ is not a max-min strategy. As we show in Proposition 8.3 below, in a zero-sum game an equilibrium strategy is the same as a max-min strategy. Therefore, unlike in a bimatrix game, $p = \frac{3}{5}$ does not need to be checked as a possible equilibrium strategy.

The max-min payoff is well defined and does not depend on what player II does, because we always take the minimum over player II's possible responses. The maximum (and thus the max-min payoff) is unique, but there may be more than one choice of the max-min strategy $\hat{x}$. An example is shown in Figure 8.3 where both L and R have the same payoff 0.6 in column w (and are therefore both best responses to w, so that this game is degenerate according to Definition 6.7). Then the lower envelope has its maximum 0.6 anywhere for $\hat{p} \in [\frac{1}{5}, \frac{2}{3}]$, which defines the possible max-min strategies $\hat{x} = (1 - \hat{p}, \hat{p})$; the endpoints of the interval are found with the difference trick for the indifferences between l and w, and w and r, respectively. On the side of player II, the unique best response against $\hat{p} = \frac{1}{2}$ (for example) is the pure strategy w. Therefore, without any further considerations, w is the unique equilibrium strategy of player II in *any* equilibrium; this is a very useful consequence of Proposition 8.3, as we will explain.

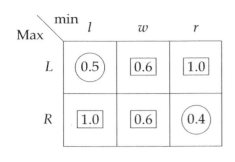

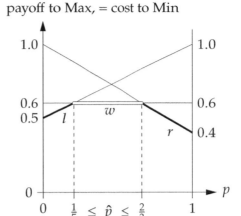

Figure 8.3 The Penalty game with the same scoring probability 0.6 when the column player chooses w. Then the game is degenerate, and the max-min strategy $(1 - \hat{p}, \hat{p})$ is not unique, $\hat{p} \in [\frac{1}{5}, \frac{2}{3}]$.

⇒ Exercise 8.1 is a variant of the Penalty game where the striker is more likely to score, and in part (b) has a third strategy M of shooting down the middle.

8.3 Max-Min and Min-Max Strategies

The important new concept for zero-sum games is that of a *max-min* strategy. This concept applies in principle to any game, where we focus on two-player games. A max-min strategy is a strategy of a player that is defined in terms of the player's *own* payoffs. It also depends on the strategy set of the player. In the following definition, we assume general (nonempty) strategy sets X and Y, which are normally, but don't have to be, the mixed-strategy sets in (6.6). A *min-max* strategy is a strategy of the *other* player.

Definition 8.1 (Max-min strategy, min-max strategy). Consider a two-player game with strategy set X for player I and Y for player II, and payoff $a(x, y)$ to player I for $x \in X$ and $y \in Y$. A *max-min strategy* $\hat{x} \in X$ of player I fulfills

$$\min_{y \in Y} a(\hat{x}, y) = \max_{x \in X} \min_{y \in Y} a(x, y), \tag{8.1}$$

assuming these minima and maxima exist, which defines the *max-min payoff* to player I. A *min-max strategy* $\hat{y} \in Y$ of player II fulfills

$$\max_{x \in X} a(x, \hat{y}) = \min_{y \in Y} \max_{x \in X} a(x, y), \tag{8.2}$$

which defines the *min-max payoff* to player I.   □

Consider first a finite game, for example the game at the top right in Figure 8.2 with the 2×3 payoff matrix A to the row player I. With the strategy sets $X = \{L, R\}$ and $Y = \{l, w, r\}$ in Definition 8.1, the minima and maxima in (8.1) and (8.2) clearly exist. Because player I cannot randomize, he has to consider the minimum payoff against L (which is 0.5) or R (which is 0.4, both indicated by the circles), so his max-min strategy is L with max-min payoff 0.5. For the min-max strategy, the column player II has to consider the maximum payoff to player I (indicated by the boxes) against l, w, and r, which is 1.0, 0.7, and 1.0, respectively, so the min-max strategy is w with min-max payoff 0.7. Here, max-min and min-max payoff are different.

We normally consider the mixed extension of a finite game. Then X and Y are the sets of mixed strategies, and the payoff to player I in Definition 8.1 is given by $a(x, y) = x^\top A y$. With the goalpost diagram in Figure 8.2, we have identified the (mixed) max-min strategy of player I as $\hat{x} = (\frac{2}{3}, \frac{1}{3})$ with max-min payoff $\frac{2}{3}$. We will shortly see that the min-max strategy player II is her equilibrium strategy $(\frac{1}{6}, \frac{5}{6}, 0)$, with the min-max payoff to player I also given by $\frac{2}{3}$.

The rationale behind a max-min strategy is that it gives a *guaranteed* (expected) payoff to the player, irrespective of what the other player does. (A max-min strategy is therefore sometimes called a "security strategy".) If player I plays x, his worst-case payoff is $\min_{y \in Y} a(x, y)$, which is maximized by player I's max-min strategy $\hat{x}$ (8.1) and gives him his max-min payoff. In a game with general payoffs, player II may not necessarily want to play the minimizing strategy y against x in $\min_{y \in Y} a(x, y)$. However, in a zero-sum game, this strategy y is player II's *best response* to x, so this is what player I should expect.

In a min-max strategy $\hat{y}$ in (8.2), player II chooses $\hat{y}$ under the assumption that player I chooses a best response x to achieve the payoff $\max_{x \in X} a(x, \hat{y})$. By playing $\hat{y}$, player II minimizes this payoff. In a game with general payoffs, this is a rather malevolent action by player II and not justified by any security considerations in terms of her own payoffs.

However, in a zero-sum game, a *min-max strategy is just a max-min strategy for the minimizer*. Namely, suppose the zero-sum game is (A, B) with $B = -A$. Then with $a(x, y) = x^\top A y$, a min-max strategy $\hat{y}$ fulfills (8.2), or equivalently

$$\min_{x \in X} x^\top B \hat{y} = \max_{y \in Y} \min_{x \in X} x^\top B y,$$

so $\hat{y}$ is a max-min strategy in terms of player II's payoff matrix B. In zero-sum games, a min-max strategy is therefore not a different concept, but merely expresses player II's preferences in terms of her cost matrix A, which is also the payoff matrix to player I and defines the matrix game.

The minima and maxima in Definition 8.1 are assumed to exist, which holds if X and Y are compact and $a(x, y)$ is continuous in x and y. This is certainly the case for our standard assumption that X and Y are the mixed-strategy sets for a matrix game A and $a(x,y) = x^{\top}Ay$. The following proposition observes that in $\max_{x \in X} \min_{y \in Y} x^{\top}Ay$ the minimum is already obtained for pure strategies j instead of mixed strategies $y \in Y$ (this has been used in considering the lower envelope in Figure 8.2, for example), and similarly that in $\min_{y \in Y} \max_{x \in X} x^{\top}Ay$ the maximum is already obtained for pure strategies i instead of mixed strategies $x \in X$.

Proposition 8.2. *For an $m \times n$ matrix game A and $x \in X$,*

$$\min_{y \in Y} x^{\top}Ay = \min\{ (x^{\top}A)_j \mid 1 \le j \le n \}. \tag{8.3}$$

For $y \in Y$,

$$\max_{x \in X} x^{\top}Ay = \max\{ (Ay)_i \mid 1 \le i \le m \}. \tag{8.4}$$

Moreover, a max-min strategy $\hat{x} \in X$ and a min-max strategy $\hat{y} \in Y$ exist.

Proof. Let $y \in Y$. The best response condition (6.10) says that a best response to y can be found among the pure strategies i of the row player, which implies (8.4). Condition (8.3) holds similarly.

The right-hand side of (8.3) is the minimum of a finite number of terms $(x^{\top}A)_j$ and continuous in x. The equal left-hand side $\min_{y \in Y} x^{\top}Ay$ is therefore also continuous in x and assumes its maximum for some $\hat{x}$ on the compact set X. Then $\hat{x}$ defines a max-min strategy. Similarly, a min-max strategy $\hat{y}$ exists. $\square$

Because max-min payoffs and min-max costs are (in expectation) "guaranteed", the first cannot be higher than the second: With a max-min strategy $\hat{x}$ and min-max strategy $\hat{y}$ as in Definition 8.1, we have

$$\begin{aligned}
\max_{x \in X} \min_{y \in Y} a(x,y) &= \min_{y \in Y} a(\hat{x}, y) \\
&\le a(\hat{x}, \hat{y}) \\
&\le \max_{x \in X} a(x, \hat{y}) = \min_{y \in Y} \max_{x \in X} a(x, y).
\end{aligned} \tag{8.5}$$

This is the near-obvious payoff inequality "maxmin $\le$ minmax", which holds for any nonempty sets X and Y, assuming that the minima and maxima exist. We have seen that for pure-strategy sets X and Y, the inequality may be strict.

If X and Y are the mixed-strategy sets and $a(x,y) = x^{\top}Ay$, the famous *minimax* (short for "maxmin = minmax") theorem of von Neumann (1928) states that this inequality holds as an equality. The following proposition states that this theorem is equivalent to $(\hat{x}, \hat{y})$ being an *equilibrium* of the game.

Proposition 8.3. *Consider an $m \times n$ matrix game A with mixed-strategy sets X and Y as in (6.6), and let $(x^*, y^*) \in X \times Y$. Then the following are equivalent:*

(a) (x^*, y^*) is an equilibrium of $(A, -A)$;

(b) x^* is a max-min strategy and y^* is a min-max strategy, and

$$\max_{x \in X} \min_{y \in Y} x^\top A y = \min_{y \in Y} \max_{x \in X} x^\top A y. \tag{8.6}$$

Proof. Suppose (b) holds. Consider a max-min strategy $\hat{x}$ as in (8.1) and a min-max strategy $\hat{y}$ as in (8.2). Then (8.6) implies

$$
\begin{aligned}
\hat{x}^\top A \hat{y} &\geq \min_{y \in Y} \hat{x}^\top A y \\
&= \max_{x \in X} \min_{y \in Y} x^\top A y \\
&= \min_{y \in Y} \max_{x \in X} x^\top A y \\
&= \max_{x \in X} x^\top A \hat{y} \geq \hat{x}^\top A \hat{y}
\end{aligned}
$$

so all these inequalities hold as equalities. Hence,

$$\max_{x \in X} x^\top A \hat{y} = \hat{x}^\top A \hat{y} = \min_{y \in Y} \hat{x}^\top A y, \tag{8.7}$$

which says that $(\hat{x}, \hat{y})$ is a equilibrium because (8.7) is equivalent to

$$x^\top A \hat{y} \leq \hat{x}^\top A \hat{y} \leq \hat{x}^\top A y \qquad \text{for all } x \in X, \, y \in Y, \tag{8.8}$$

where the left inequality in (8.8) states that $\hat{x}$ is a best response to $\hat{y}$ and the right inequality that $\hat{y}$ is a best response to $\hat{x}$. This shows (a).

Conversely, suppose (a) holds, that is, as in (8.7),

$$\max_{x \in X} x^\top A y^* = x^{*\top} A y^* = \min_{y \in Y} x^{*\top} A y.$$

Using (8.5), this implies

$$
\begin{aligned}
x^{*\top} A y^* &= \min_{y \in Y} x^{*\top} A y \\
&\leq \max_{x \in X} \min_{y \in Y} x^\top A y \\
&\leq \min_{y \in Y} \max_{x \in X} x^\top A y \\
&\leq \max_{x \in X} x^\top A y^* = x^{*\top} A y^*
\end{aligned}
$$

so we also have equalities throughout, which state that x^* is a max-min strategy and y^* is a min-max strategy, and (8.6) holds, which shows (b). ☐

The minimax theorem is therefore a consequence of the Theorem 6.5 of Nash (1951).

Theorem 8.4 (The minimax theorem of von Neumann, 1928). *In a matrix game with payoff matrix A to the maximizing row player I and mixed-strategy sets X and Y for players I and II,*

$$\max_{x \in X} \min_{y \in Y} x^\top A y = v = \min_{y \in Y} \max_{x \in X} x^\top A y \qquad (8.9)$$

where v is the unique max-min payoff to player I and min-max cost to player II, called the value *of the game.*

Nash's theorem is proved with the help of Brouwer's fixed-point theorem, which has a complex proof. We give a short self-contained proof of the minimax Theorem 8.4 in the next Section 8.4.

Equilibrium strategies for a two-player zero-sum game have the following very desirable properties, which usually do not hold in games with general payoffs:

- The equilibrium payoff to the players is unique, and called the *value* of the game (as payoff v in (8.9) to the maximizer and cost to the minimizer).

- An equilibrium strategy is the same as a max-min strategy (for the maximizer, and min-max strategy for the minimizer). It is therefore *independent* of what the other player does. For that reason it is also called an *optimal strategy* of the player.

"Optimal strategy" assumes that players play as a well as they can. Optimal strategies are often rather "defensive" strategies that give the same expected payoff to many responses of the opponent. For example, a symmetric zero-sum game necessarily has value zero (see Exercise 8.3(c)). Hence, an optimal strategy of the maximizer will guarantee him at least payoff zero for all pure responses of the opponent, even if the opponent does not play optimally. Bad play of the opponent is thereby not exploited, but it can only help the player who plays optimally.

Because an optimal strategy does not depend on the other player, we also have:

- In a zero-sum game, equilibrium strategies are *exchangeable*: If (x, y) and $(\bar{x}, \bar{y})$ are two equilibria of a zero-sum game, then $(\bar{x}, y)$ and $(x, \bar{y})$ are also equilibria.

- As stated in Proposition 8.3, in a zero-sum game an equilibrium strategy is the same as a max-min (respectively, min-max) strategy. For example, if the zero-sum game has an equilibrium with a unique pure best response, that pure strategy is also the unique max-min strategy of the maximizer or min-max strategy of the minimizer. The reason is that if the min-max strategy (say) was not unique, it would be a non-unique strategy in *any* equilibrium. An example is the game in Figure 8.3: The max-min strategies are given by $(1 - \hat{p}, \hat{p})$ for $p \in [\frac{1}{5}, \frac{2}{3}]$; for any value of $\hat{p}$ in the interior of that interval (for example $\hat{p} = \frac{1}{2}$), the *unique* best response of player II is the pure strategy w, which is therefore also the unique equilibrium strategy of player II and min-max strategy.

Further properties are discussed in Section 8.5, namely the convexity of the set of optimal strategies (Proposition 8.6), and that weakly dominated strategies can be removed without affecting the value of the game (Proposition 8.7); however, this removal may affect the set of optimal strategies.

8.4   A Short Proof of the Minimax Theorem

The minimax Theorem 8.4 of von Neumann (1928) is, via Proposition 8.3, a special case of Nash's Theorem 6.5 on the existence of an equilibrium in a finite game. However, von Neumann's theorem predates Nash's result, and can be proved on its own. The following proof by induction is particularly simple.

Theorem 8.5. *For an $m \times n$ matrix game A with mixed-strategy sets X and Y in (6.6) there are $x \in X$, $y \in Y$ and $v \in \mathbb{R}$ so that*

$$Ay \leq \mathbf{1}v, \qquad x^\top A \geq v\mathbf{1}^\top.$$

Then v is the unique value of the game and (8.9) holds, and x and y are optimal strategies of the players.

Proof. Consider the two optimization problems (with $v, u \in \mathbb{R}$)

$$\underset{y,v}{\text{minimize}}\ v \quad \text{subject to}\ \ Ay \leq \mathbf{1}v, \quad y \in Y. \tag{8.10}$$

and

$$\underset{x,u}{\text{maximize}}\ u \quad \text{subject to}\ \ x^\top A \geq u\mathbf{1}^\top, \quad x \in X. \tag{8.11}$$

The inequalities $Ay \leq \mathbf{1}v$ state that by playing y player II pays at most v in every row of the game. Similarly, $x^\top A \geq u\mathbf{1}^\top$ states that by playing x player I gets at least u in every column of the game.

Let y, v be optimal for (8.10). Then at least one of the m inequalities $Ay \leq \mathbf{1}v$ holds as equality because if all inequalities were strict then v could be reduced further. Hence, v is the largest entry $(Ay)_i$ of Ay (as in any pure best response i to y) which by (8.4) fulfills

$$v = \max_{1 \leq i \leq m}(Ay)_i = \max_{y \in Y} x^\top Ay,$$

and an optimal choice of y in (8.10) is a min-max strategy. Similarly, for an optimal pair x, u in (8.11) we have by (8.3)

$$u = \min_{1 \leq j \leq n}(x^\top A)_j = \min_{y \in Y} x^\top Ay,$$

where x is a max-min strategy.

Consider *feasible* (not necessarily optimal) solutions y, v, x, u to the problems (8.10) and (8.11), that is,

$$Ay \leq \mathbf{1}v, \qquad y \in Y, \qquad x^\top A \geq u\mathbf{1}^\top, \qquad x \in X. \tag{8.12}$$

Then

$$u = u\mathbf{1}^\top y \leq x^\top Ay \leq x^\top \mathbf{1}v = v. \tag{8.13}$$

This implies that the objective functions in (8.10) and (8.11) are bounded and (without changing their optimum values) can be restricted to nonempty compact domains, namely to $(y, v) \in Y \times [\bar{u}, \bar{v}]$ and to $(x, u) \in X \times [\bar{u}, \bar{v}]$ for some $\bar{y}, \bar{v}, \bar{x}, \bar{u}$ that fulfill (8.12). Hence, the optimum values v and u in (8.10) and (8.11) exist (and are unique). The claim is that $u = v$, which proves "maxmin = minmax" and (8.9).

We prove this claim by induction on $m + n$. It is trivial for $m + n = 2$. Consider a matrix A of any dimension $m \times n$ and let y, v and x, u be optimal solutions to (8.10) and (8.11), which are feasible, that is, (8.12) holds. If all inequalities in (8.12) hold as equalities, that is, $Ay = \mathbf{1}v$ and $u\mathbf{1}^\top = x^\top A$, then we also have equality in (8.13) and $u = v$ as claimed. Hence, we can assume that at least one inequality in $Ay \leq \mathbf{1}v$ is strict (or in $x^\top A \geq u\mathbf{1}^\top$, where the same reasoning applies). Suppose this is the kth row, that is, with matrix entries a_{ij} of A,

$$\sum_{j=1}^{n} a_{kj}y_j < v. \tag{8.14}$$

Let $\bar{A}$ be the matrix A with the kth row deleted (clearly $m > 1$). Let $\bar{y}, \bar{v}$ be the optimal solution to (8.10) with $\bar{A}$ instead of A, that is, to

$$\underset{\bar{y}, \bar{v}}{\text{minimize }} \bar{v} \quad \text{subject to} \quad \bar{A}\bar{y} \leq \mathbf{1}\bar{v}, \quad \bar{y} \in Y. \tag{8.15}$$

The maximization problem that corresponds to (8.11) with $\bar{A}$ instead of A can be written as

$$\underset{\bar{x}, \bar{u}}{\text{maximize }} \bar{u} \quad \text{subject to} \quad \bar{x}^\top A \geq \bar{u}\mathbf{1}^\top, \quad \bar{x} \in X, \quad \bar{x}_k = 0. \tag{8.16}$$

Let $\bar{x}, \bar{u}$ be an optimal solution to (8.16). Then

$$\bar{u} \leq u, \qquad \bar{v} \leq v, \tag{8.17}$$

because $\bar{x}, \bar{u}$ is feasible for (8.11) (so $\bar{u} \leq u$ for the maximum u in (8.11)) and because y, v is feasible for (8.15). Intuitively, (8.17) holds because the game $\bar{A}$ is at least as favorable for player II as game A. By inductive hypothesis, $\bar{u} = \bar{v}$.

We claim (this is the main point) that $\bar{v} = v$. Namely, if $\bar{v} < v$, let $\varepsilon \in (0, 1]$, so $v(1 - \varepsilon) + \bar{v}\varepsilon = v - \varepsilon(v - \bar{v}) < v$. Consider the strategy $y' = y(1 - \varepsilon) + \bar{y}\varepsilon$, which belongs to Y because Y is convex. Because $\bar{A}y \leq \mathbf{1}v$ and by (8.15), y' fulfills

$$\bar{A}y' = \bar{A}(y(1 - \varepsilon) + \bar{y}\varepsilon) \leq \mathbf{1}(v(1 - \varepsilon) + \bar{v}\varepsilon) < \mathbf{1}v.$$

Similarly, (8.14) implies for sufficiently small ε

$$\sum_{j=1}^{n} a_{kj} y_j' = \sum_{j=1}^{n} a_{kj} y_j + \varepsilon \sum_{j=1}^{n} a_{kj} (\bar{y}_j - y_j) < v$$

which implies that there are $y' \in Y$ and $v' < v$ so that $Ay' \leq 1v'$, in contradiction to the minimality of v in (8.10). This shows $v = \bar{v}$.

With (8.17) this gives $\bar{u} \leq u \leq v = \bar{v} = \bar{u}$ and therefore $u = v$, which was to be shown. $\square$

The idea behind this induction proof can be summarized as follows: If the players have optimal strategies that equalize the payoffs v and u for all rows and columns, then $u = v$. Otherwise (if needed by exchanging the players), there is at least one row with lower payoff than v, which will not be chosen by the row player, and by omitting this row from the game the theorem holds by the inductive hypothesis.

The optimization problem (8.10) is a special case of a *linear program* which asks to optimize a linear function subject to linear inequalities and equations. Here the variables are $(y, v) \in \mathbb{R}^n \times \mathbb{R}$ with the linear objective function v, to be minimized, and the linear constraints are $Ay - 1v \leq \mathbf{0}$, $y \geq \mathbf{0}$, $\mathbf{1}^\top y = 1$. The maximization problem (8.11) is the so-called *dual* linear program to (8.10), which uses the same data of the problem but processed "vertically" rather than "horizontally", that is, using the transposed matrix and with the roles of right-hand side and objective function exchanged (we omit the precise definition here; see the references in Section 8.6).

The inequality (8.13) is a special case of the *weak duality* of linear programming and, like (8.5), nearly immediate. The *strong duality* theorem of linear programming states that if both the original linear program and its dual are feasible, then their optimal values are not just mutual bounds but in fact *equal*. This immediately implies the minimax theorem.

Linear programming duality is a beautiful and very useful theory, for which the proof of Theorem 8.5 is a nice starting point. Moreover, algorithms for solving linear programs (which now work for hundreds of thousands, even millions, of variables and constraints) can be used to find optimal strategies of zero-sum games.

8.5   Further Notes on Zero-Sum Games

A max-min strategy, and the max-min payoff, only depends on the player's *own* payoffs. It can therefore also be defined for a game with general payoffs which are not necessarily zero-sum, and even for an N-player game. In that case, a max-min strategy represents a rather pessimistic view of the world: Player I chooses a strategy $\hat{x}$ that maximizes his payoff against an opponent (or opponents) who only want to harm player I. (In a zero-sum game, this coincides with the opponent's own objectives.) A min-max strategy in (8.2) can also be defined for a general game, with a different interpretation: The opponent chooses a strategy $\hat{y}$ with the intent to punish player I, who can defend himself against this by subsequently choosing x. The min-max payoff is the maximum punishment that can be inflicted on player I that way. This also makes sense in a game with more than two players, where the opponents agree on a partial mixed-strategy profile $\hat{y}$ that maximally punishes player I. (Such "punishment strategies" are used to enforce certain equilibria in repeated games, which we do not cover in this book.)

Here, we consider max-min and min-max strategies only for zero-sum games, which define the same concept and are based on the rational assumption that the opponent plays a best response, against which the player wants to play optimally.

We state some further observations on the max-min strategies of the maximizer and min-max strategies of the minimizer in a zero-sum game.

Proposition 8.6. *In a zero-sum game, the set of optimal strategies of each player is convex.*

Proof. This follows from Propositions 8.3 and 6.9.   □

A *weakly dominated* strategy in a zero-sum game can be disregarded without affecting the value of the game.

Proposition 8.7. *Consider two matrix games A and A', where A' is obtained from A by removing a weakly dominated strategy of one of the players. Then any (mixed) equilibrium of A' is an equilibrium of A, and A and A' have the same game value.*

Proof. Consider a mixed equilibrium of A', which defines a mixed strategy pair in A. By the best-response condition, it is also an equilibrium in A if there is no profitable deviation to a pure strategy. The only profitable deviation not available in A' would be to the weakly dominated strategy, but then deviating to the strategy that dominates it would also be profitable and available in A'. Hence, the equilibrium of A' is also an equilibrium of A (Proposition 3.3(a) is similar but about pure equilibria). By Proposition 8.3, the equilibrium payoff in a zero-sum game is the value of the game, which is therefore the same in both games.   □

The following game, with its pure best responses shown on the right,

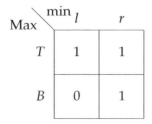   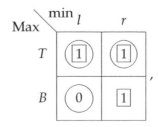

has two pure equilibria (T, l) and (T, r). Hence, any convex combination of l and r (which is an arbitrary mixed strategy of the minimizer) is a min-max strategy. The only max-min strategy is T, as the unique best response to l. The game has value 1. This is also the case when omitting the weakly dominated strategies B and r from the game. However, r is thereby lost as a possible min-max strategy. Consequently, if we are only interested in the value of a zero-sum game (which is often the case), then we can remove weakly dominated strategies (and iterate this), but the set of optimal strategies may become smaller that way.

8.6   Further Reading

The history of the minimax theorem is discussed in von Neumann and Fréchet (1953). In earlier articles in the same issue of *Econometrica*, Fréchet explains how the mathematician Emile Borel first suggested the use of active randomizations (that is, mixed strategies) in games, and posed this theorem (for symmetric zero-sum games) as an open problem. The theorem was first proved by von Neumann (1928), without knowledge of Borel's papers and question. Its proof was later substantially simplified.

The short inductive proof of Theorem 8.5 is due to Loomis (1946) (for a more general statement, which we have specialized to zero-sum games). It is very close to the proof by Owen (1967) who removes only one strict inequality to apply the induction hypothesis, not all such inequalities like Loomis (1946).

A geometric proof of the minimax theorem, which is also used in the book of von Neumann and Morgenstern (1947), is based on the theorem of the separating hyperplane (see Corollary 7.16 and Figure 7.13). This theorem applies to the following situation, where it is known as *Farkas's Lemma*: Let $A \in \mathbb{R}^{d \times n}$ and $x \in \mathbb{R}^d$ and assume that $x \notin D = \{Ay \mid y \in \mathbb{R}^n, y \geq \mathbf{0}\}$. The lemma states that then there is some $v \in \mathbb{R}^d$ so that $v^\top A \geq \mathbf{0}^\top$ and $v^\top x < 0$. The set D is the "convex cone" of nonnegative linear combinations of the columns of A, and if the vector x is not in that cone then there is a "separating hyperplane" with normal vector v so that all columns of A are on one side of the hyperplane ($v^\top A \geq \mathbf{0}^\top$) and x is strictly on the other side ($v^\top x < 0$). This is a consequence of Corollary 7.16 if one proves (which is the only missing piece) that D is closed. Because D is closed, one can choose c as the closest point to x in D and let $v = c - x$ (which has opposite sign to v as described following Corollary 7.16). Then $D \subseteq \{z \in \mathbb{R}^d \mid v^\top z \geq \alpha\}$ where $\alpha = v^\top c = -(x-c)^\top c = 0$ because, if not already $c = \mathbf{0}$, then similarly to Lemma 7.17 one can see that $x - c$ is orthogonal to the ray $\{c\beta \mid \beta \geq 0\}$ which is a subset of D; this also implies $v^\top(x - c) = -\|v\|^2 < 0$ and thus $v^\top x < v^\top c = 0$.

The strong duality theorem of linear programming, and therefore the minimax theorem, can be proved with some further algebraic manipulation from Farkas's Lemma (see Gale, 1960, p.78f). A classic text on linear programming is Dantzig

(1963). Accessible introductions are Chvátal (1983) and, in particular for the geometric aspects, Matoušek and Gärtner (2007). A scholarly book with superb historical references is Schrijver (1986). The minimax theorem is easy to obtain from linear programming duality; the converse direction has been studied by Adler (2013).

8.7 Exercises for Chapter 8

Exercise 8.1. Consider the following Penalty game. It gives the probability of scoring a goal when the row player (the striker) adopts one of the strategies L (shoot left), R (shoot right), and the column player (the goalkeeper) uses the strategies l (jump left), w (wait then jump), r (jump right). The row player is interested in maximizing and the column player in minimizing the probability of scoring a goal.

Max \ min	l	w	r
L	0.6	0.7	1.0
R	1.0	0.8	0.7

(a) Find all equilibria of this game in mixed (including pure) strategies, and their equilibrium payoffs. Why is the payoff unique?

(b) Now suppose that player I has an additional strategy M (shoot down the middle), so that the payoff matrix is

Max \ min	l	w	r
L	0.6	0.7	1.0
R	1.0	0.8	0.7
M	1.0	0.0	1.0

Find an equilibrium of this game, and the equilibrium payoff. [*Hint: The result from* (a) *will be useful.*]

Exercise 8.2.

(a) Consider a $2 \times n$ zero-sum game where the best response to every pure strategy is unique. Suppose that the top row and leftmost column define a pure-strategy equilibrium of the game. Show that this equilibrium can be found by iterated elimination of strictly dominated strategies. [*Hint: Consider first the case n = 2.*]

(b) Does the claim in (a) still hold if some pure strategy has more than one best response (and the game is therefore degenerate)?

(c) Give an example of a 3×3 zero-sum game, where the best response to every pure strategy is unique, which has a pure-strategy equilibrium that *cannot* be found by iterated elimination of strictly dominated strategies.

Exercise 8.3. Consider the following simultaneous game between two players: Each player chooses an integer between 1 and 5 (e.g., some number of fingers shown with your hand), and the higher number wins, except when it is just one larger than the lower number, in which case the lower number wins. So 4 beats 2 and 5 but it loses against 3. Equal numbers are a draw.

(a) Write down the matrix of payoffs to the row player of this zero-sum game. A winning player gets payoff 1, a draw gives payoff 0.

(b) Show all pairs of strategies of the row player where one strategy weakly or strictly dominates the other, and indicate the type of domination.

(c) Argue carefully, without calculations, why the value of this game has to be zero.

(d) Find an equilibrium of this game in mixed (including pure) strategies, and explain why it is an equilibrium.

Exercise 8.4. Consider the following bimatrix game.

	l	c	r
T	0 / 2	2 / 2	3 / 1
B	5 / 0	4 / 4	2 / 3

(a) Find an equilibrium of this game in mixed (including pure) strategies.

(b) Find a max-min strategy of player I, which may be a mixed strategy, and the resulting max-min payoff to player I. [*Hint: Consider the game as a zero-sum game with payoffs to player I.*]

(c) Find a max-min strategy of player II, which may be a mixed strategy, and the resulting max-min payoff to player II. [*Hint: Consider the game as a zero-sum game with payoffs to player II; note the reversed roles of maximizer and minimizer.*]

(d) Compare the max-min payoffs to player I and II from (b) and (c) with their payoffs in any equilibrium from (a).

9

Geometry of Equilibria in Bimatrix Games

This chapter is about the geometric structure of equilibria in two-player games in strategic form. It shows how to quickly identify equilibria with qualitative "best-response diagrams". These diagrams (for example, Figure 9.4) are subdivisions of the two mixed-strategy simplices into labeled "best-response regions", and suitable labels of the outsides of each simplex, so that an equilibrium is a "completely labeled" pair (x, y) of points. The power of this method is that it easily allows studying 3×3 games, which have a lot more variety than the rather restricted 2×2 games and their intersecting best-response lines as in Figure 6.8.

Best-response diagrams have been introduced by Shapley (1974) to illustrate the algorithm of Lemke and Howson (1964) that finds one equilibrium of a bimatrix game. Moreover, the algorithm shows that every nondegenerate bimatrix game has an *odd* number of equilibria. In this chapter, we explain the best-response regions in Section 9.2 and the principle behind the Lemke–Howson algorithm in Section 9.3. Even a reader who is not particularly interested in the computation of equilibria should look at these two sections, which are one of the things in this book that every game theorist should know.

The remainder of this chapter gives a more detailed introduction to the computation of equilibria in two-player games based on this geometric approach. In the introductory 3×2 example (9.1), the mixed strategy simplex X of player I is a triangle, and the mixed strategy simplex Y of player II is a line segment, where one would normally apply the upper envelope method (see Section 6.7). Section 9.4 explains the best-response diagrams for 3×3 games, where they are subdivided triangles. Section 9.5 explains strategic equivalence of games, which leaves best responses and hence the diagrams unchanged, which is generally useful.

While the partition of X and Y into best-response regions is convenient to draw in low dimension, it is in general dimension better to keep the payoff to the other player as an additional real variable, as described in Section 9.6. The resulting "best-response" polyhedra represent the upper envelope of expected payoffs in general dimension. The mathematically simpler, equivalent "best-response polytopes" are the basis for an algebraic implementation of the Lemke–Howson algorithm known as "complementary pivoting", described in Section 9.7. In the last Section 9.8, we describe how to extend complementary pivoting to degenerate games. This completes an alternative proof of the existence of an equilibrium in a two-player game that is independent of Brouwer's fixed-point theorem.

9.1   Prerequisites and Learning Outcomes

This chapter is a continuation of Chapter 6, and assumes the knowledge of the upper-envelope (also known as "goalpost") method and of degenerate games. All geometric definitions will be recalled here. For example, the concept of a simplex has been defined in Section 7.3, but we will only consider the mixed-strategy simplices as in Figure 6.5. Undirected graphs have been defined as background material early in Chapter 7.

After studying this chapter, you should be able to:

- draw best-response diagrams for bimatrix games with up to three strategies per player;

- identify the equilibria of the game with the help of the labels used in these diagrams, and explain the connection of labels and the best-response condition;

- show the computation of one equilibrium with the Lemke–Howson algorithm using the diagram;

- explain strategic equivalence, for two- and three-player games;

- see the connection between best-response polyhedra, the upper envelope of expected payoffs, and the best response-regions;

- be able to perform pivoting on small dictionaries with the complementary pivoting rule.

9.2   Labeled Best-Response Regions

Our leading example will be the 3×2 bimatrix game (A, B) with

$$A = \begin{pmatrix} 2 & 0 \\ 1 & 2 \\ 0 & 3 \end{pmatrix}, \qquad B = \begin{pmatrix} 2 & 0 \\ 0 & 1 \\ 4 & 1 \end{pmatrix}. \tag{9.1}$$

which is very similar to the right game Figure 6.7 and shown in Figure 9.1.

For an $m \times n$ bimatrix game, we number the pure strategies of the two players consecutively,

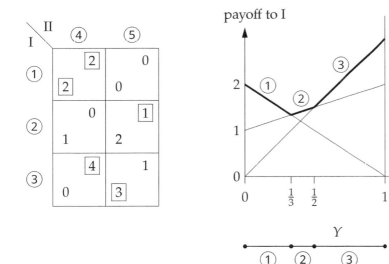

Figure 9.1 Left: The game in (9.1) with pure strategies numbered as in (9.2). Right: The upper envelope of expected payoffs to player I for a mixed strategy $y \in Y$ of player II. The mixed strategy set Y (bottom right) is subdivided into best-response regions, which are labeled with player I's pure best responses.

$$\begin{array}{lll} \text{pure strategies of player I (rows)} & : & 1, \ldots, m, \\ \text{pure strategies of player II (columns)} & : & m+1, \ldots, m+n. \end{array} \qquad (9.2)$$

In drawings we typically show them as circled numbers, here ①, ②, ③ for player I and ④, ⑤ for player II. In the text, we usually omit the circles.

The mixed strategy sets X and Y of the two players are defined in (6.6). Using (9.2), we subdivide them into *best-response regions* of the other player according to

$$\begin{array}{lll} Y(i) = \{y \in Y \mid i \text{ is a best response to } y\} & (1 \le i \le m), \\ X(j) = \{x \in X \mid j \text{ is a best response to } x\} & (m+1 \le j \le m+n). \end{array} \qquad (9.3)$$

The best-response regions are not disjoint because a mixed strategy may have more than pure best response, but this typically happens only at the "boundary" of a best-response region. Every mixed strategy has at least one best response, so the union of the best-response regions is the respective set of mixed strategies. In Figure 9.1, the subdivision of Y can be seen from the upper envelope diagram:

$$\begin{array}{l} Y(1) = \{(y_1, y_2) \in Y \mid 0 \le y_2 \le \tfrac{1}{3}\}, \\ Y(2) = \{(y_1, y_2) \in Y \mid \tfrac{1}{3} \le y_2 \le \tfrac{1}{2}\}, \\ Y(3) = \{(y_1, y_2) \in Y \mid \tfrac{1}{2} \le y_2 \le 1\}. \end{array}$$

Figure 9.2 demonstrates the subdivision of X into best-response regions. The strategy simplex X is a triangle with the unit vectors e_1, e_2, and e_3 as vertices,

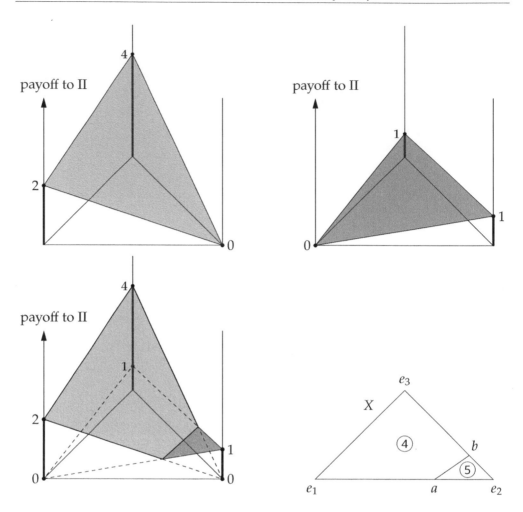

Figure 9.2 Expected payoffs to player II as a function of the mixed strategy of player I for the 3×2 game in Figure 9.2. Top left: plane of expected payoffs for the left column ④, with payoffs $2, 0, 4$ against the pure strategies e_1, e_2, e_3. Top right: plane of expected payoffs for the right column ⑤, with payoffs $0, 1, 1$. Bottom left: *upper envelope* (maximum) of expected payoffs. Bottom right: Resulting subdivision of X into best-response regions $X(4)$ and $X(5)$.

which are the pure strategies of player I. A "goalpost" is erected on each vertex, shown in the perspective drawings in Figure 9.2. In the top-left diagram, the heights $2, 0,$ and 4 are marked on these posts, which are the payoffs to player II for her left column ④. The resulting expected payoffs define a plane above the triangle through these height markings, which defines the expected payoff to player II for this column as a function of $x = (x_1, x_2, x_3)$, because the expected payoff is just the convex combination of the corresponding "heights", located above the point x in the triangle X below. Similarly, the top-right diagram shows for each goalpost the expected payoffs $0, 1,$ and 1 for the right column ⑤ of player II, which span the plane for the expected payoff. The bottom-left diagram shows the upper envelope as the maximum of the two expected-payoff planes.

Finally, the bottom-right diagram in Figure 9.2 shows the "projection" of the upper envelope diagram onto X, by ignoring the vertical coordinate, to represent the subdivision of X into the best-response regions $X(4)$ and $X(5)$. Against (x_1, x_2, x_3), the expected payoffs for the left and right column are $2x_1 + 4x_3$ and $x_2 + x_3$, respectively, so the equation

$$2x_1 + 4x_3 = x_2 + x_3 \quad \text{or equivalently} \quad 2x_1 + 3x_3 = x_2 \qquad (9.4)$$

shows where both columns have equal payoff. This defines a line segment in the triangle that belongs to both $X(4)$ and $X(5)$. The endpoints a and b of this line segment, also shown in the diagram, are obtained by setting in equation (9.4) either $x_3 = 0$, which gives $a = (\frac{1}{3}, \frac{2}{3}, 0)$, or $x_1 = 0$, which gives $b = (0, \frac{3}{4}, \frac{1}{4})$. The equation holds also for all convex combinations of a and b, which define the line segment itself.

In Figures 9.1 and 9.2, the best-response regions $Y(i)$ and $X(j)$ in (9.3) are labeled with the respective pure strategies i of player I and j of player II. We now introduce a second set of labels for *unplayed own strategies* via

$$\begin{aligned}
X(i) &= \{ \ (x_1, \ldots, x_m) \in X \mid \quad x_i = 0 \} \quad &(1 \le i \le m), \\
Y(j) &= \{ \ (y_1, \ldots, y_n) \in Y \mid y_{j-m} = 0 \} \quad &(m+1 \le j \le m+n).
\end{aligned} \qquad (9.5)$$

We refer to y_{j-m} because j denotes a pure strategy in $\{m+1, \ldots, m+n\}$ of player II, but the components of y still have subscripts $1, \ldots, n$. Similarly, the pure strategies of player II, which are the vertices of Y, are the unit vectors $e_1, \ldots, e_n$ in $\mathbb{R}^n$.

Definition 9.1. Consider an $m \times n$ bimatrix game (A, B) with mixed strategy sets X and Y in (6.6). Let $k \in \{1, \ldots, m+n\}$ be any pure strategy of player I or II with $X(k)$ and $Y(k)$ defined as in (9.3) and (9.5). Then $x \in X$ is said to have *label* k if $x \in X(k)$, and $y \in Y$ is said to have label k if $y \in Y(k)$. If every $k \in \{1, \ldots, m+n\}$ is a label of x or y (or both), then (x, y) is called *completely labeled*. The sets X and Y annotated with these labels define the *best-response diagram* for the game.   □

For the game (9.1), the labels for the unplayed own strategies as defined in (9.5) are shown in Figure 9.3. For example, $X(1)$ is the side of the triangle given by the points (x_1, x_2, x_3) in X where $x_1 = 0$, which are all the convex combinations of the unit vectors $e_2 = (0, 1, 0)$ and $e_3 = (0, 0, 1)$. In general, $X(i)$ is the facet of the triangle that is *opposite* to the vertex given by the unit vector e_i, for $1 \le i \le m$. The mixed strategy set Y of player II is a line segment with endpoints $e_1 = (1, 0)$ (which is the left column ④ of the game) and $e_2 = (0, 1)$ (which is the right column ⑤). Each "side" or facet $Y(4)$ or $Y(5)$ of this low-dimensional simplex Y is just one of these endpoints, where e_1 has label 5 because player II's second strategy 5 is played with probability zero (and this is the only element of Y with that property), and the other endpoint e_2 has label 4 because player II's first strategy 4 has probability zero.

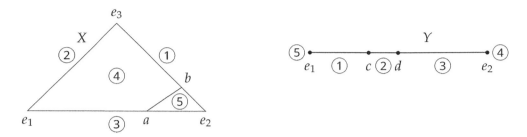

Figure 9.3 The mixed strategy simplices X and Y for the game in Figure 9.1 with best-response regions labeled with pure strategies of the other player, and sides labeled with *unplayed* own strategies. This defines the *best-response diagram* of the game.

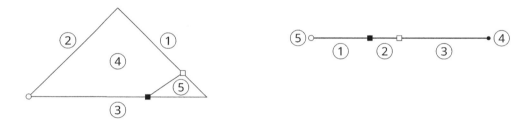

Figure 9.4 The same picture as Figure 9.3 with all annotations omitted except for the labels, which fully define the best-response diagram. The equilibria as completely labeled vertex pairs are shown with symbols: a white circle for the pure equilibrium (e_1, e_1), a black square for (a, c), and a white square for (b, d).

Any mixed strategy x or y may have several labels according to Definition 9.1. The inclusion of the *unplayed* own strategies among these labels is needed for the following, very useful theorem.

Theorem 9.2. *Under the assumptions of Definition 9.1, a mixed strategy pair (x, y) is an equilibrium of (A, B) if and only if (x, y) is completely labeled.*

Proof. The theorem is essentially another way of stating the best-response condition Proposition 6.1. Let $(x, y) \in X \times Y$, let u be the best-response payoff to player I against y, and let v be the best-response payoff to player II against x, as defined in (6.9). Suppose (x, y) is an equilibrium and therefore x a best response to y as in (6.11), and y a best response to x as in (6.12). These conditions can be rewritten as

$$
\begin{aligned}
x_i = 0 \quad &\text{or} \quad (Ay)_i = u \quad (1 \leq i \leq m), \\
y_j = 0 \quad &\text{or} \quad (x^\top B)_j = v \quad (1 \leq j \leq n),
\end{aligned}
\tag{9.6}
$$

where the first condition says that x has label i (because $x \in X(i)$) or y has label i (because i is a best response to y and therefore $y \in Y(i)$), and the second condition that y has label $m + j$ or x has label $m + j$ (because $m + j$ is a best response to x, with the numbering of pure strategies in (9.2)). In other words, (9.6) states that every label $i \in \{1, \ldots, m\}$ or $m + j \in \{m + 1, \ldots, m + n\}$ appears as a label of x or y, and (x, y) is therefore completely labeled.

Conversely, if (x, y) is completely labeled, then (9.6) holds and x and y are mutual best responses, so that (x, y) is an equilibrium, as claimed.   □

Another way of quickly seeing Theorem 9.2 is that a *missing label* from (x, y) would be a pure strategy of a player that is played with positive probability (because it is not an unplayed pure strategy as in (9.5)) and is also not a best response, which is precisely not allowed according to the best-response condition, and violates (9.6). The players' pure strategies are numbered separately as in (9.2) so that they can appear as labels of either x or y, and in order to distinguish the two sets of conditions in (9.6).

As Figure 9.3 shows, only the points e_1, e_2, e_3, a, b in X have three labels (which is the maximum number of labels of a point in X), and only the points e_1, e_2, c, d in Y have two labels (which is the maximum number of labels of a point in Y). Because an equilibrium (x, y) needs all five labels for the players' pure strategies, one can quickly identify the equilibria of the game. These are:

- The pure equilibrium (e_1, e_2) with labels 2, 3, 4 in X (which means 2 and 3 are not played and 4 is a best response) and 5, 1 in Y (which means 5 is not played and 1 is a best response). This equilibrium is marked by white circles in Figure 9.4.

- The mixed equilibrium (a, c) with labels 3, 4, 5 for $a = (\frac{1}{3}, \frac{2}{3}, 0)$ in X (which means 3 is not played and 4 and 5 are best responses) and 1, 2 for $c = (\frac{2}{3}, \frac{1}{3})$ in Y (which means both 1 and 2 are best responses). This is marked by black squares in Figure 9.4.

- The mixed equilibrium (b, d) with labels 1, 4, 5 for $b = (0, \frac{3}{4}, \frac{1}{4})$ in X (which means 1 is not played and 4 and 5 are best responses) and 2, 3 for $c = (\frac{1}{2}, \frac{1}{2})$ in Y (which means both 2 and 3 are best responses). This is marked by white squares in Figure 9.4.

The quickest way to find these equilibria as completely labeled pairs of points is to check the points in Y that have two labels, where the three other labels required to complete the full set of labels are unique, and may or may not appear as labels of points in X. The three points e_1, c, d in Y are part of an equilibrium, but e_2 is not because it has labels 3 and 4 and the required three other labels $1, 2, 5$ do not appear in X, for example because the sets $X(2)$ and $X(5)$ have no point in common. This suffices to find all equilibria. A similar, slightly longer method is to check e_2

in X with labels $1, 3, 5$ where no point in Y has the other two labels 2 and 4, and e_3 in X with labels $1, 2, 4$ where no point in Y has the other two labels 3 and 5.

The advantage of the best-response diagram for the game is that it is a *qualitative* diagram that allows identifying the equilibria, in a much more informative way than the payoffs do. In Figure 9.4 we have only kept the labels as annotations, which nevertheless encode all the necessary information. In particular, X is the strategy set of player I because its sides are labeled with the pure strategies of player I, and the inside labels are the pure best responses of player II; the reverse holds for Y. The vertices of X are unit vectors that are identified by the incident labels for the unplayed pure strategies; for example, $e_1 = (1, 0, 0)$ is the vertex with labels 2 and 3. In Y this holds for $e_1 = (1, 0)$ as the first strategy of player II where the unplayed pure strategy is $5 = m + 2$ as the label of that endpoint of Y, and for $e_2 = (0, 1)$ as the second endpoint of Y with label $4 = m + 1$. The equilibria themselves are identified by their complementary sets of labels.

9.3   The Lemke–Howson Algorithm

Identifying equilibria as completely labeled pairs (x, y) according to Theorem 9.2 does not yet prove that there is such a pair. Figure 9.5 is an extension of Figure 9.3 that illustrates such an existence proof, which is the topic of this section. In general, it provides an algorithm for finding an equilibrium of a nondegenerate $m \times n$ bimatrix game, due to Lemke and Howson (1964). In Section 9.7 below, we give a more specific (and more rigorous) description of this algorithm that can also be applied to degenerate games. This provides a new, elementary proof of the existence of an equilibrium in a two-player game, which does not require Brouwer's fixed point theorem. Nevertheless, this proof will have some similarity to the techniques used for proving the fixed point theorem, in particular the "parity argument" used in the proof of Sperner's lemma.

The mixed-strategy set X is a subset of $\mathbb{R}^m$, and Y is a subset of $\mathbb{R}^n$. Neither X nor Y contains the origin $\mathbf{0}$ of $\mathbb{R}^m$ or $\mathbb{R}^n$, respectively. We now add $\mathbf{0}$ to X and connect it with a line segment to each vertex $e_1, \ldots, e_m$ of X, and denote this extension of X by $\tilde{X}$. We similarly add $\mathbf{0}$ to Y and call the resulting extension $\tilde{Y}$. Figure 9.5 shows $\tilde{X}$ and $\tilde{Y}$ for our example (9.1).

Then $\mathbf{0}$ in $\tilde{X}$ has all labels $1, \ldots, m$, and $\mathbf{0}$ in $\tilde{Y}$ has all labels $m + 1, \ldots, m + n$, according to (9.5). Because the pair $(\mathbf{0}, \mathbf{0})$ is therefore completely labeled, we call it the *artificial equilibrium* of $\tilde{X} \times \tilde{Y}$. It does not represent a mixed strategy pair.

We now consider $\tilde{X}$ and $\tilde{Y}$ as *graphs*, rather like they are drawn in Figure 9.5. The nodes of the graph $\tilde{X}$ are the points that have m labels, and they are connected by an edge if they share $m - 1$ labels, which are the labels of that edge. For example, the nodes $\mathbf{0}$ (with labels $1, 2, 3$) and e_2 (with labels $1, 3, 5$) are connected by the

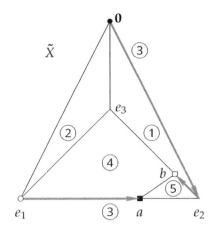
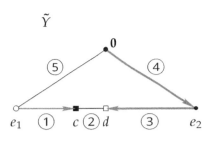

Figure 9.5   The labeled mixed-strategy sets X and Y in Figure 9.3, each extended by an extra point **0** with all components zero, so that $(\mathbf{0},\mathbf{0})$ is a completely labeled "artificial equilibrium". The gray arrows show the two Lemke–Howson paths for the missing label 2. The end nodes of these paths are the artificial equilibrium and the equilibria of the game, so the number of equilibria is *odd*.

edge that has labels 1 and 3. The nodes of the graph $\tilde{Y}$ are the points that have n labels, and are connected by an edge if they have $n-1$ labels in common (which in Figure 9.5 is just a single label because $n-1=1$), which are the labels of that edge.

Next, we choose a label in $\{1, \ldots, m+n\}$ that is allowed to be *missing*, which is kept fixed throughout. We consider all pairs (x, y) in $\tilde{X} \times \tilde{Y}$ that have all labels except possibly this missing label, and call such a pair (x, y) *almost completely labeled*. The almost completely labeled pairs (x, y) belong to two types of sets, which are either of the form $\{x\} \times E$ where x is a node of $\tilde{X}$ and E is an edge of $\tilde{Y}$, with m labels for x and $n-1$ labels for $y \in E$, except for the endpoints of E which have n labels; or these sets are of the form $E \times \{y\}$ where E is an edge of $\tilde{X}$, with $m-1$ labels for $x \in E$, except for the endpoints of E which have m labels, and a node y of $\tilde{Y}$. (Technically, these sets are the edges of the *product graph* of the two graphs $\tilde{X}$ and $\tilde{Y}$.)

In Figure 9.5, the missing label is chosen to be 2. Any completely labeled pair (x, y) is by definition also almost completely labeled. It serves as the unique starting pair of a *path* of almost completely labeled pairs (x, y). We start with the artificial equilibrium $(\mathbf{0}, \mathbf{0})$. With label 2 allowed to be missing, we *drop* label 2, which means changing x along the unique edge of $\tilde{X}$ that connects $\mathbf{0}$ to e_2 (shown by an arrow in $\tilde{X}$), while keeping $y = \mathbf{0}$ fixed. The endpoint $x = e_2$ of that arrow in $\tilde{X}$ has a new label 5 which is *picked up*. Because x has three labels $1, 3, 5$ and $y = \mathbf{0}$ has two labels $4, 5$ but label 2 is missing, the label 5 that has just been picked up in $\tilde{X}$ is now *duplicate*. Because y no longer needs to have the duplicate label 5, the next step is to drop label 5 in $\tilde{Y}$, that is, change y from $\mathbf{0}$ to e_2 along the edge which has only label 4. The endpoint e_2 of that edge of $\tilde{Y}$ has labels 4 and 3, where label 3 has been picked up. Therefore, the current pair $(x, y) = (e_2, e_2)$ has the duplicate label 3. Correspondingly, we can now drop label 3 in $\tilde{X}$, that is, move x along the edge with labels 1 and 5 to point b, where label 4 is picked up. At the current pair $(x, y) = (b, e_2)$, label 4 is duplicate. Next, we drop label 4 in $\tilde{Y}$ by moving y along the edge with label 3 to reach point d, where label 2 is picked up. Because 2 is the missing label, the reached pair $(x, y) = (b, d)$ is completely labeled. This is the equilibrium that is found at the end of the path. In summary, the path is given by the sequence of points in $\tilde{X} \times \tilde{Y}$ with corresponding label sets

$$
\begin{array}{lllll}
(x, y) & & \text{labels} & \text{drop label} & \\
\hline
(\mathbf{0}, \mathbf{0}) & \bullet & 123,\ 45 & 2 \text{ in } \tilde{X} & \\
(e_2, \mathbf{0}) & & 135,\ 45 & 5 \text{ in } \tilde{Y} & \\
(e_2, e_2) & & 135,\ 43 & 3 \text{ in } \tilde{X} & (9.7) \\
(b, e_2) & & 145,\ 43 & 4 \text{ in } \tilde{Y} & \\
(b, d) & \square & 145,\ 23 & (\text{found } 2 \text{ in } \tilde{Y}). &
\end{array}
$$

Hence, by starting from the artificial equilibrium $(\mathbf{0}, \mathbf{0})$, the sequence of almost completely labeled node pairs of $\tilde{X} \times \tilde{Y}$ terminates at an actual equilibrium of the game. In general, this defines an algorithm for finding an equilibrium of a bimatrix game, due to Lemke and Howson (1964). The path in (9.7) is therefore also called the *Lemke–Howson path* for missing label 2. Note that it alternates between $\tilde{X}$ and $\tilde{Y}$ by keeping in each step one of the components x or y in (x, y) fixed, and changing the other component by traversing an edge in the respective graph $\tilde{Y}$ or $\tilde{X}$.

A Lemke–Howson path can be started at any completely labeled pair (x, y), not just at the artificial equilibrium. For example, if started from (b, d) in Figure 9.5, then the path (for missing label 2) just traverses the path (9.7) backwards, which will then go back to $(\mathbf{0}, \mathbf{0})$. However, the game has further equilibria, for example the pure-strategy equilibrium (e_1, e_1), indicated by the white circles. The Lemke–Howson path for missing label 2 that starts at this equilibrium is also shown in Figure 9.5 and consists of only two steps, which end at the third equilibrium (a, c):

$$
\begin{array}{lllll}
(x, y) & & \text{labels} & \text{drop label} & \\
\hline
(e_1, e_1) & \circ & 234,\ 15 & 2 \text{ in } \tilde{X} & \\
(a, e_1) & & 345,\ 15 & 5 \text{ in } \tilde{Y} & (9.8) \\
(a, c) & \blacksquare & 345,\ 12 & (\text{found } 2 \text{ in } \tilde{Y}). &
\end{array}
$$

As a consequence, for a given missing label, the equilibria of the game plus the artificial equilibrium are the end nodes of the Lemke–Howson paths. Because each path has two end nodes, their total number is even and therefore the number of equilibria of the game is *odd*. This is a *parity argument* similar to that used for Sperner's lemma in Section 7.4.

In order to make this parity argument for a general bimatrix game, we have to assume that the game is nondegenerate (see Definition 6.7) because a degenerate game does not necessarily have a odd number of equilibria; for example, the degenerate game in Figure 6.12 has an infinite set of equilibria (which is very well understood with the help of the best-response diagram, see Exercise 9.2).

Lemma 9.3. *An $m \times n$ bimatrix game as in Definition 9.1 is nondegenerate if and only if no point in X has more than m labels, and no point in Y has more than n labels.*

Proof. Let $x = (x_1, \ldots, x_m) \in X$ and consider the sets $I = \{i \mid x_i = 0,\ 1 \le i \le m\}$ and $J = \{j \mid j$ is a best response to $x,\ m + 1 \le j \le m + n\}$, so that $I \cup J$ is the set of labels of x. The number of pure strategies played with positive probability by x is $m - |I|$, and nondegeneracy requires $|J| \le m - |I|$ or equivalently $|I| + |J| \le m$, that is, x has no more than m labels. Similarly, in a nondegenerate game no $y \in Y$ has more than n labels. In a degenerate game, for some x or y this fails to hold.   □

For a nondegenerate $m \times n$ game, the graphs $\tilde{X}$ and $\tilde{Y}$ have the following properties (which we state without proof, because we will give an alternative, more rigorous description of the Lemke–Howson method in Section 9.7 below):

- The nodes of $\tilde{X}$ are the finitely many points in X that have m labels, plus $\mathbf{0} \in \mathbb{R}^m$. The nodes of $\tilde{Y}$ are the finitely many points in Y that have n labels, plus $\mathbf{0} \in \mathbb{R}^n$.

- Any two nodes of $\tilde{X}$ are joined by an *edge* if they have $m - 1$ labels in common, which are the labels of that edge. (As a geometric object, the edge is the set of points that have at least these labels; as an edge of the graph $\tilde{X}$, the edge is just a pair $\{x, x'\}$ of nodes of $\tilde{X}$.) Any two nodes of $\tilde{Y}$ are joined by an edge if they have $n - 1$ labels in common, which are the labels of that edge of $\tilde{Y}$.

- For any node x in $\tilde{X}$ and a label k of x, there is a unique edge of $\tilde{X}$ with endpoint x that has all the labels of x except k. We say this edge is traversed from x by *dropping label* k (the other endpoint x' of this edge $\{x, x'\}$ has some other label than k that is *picked up*, as described for Figure 9.5 above). The same holds for any node in $\tilde{Y}$.

Let $\ell \in \{1, \ldots, m + n\}$. Using the above properties, we define the *LH graph for missing label ℓ* as a certain subset of the "product graph" of $\tilde{X}$ and $\tilde{Y}$. The nodes of the LH graph are the node pairs (x, y) in $\tilde{X} \times \tilde{Y}$ that are almost completely labeled, that is, every label in $\{1, \ldots, m + n\} - \{\ell\}$ appears as a label of x or y. Any edge of the LH graph is of the form $\{(x, y), (x, y')\}$ where $\{y, y'\}$ is an edge of $\tilde{Y}$, or of the form $\{(x, y), (x', y)\}$ where $\{x, x'\}$ is an edge of $\tilde{X}$. If (x, y) is completely labeled, then it has a unique neighbor in the LH graph reached by *dropping label ℓ*, that is, by following the edge which does not have label ℓ (via the edge in $\tilde{X}$ if $\ell \in \{1, \ldots, m\}$, and the edge in $\tilde{Y}$ otherwise). If (x, y) is almost completely labeled and does not have label ℓ, then there is exactly one *duplicate label k* that appears as a label of both x and y. Such a node in the LH graph has exactly two neighbors, reached by dropping the duplicate label k in either $\tilde{X}$ or $\tilde{Y}$.

Because every node (x, y) in the LH graph has degree one or two, the graph is a collection of paths and cycles. The end nodes of paths are the degree-one nodes, whose number is even. One of these end nodes is the artificial equilibrium $(0, 0)$, and every other end node is an equilibrium of the game. This completes the parity argument. We summarize it as a theorem (whose complete proof requires proving the above bullet points).

Theorem 9.4. *A nondegenerate bimatrix game has an odd number of equilibria.*

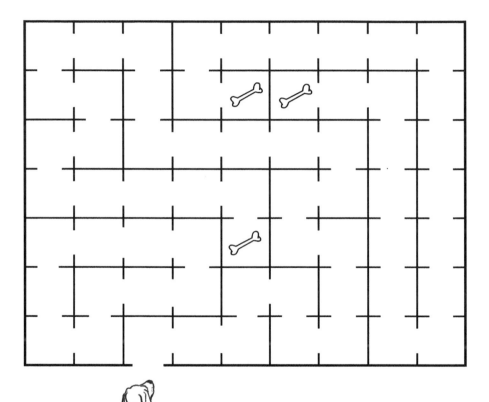

Figure 9.6 Floorplan of a castle with one outside door and each room having only one or two doors. Each room with only one door has a bone in it. There is an odd number of them.

A way to remember the parity argument is the story of a castle whose rooms create a maze (which is not really a maze because one cannot make a mistake when entering it), shown in Figure 9.6. Every room in the castle has only one or two doors in it. Only one door leads outside, every other door leads to another room. Every room that has only one door has a bone in it. A hungry dog arrives at the castle, and easily finds a bone, as follows. The dog enters the castle from the outside door and enters a room that either has a bone in it, or is a room with two doors, in which case the dog goes through the other door of that room, and continues in that manner. In that way the dog passes through a number of rooms in the castle, but cannot enter a room again, because that room would have needed a third door to enter it. The dog can also not get out of the castle because it has only one door to the outside, which is the one the dog entered. So the sequence of rooms (the castle has only a finite number of rooms) must end somewhere, which can only be a room with only one door that contains a bone. In this analogy, a room in a castle is a node of the LH graph, and a door that connects two rooms is an edge of the LH graph. The outside of the castle is the artificial equilibrium. Every room with only one door (and a bone in it) is an equilibrium. Moreover, by Theorem 9.4, the number of bones is odd, because starting from a bone either leads to the outside or to another bone. In addition, there may be cycles of rooms that all have two doors, like at the bottom left in Figure 9.6.

9.4   Using Best-Response Diagrams

In this section, we describe the use of best-response diagrams for 3×3 games. For 2×2 games, it is possible to draw all mixed strategy pairs simultaneously in a square that represents $X \times Y$, like in Figures 6.4 and 6.8, so that an equilibrium is identified as a single point in the square. In a 3×3 game, the best-response diagram is given by two triangles that represent X and Y, and each equilibrium has to be identified by a separate symbol that identifies the equilibrium strategies in X and Y, like for the 3×2 game (9.1) in Figure 9.4. However, 3×3 games have a lot more variety than 2×2 games. It is therefore useful that they can be analyzed with the help of two-dimensional diagrams.

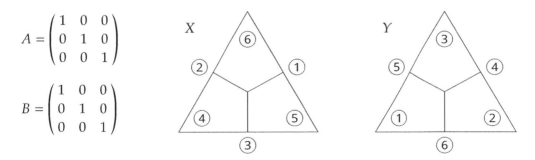

Figure 9.7 The 3×3 coordination game (A, B) and its best-response diagram. Every point in X with three labels together with the same point in Y is an equilibrium, so the game has seven equilibria.

In a 3×3 game, the rows (strategies of player I) have labels $1, 2, 3$, and the columns (strategies of player II) have labels $4, 5, 6$. Each strategy set X and Y is a triangle, with (up to) three best-response regions.

Figure 9.7 shows the game (A, B) where A and B are the identity matrix. This *coordination game* is a symmetric game where both players get payoff one if they choose the same row and column, and zero otherwise. In addition to the pure strategy equilibria (e_1, e_1), (e_2, e_2), and (e_3, e_3), the game has many mixed equilibria, because any subset of pure strategies, chosen equally on both sides, with a uniformly at random chosen strategy from that subset, defines a mixed equilibrium (x, y). Here, these are the mixed equilibrium strategies $x = y = (\frac{1}{2}, \frac{1}{2}, 0)$, $(\frac{1}{2}, 0, \frac{1}{2})$, $(0, \frac{1}{2}, \frac{1}{2})$, and the completely mixed equilibrium where $x = y = (\frac{1}{3}, \frac{1}{3}, \frac{1}{3})$. These are seven equilibria in total, as can be seen from the labels in Figure 9.7. The best-response regions have all the same shape; for example, $X(4) = \{(x_1, x_2, x_3) \in X \mid x_1 \geq x_2, \ x_1 \geq x_3\}$.

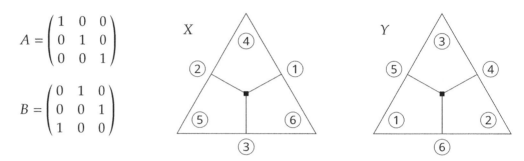

$$A = \begin{pmatrix} 1 & 0 & 0 \\ 0 & 1 & 0 \\ 0 & 0 & 1 \end{pmatrix}$$

$$B = \begin{pmatrix} 0 & 1 & 0 \\ 0 & 0 & 1 \\ 1 & 0 & 0 \end{pmatrix}$$

Figure 9.8 A game (A, B) with a full cycle (9.9) of pure best responses. Its only equilibrium is completely mixed, marked by a black square.

Figure 9.8 shows a game (A, B) where A is the identity matrix and B is the identity matrix with permuted columns. The best-response regions in X therefore have the same shape but different labels. As the best-response diagram and the payoff matrices show, there is a full cycle of pure best responses

$$① \to ⑤ \to ② \to ⑥ \to ③ \to ④ \to ① \tag{9.9}$$

in this game. There is no completely labeled pair of mixed strategies (x, y) except for the completely mixed equilibrium where $x = y = (\frac{1}{3}, \frac{1}{3}, \frac{1}{3})$. This can be seen from the best-response diagram because there is clearly no pure equilibrium by (9.9), and the point $x = (\frac{1}{2}, \frac{1}{2}, 0)$ in X with labels $3, 5, 6$ requires some y with labels $1, 2, 4$ to be part of an equilibrium, but no point in Y has labels 1 and 4; similarly, labels $1, 4, 6$ in X require labels $2, 3, 5$ in Y but no point in Y has labels 2 and 5, and labels $2, 4, 5$ in X require labels $1, 3, 6$ in Y but no point in Y has labels 3 and 6.

This raises the general question if the cycle (9.9) of pure best responses (for a nondegenerate game) implies that the game has a unique mixed equilibrium (which is true for 2×2 games). That question is easily answered with the help of the best-response diagrams. The pure best responses in the corners of the triangle are determined by (9.9). We start with Figure 9.9, where we first keep A and therefore the best-response regions $Y(1)$, $Y(2)$, and $Y(3)$ fixed. Can $y = (\frac{1}{2}, \frac{1}{2}, 0)$ in Y with labels $1, 2, 6$ be part of an equilibrium (x, y)? This requires x to have labels $3, 4, 5$, which is indeed possible if, along the bottom edge of X with label 3, label 4 appears between labels 5 and 6. In the resulting best-response diagram, $x = (\frac{2}{3}, \frac{1}{3}, 0)$ has these labels $3, 4, 5$, and the equilibrium strategies x and y are marked with a black square. The best-response diagram shows that this is the only equilibrium. Hence, the game has indeed only one equilibrium, although it is not completely mixed.

$$A = \begin{pmatrix} 1 & 0 & 0 \\ 0 & 1 & 0 \\ 0 & 0 & 1 \end{pmatrix}$$

$B = ?$

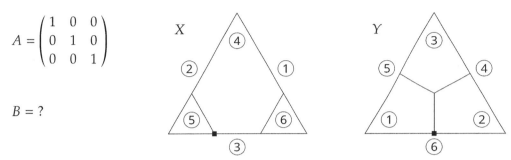

Figure 9.9 Best-response diagram for a game with a full cycle (9.9) of pure best responses and a unique equilibrium (x, y) for $x = (\frac{2}{3}, \frac{1}{3}, 0)$ and $y = (\frac{1}{2}, \frac{1}{2}, 0)$. The payoff matrix B is constructed in Figure 9.10.

We have drawn the best-response diagram first. The construction of a payoff matrix B to get this subdivision of X into best-response regions is shown in Figure 9.10.

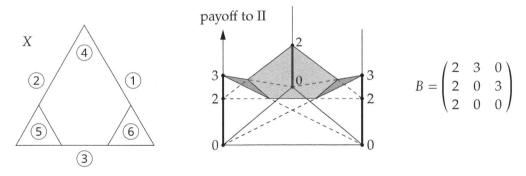

$$B = \begin{pmatrix} 2 & 3 & 0 \\ 2 & 0 & 3 \\ 2 & 0 & 0 \end{pmatrix}$$

Figure 9.10 Construction of the game matrix B from the best-response diagram on the left via the upper envelope of suitable payoffs. The first column ④ defines a horizontal plane of constant payoffs 2, and the second and third columns ⑤ and ⑥ have higher payoff 3 in the corners e_1 and e_2, respectively, and 0 otherwise.

Figure 9.11 shows a game where the best-response regions $Y(1)$, $Y(2)$, $Y(3)$, and the payoff matrix A, are also changed compared to Figure 9.10. The game still has the full cycle (9.9) of pure best responses, but it has three equilibria. Hence, condition (9.9) does not imply a unique equilibrium. The counterexample was easy to construct with the help of the best-response diagrams.

$$A = \begin{pmatrix} 3 & 0 & 0 \\ 2 & 2 & 2 \\ 0 & 0 & 3 \end{pmatrix}$$

$$B = \begin{pmatrix} 2 & 3 & 0 \\ 2 & 0 & 3 \\ 2 & 0 & 0 \end{pmatrix}$$

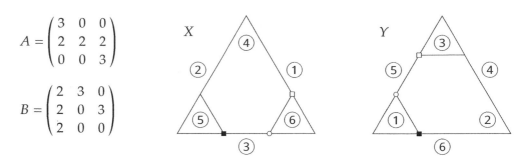

Figure 9.11   A game with a full cycle (9.9) of pure best responses and multiple equilibria.

Figure 9.12 shows best-response regions where the subdivision of X is qualitatively at the transition between Figures 9.8 and 9.9. According to Lemma 9.3, this is a degenerate game because $x = (\frac{1}{2}, \frac{1}{2}, 0)$ has three pure best responses and four labels $3, 4, 5, 6$. Consequently, only the two labels 1 and 2 of a point y are needed for an equilibrium (x, y). These two labels define $Y(1) \cap Y(2)$, which is the line segment with endpoints $(\frac{1}{2}, \frac{1}{2}, 0)$ and $(\frac{1}{3}, \frac{1}{3}, \frac{1}{3})$ in Y. Any y in that line segment is part of an equilibrium (x, y), so this game has an infinite set of equilibria.

$$A = \begin{pmatrix} 1 & 0 & 0 \\ 0 & 1 & 0 \\ 0 & 0 & 1 \end{pmatrix}$$

$$B = \begin{pmatrix} 1 & 2 & 0 \\ 1 & 0 & 2 \\ 1 & 0 & 0 \end{pmatrix}$$

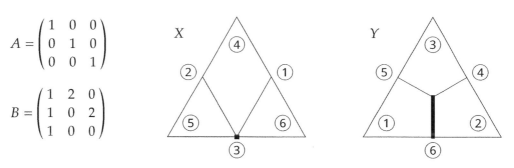

Figure 9.12   A degenerate game with a line segment of multiple equilibria.

For a nondegenerate game, Proposition 6.8 states that the equilibrium strategies in an equilibrium (x, y) mix the same number of pure strategies. In terms of the best-response diagram, this means that the smallest face of X that contains x has the same dimension as the smallest face of Y that contains y. This face is either a vertex if (x, y) is a pure-strategy equilibrium, an edge if x and y mix only two pure strategies, or a triangle if they mix three strategies (and a face of higher dimension for larger games). This is the case in the diagrams seen so far, but fails for the degenerate game in Figure 9.12 where the equilibrium strategy $x = (\frac{1}{2}, \frac{1}{2}, 0)$ on the lower edge of X and any point $y = (\alpha, \frac{1}{2} - \frac{\alpha}{2}, \frac{1}{2} - \frac{\alpha}{2})$ for $0 < \alpha \leq \frac{1}{3}$ in the interior of the triangle Y form an equilibrium (x, y).

9.5   Strategic Equivalence

When constructing the best-response diagram of a game, it is often simpler to consider another game that has the same diagram. Two such games are called *strategically equivalent*, which we define for N-player games in general.

Definition 9.5. Consider an N-player game G in strategic form with finite strategy set S_i for player $i = 1, \ldots, N$, payoff $u_i(s)$ for each strategy profile s, mixed strategies $\sigma_i \in X_i = \Delta(S_i)$, and partial profiles of mixed strategies $\sigma_{-i} \in X_{-i} = \times_{k \neq i} X_k$. Let $\hat{G}$ be another game with the same strategy sets, and payoff function $\hat{u}_i$ for each player i. Then G and $\hat{G}$ are called *strategically equivalent* if for all $i = 1, \ldots, N$ and all $\sigma_{-i} \in X_{-i}$, any strategy $s_i \in S_i$ is a best response to σ_{-i} in G if and only if s_i is a best response to σ_{-i} in $\hat{G}$.   □

Hence, strategic equivalence of two $m \times n$ bimatrix games means that any row i is a best response to a mixed strategy y of player II in one game whenever it is a best response to y in the other game, and similarly any column j is a best response to a mixed strategy x of player I in one game whenever it is in the other game. This is equivalent to saying that the two games have the same best-response regions $Y(i)$ and $X(j)$ in (9.3), that is, the same best-response diagram.

As Figures 6.10 and 9.1 show, the game on the right in Figure 6.7 and in Figure 9.1 are strategically equivalent. The two games have the same payoff matrix B for player II, and the payoff matrix A in the second game is obtained from the first by adding the constant 1 to each entry in the first column of A. This is generally a way to obtain a strategically equivalent game. Consider a bimatrix game (A, B), and add a constant α to each entry in column j of A. Then if player II plays the mixed strategy y, with probability y_j for column j, then the expected payoff in row i is changed from $(Ay)_i$ to $(Ay)_i + \alpha y_j$. That is, all these expected payoffs change by the *same amount* αy_j. Although this depends on y, it does not change which rows i have maximal payoff, and therefore it does not change player I's best responses to y. We call such a payoff change of adding a constant to some column of the row player's payoffs a *column shift* of the game.

We use "column shift" as a general term in an N-player game, by assuming that the pure strategies of the player under consideration are temporarily arranged as rows, so that any partial profile of strategies of the other $N - 1$ players is given by a column. For example, the three-player game in Exercise 3.2 has as rows the strategies of player I and as columns the partial pure strategy profiles of players II and III. However, one should remember that therefore in a bimatrix game (A, B), a "column shift" for player II means adding a constant to some *row* of player II's payoff matrix B.

Definition 9.6. Consider an N-player game G as in Definition 9.5. Let $i = 1, \ldots, N$ be a player, let $a_i(s_{-i}) \in \mathbb{R}$ be a separate constant for each partial profile s_{-i} of pure strategies of the other $N - 1$ players, and let

$$\hat{u}_i(s_i , s_{-i}) = u_i(s_i , s_{-i}) + a_i(s_{-i})$$

be a new utility function $\hat{u}_i$ for player i. This is called a *column shift* for player i. $\square$

Proposition 9.7. *Consider an N-player game G as in Definition 9.6, with a new utility function $\hat{u}_i$ for player i obtained by a column shift for player i. Then the game $\hat{G}$ with $\hat{u}_i$ instead of u_i is strategically equivalent to G.*

Proof. With the notation in Definition 9.6, let $s_i \in S_i$ be a strategy of player i and let σ_{-i} be a partial profile of mixed strategies of the other players. Then the expected payoff to player i is $U_i(s_i , \sigma_{-i})$ in G, and $U_i(s_i , \sigma_{-i}) + a_i(s_{-i}) \prod_{k=1, \ k \neq i}^{N} \sigma_k(s_k)$ in $\hat{G}$. The added term does not depend on s_i. Hence, the maximum over $s_i \in S_i$ for determining the best response of player i to σ_{-i} is not affected by the column shift. That is, the best responses in G and $\hat{G}$ to σ_{-i} are the same, as claimed. $\square$

A column shift may mean adding the *same* constant to all "columns" and hence to all payoffs. This just changes the origin of the utility function of the player and always represents the same preference. In addition, the player's payoffs may be *scaled* by multiplying them with a positive constant (which cannot be varied with the column), which trivially creates a strategically equivalent game. However, two games may be strategically equivalent even if one game cannot be obtained from the other by scaling and column shifts. Consider the 3×2 games (A, B) and $(\hat{A}, B)$ with

$$A = \begin{pmatrix} 2 & 0 \\ 1 & 2 \\ 0 & 3 \end{pmatrix}, \qquad \hat{A} = \begin{pmatrix} 2 & 0 \\ 1 & 2 \\ \frac{1}{2} & \frac{5}{2} \end{pmatrix} \tag{9.10}$$

and any B, for example as in (9.1). The upper envelope diagrams for the two games in Figure 9.13 show that the two games are strategically equivalent because they have the same best-response regions $Y(1), Y(2), Y(3)$, in particular the indifference between the two rows 2 and 3, which are both best responses when $y = (\frac{1}{2}, \frac{1}{2})$. However, $\hat{A}$ is not obtained from A by scaling and column shifts: The first two entries of the first column of A and $\hat{A}$ are the same, 2 and 1, and fix the "utility scale" for the payoffs in the first column, so that no scaling or additive shift can be applied to that column, but the third entry in that column is different for A and $\hat{A}$.

Note that we have defined strategic equivalence only in terms of *best responses*, which are relevant for determining an equilibrium. The rows 1 and 3 for A and $\hat{A}$ in (9.10) do not have the same point of indifference, which occurs for $y = (\frac{3}{5}, \frac{2}{5})$ with A when both rows have payoff $\frac{6}{5}$, and for $y = (\frac{5}{8}, \frac{3}{8})$ with $\hat{A}$ when both rows have payoff $\frac{5}{4}$. However, in both cases rows 1 and 3 are not on the upper envelope because row 2 has a higher payoff, as Figure 9.13 shows.

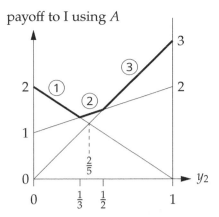
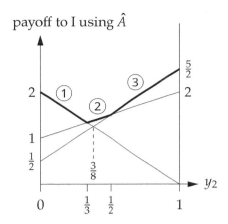

Figure 9.13 Upper envelopes of expected payoffs to player I against (y_1, y_2) for the payoff matrices A and $\hat{A}$ in (9.10), which define strategically equivalent but not strongly strategically equivalent games (A, B) and $(\hat{A}, B)$.

A stronger definition (with the notation in Definition 9.5) is to say that G and $\hat{G}$ are *strongly strategically equivalent* if for all players $i = 1, \ldots, N$, all $\sigma_{-i} \in X_{-i}$, and any two strategies s_i and s_i' in S_i we have

$$u_i(s_i, \sigma_{-i}) \geq u_i(s_i', \sigma_{-i}) \quad \Leftrightarrow \quad \hat{u}_i(s_i, \sigma_{-i}) \geq \hat{u}_i(s_i', \sigma_{-i}). \qquad (9.11)$$

Clearly, strong strategic equivalence implies strategic equivalence, and is implied by any column shift because the proof of Proposition 9.6 still applies. However, the converse does not hold, as the example (9.10) shows, because (9.11) is violated for i as the row player, $\sigma_{-i} = y = (0.61, 0.39)$, and rows 1 and 3 for s_i and s_i'.

As an example of using column shifts to simplify the drawing of best-response diagrams, we consider the 3×3 game (A, B) with

$$A = \begin{pmatrix} 0 & 3 & 0 \\ 1 & 0 & 1 \\ -3 & 4 & 5 \end{pmatrix}, \qquad B = \begin{pmatrix} 0 & 1 & -2 \\ 2 & 0 & 3 \\ 2 & 1 & 0 \end{pmatrix}. \qquad (9.12)$$

We add the constant 3 to the first and third column of A, and the constant 2 to the first row of B. This produces a new game $(\hat{A}, \hat{B})$ that is strategically equivalent to (A, B) by Proposition 9.7,

$$\hat{A} = \begin{pmatrix} 3 & 3 & 3 \\ 4 & 0 & 4 \\ 0 & 4 & 8 \end{pmatrix}, \qquad \hat{B} = \begin{pmatrix} 2 & 3 & 0 \\ 2 & 0 & 3 \\ 2 & 1 & 0 \end{pmatrix}. \qquad (9.13)$$

The game $(\hat{A}, \hat{B})$, and hence the game (A, B), has the best-response diagram shown in Figure 9.14. These diagrams can be relatively easily visualized via an upper envelope of three planes of expected payoffs, spanned across three goalposts on the corners of each triangle, because the plane for the first strategy has goalposts at equal heights and is therefore horizontal. This has been shown in Figure 9.10, which produces almost the same subdivision of X because the matrix B there is almost the same as $\hat{B}$.

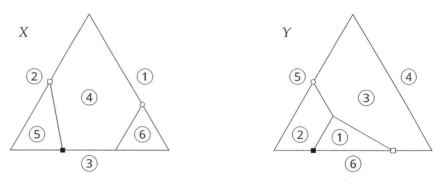

Figure 9.14 Best-response diagram for the games in (9.12) and (9.13). The games have the same three equilibria, namely $\left(\left(\frac{2}{3}, \frac{1}{3}, 0\right), \left(\frac{3}{4}, \frac{1}{4}, 0\right)\right)$ (black square), $\left(\left(\frac{1}{2}, 0, \frac{1}{2}\right), \left(\frac{1}{4}, \frac{3}{4}, 0\right)\right)$ (white square), and $\left(\left(0, \frac{2}{3}, \frac{1}{3}\right), \left(\frac{1}{2}, 0, \frac{1}{2}\right)\right)$ (white circle).

The subdivision of Y into its best-response regions $Y(1)$, $Y(2)$, $Y(3)$ can be obtained similarly to Figure 9.10 by considering an upper envelope of three planes of expected payoffs. We also describe another method that can be applied directly to the triangle Y, without considering a third dimension. For the original game (A, B), the best-response regions are defined by comparing the expected payoffs in each row:

$$
\begin{aligned}
Y(1) &= \{(y_1, y_2, y_3) \in Y \mid 3y_2 \geq y_1 + y_3, \\
&\qquad\qquad\qquad\qquad 3y_2 \geq -3y_1 + 4y_2 + 5y_3 \}, \\
Y(2) &= \{(y_1, y_2, y_3) \in Y \mid y_1 + y_3 \geq 3y_2, \\
&\qquad\qquad\qquad\qquad y_1 + y_3 \geq -3y_1 + 4y_2 + 5y_3 \}, \\
Y(3) &= \{(y_1, y_2, y_3) \in Y \mid -3y_1 + 4y_2 + 5y_3 \geq 3y_2, \\
&\qquad\qquad\qquad\qquad -3y_1 + 4y_2 + 5y_3 \geq y_1 + y_3 \}.
\end{aligned}
\tag{9.14}
$$

For the game $(\hat{A}, \hat{B})$, these conditions are

$$
\begin{aligned}
Y(1) &= \{(y_1, y_2, y_3) \in Y \mid 3y_1 + 3y_2 + 3y_3 \geq 4y_1 + 4y_3, \\
&\qquad\qquad\qquad\qquad 3y_1 + 3y_2 + 3y_3 \geq 4y_2 + 8y_3 \}, \\
Y(2) &= \{(y_1, y_2, y_3) \in Y \mid 4y_1 + 4y_3 \geq 3y_1 + 3y_2 + 3y_3, \\
&\qquad\qquad\qquad\qquad 4y_1 + 4y_3 \geq 4y_2 + 8y_3 \}, \\
Y(3) &= \{(y_1, y_2, y_3) \in Y \mid 4y_2 + 8y_3 \geq 3y_1 + 3y_2 + 3y_3, \\
&\qquad\qquad\qquad\qquad 4y_2 + 8y_3 \geq 4y_1 + 4y_3 \}.
\end{aligned}
\tag{9.15}
$$

In agreement with Proposition 9.7, these two sets of conditions are completely equivalent, because the term $3y_1 + 3y_3$ is added on both sides of each inequality;

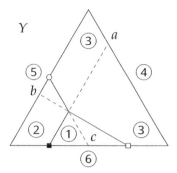

Figure 9.15 Construction of the subdivision of Y in the plane, without an upper envelope in three dimensions.

this corresponds to the column shift of adding the constant 3 to the first and third column of A to obtain $\hat{A}$.

We now consider the *indifferences* between each pair of rows, and write them as follows:

$$
\begin{array}{ll}
①\sim② : & 3y_2 = y_1 + y_3\,, \\
①\sim③ : & 3y_1 = y_2 + 5y_3\,, \\
②\sim③ : & 4y_2 + 4y_3 = 4y_1\,.
\end{array}
\qquad (9.16)
$$

The equations in (9.16) are obtained by converting an inequality in (9.14) or (9.15) into an equality and then *sorting* the coefficients of y_1, y_2, and y_3 so that each variable appears only on one side of the equation with a *positive* coefficient (or not at all, if the variable has the same coefficient on both sides, which is not the case here). This has already been done in (9.4) for the game (9.1), where it conveniently identified the common line segment of the two best-response regions $X(4)$ and $X(5)$.

In the present case, the three best-response regions $Y(1)$, $Y(2)$, and $Y(3)$ can be found by extending that approach, which is illustrated in Figure 9.15. The first equation in (9.16) shows where rows 1 and 2 have the same expected payoff. All three variables y_1, y_2, y_3 appear in this equation, with y_1 on the left-hand side and y_2 and y_3 on the right-hand side. Setting $y_3 = 0$ then gives the solution $y = (\frac{3}{4}, \frac{1}{4}, 0)$, which is shown with a black square in Figure 9.15, and setting $y_1 = 0$ gives another solution $y = (0, \frac{1}{4}, \frac{3}{4})$, shown as point a in the figure. The line segment that connects these two points separates $Y(1)$ and $Y(2)$. In order to decide on which side of this line segment the labels 1 and 2 should be drawn, it is useful to identify the best responses to the pure strategies of player II, which are row 2 against e_1 and row 3 against both e_2 and e_3, shown by the labels in the corner of the triangle.

Similarly, the "extreme" solutions to the second equation in (9.16) are $y = (\frac{1}{4}, \frac{3}{4}, 0)$ when $y_3 = 0$, shown as a white square in Figure 9.15, and $b = (\frac{5}{8}, 0, \frac{3}{8})$ when $y_2 = 0$. For the third equation in (9.16), they are $c = (\frac{1}{2}, \frac{1}{2}, 0)$ and $(\frac{1}{2}, 0, \frac{1}{2})$, shown as a white circle (which shows that $y_1 = \frac{1}{2}$ in any solution to this equation). The three best-response regions have a common point in the interior of Y, which can be computed (although its precise location is not relevant for the qualitative shape of the best-response diagram) as $(\frac{1}{2}, \frac{1}{4}, \frac{1}{4})$.

In order to identify the equilibria of the game (9.12), the best-response diagram in Figure 9.14 need not necessarily be constructed with full precision, only to show where the best-response regions meet in each triangle to identify the points that have three labels. The three equilibria are shown with symbols as before.

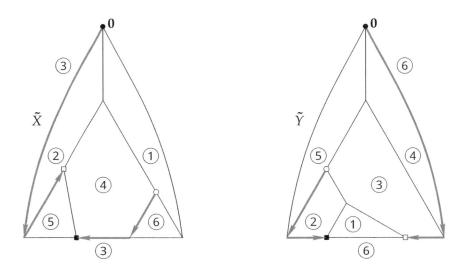

Figure 9.16   Lemke–Howson paths for the game (9.12), shown in (9.17) and (9.18), for missing label 1.

Figure 9.16 shows the Lemke–Howson paths for missing label 1 for the game (9.12) using the best-response diagram in Figure 9.14. As before, the graphs $\tilde{X}$ and $\tilde{Y}$ are created by adding the point $\mathbf{0}$ to X and Y, which is connected to the vertices of X and Y, respectively, by an edge. The paths are as follows, where we only identify the pairs (x, y) in $\tilde{X} \times \tilde{Y}$ by their sets of labels, and equilibria by their symbols, so as not to clutter the diagram:

(x, y)	labels	drop label	
•	123, 456	1 in $\tilde{X}$	
	235, 456	5 in $\tilde{Y}$	
	235, 356	3 in $\tilde{X}$	(9.17)
	245, 356	4 in $\tilde{Y}$	
□	245, 136	(found 1 in $\tilde{Y}$)	

and

$$
\begin{array}{c|cc|c}
(x,y) & \text{labels} & & \text{drop label} \\
\hline
\circ & 146, & 235 & 1 \text{ in } \tilde{X} \\
 & 346, & 235 & 3 \text{ in } \tilde{Y} \\
 & 346, & 256 & 6 \text{ in } \tilde{X} \\
 & 345, & 256 & 5 \text{ in } \tilde{Y} \\
\blacksquare & 345, & 126 & (\text{found } 1 \text{ in } \tilde{Y}).
\end{array}
\tag{9.18}
$$

It is a useful exercise to find the Lemke–Howson paths in Figure 9.16 for each missing label. For example, the equilibrium $((0, \frac{2}{3}, \frac{1}{3}), (\frac{1}{2}, 0, \frac{1}{2}))$ marked with a white circle is found from the artificial equilibrium $(\mathbf{0}, \mathbf{0})$ for missing label 3, so this is different from the equilibrium found for missing label 1. The LH path that starts at the white circle for missing label 1 again is shown in (9.18) and finds the third equilibrium. By considering all possible missing labels, one can construct a "network" of LH paths that can be traversed back and forth. However, this network may be disconnected, so that not all equilibria are found when starting from the artificial equilibrium; see Exercise 9.1.

9.6 Best-Response Polyhedra and Polytopes

As shown in the previous two sections, best-response diagrams help visualize the structure of equilibria in 3×3 games. They are also a good basis for general algorithms to find equilibria in larger bimatrix games. These algorithms, including the Lemke–Howson algorithm, are based on the upper envelope diagram, which has a general definition in terms of suitable "best-response polyhedra", as we explain in this section.

As mentioned earlier, a *hyperplane* H in $\mathbb{R}^d$ is a set $H = \{z \in \mathbb{R}^d \mid v^\top z = \alpha\}$ for a nonzero *normal vector* v in $\mathbb{R}^d$ and a real number α. The corresponding *halfspace* is defined as $\{z \in \mathbb{R}^d \mid v^\top z \le \alpha\}$. A *polyhedron* is the intersection of a finite number of halfspaces. If the polyhedron S is the intersection of n halfspaces $\{z \in \mathbb{R}^d \mid c_i^\top z \le q_i\}$ with $c_i \in \mathbb{R}^d$ and $q_i \in \mathbb{R}$ for $i = 1, \ldots, n$, then it can be written as

$$
S = \{z \in \mathbb{R}^d \mid Cz \le q\}
\tag{9.19}
$$

for the $n \times d$ matrix C and $q \in \mathbb{R}^n$ where $C^\top = [c_1 \cdots c_n]$, that is, $c_i^\top$ is the ith row of C.

If $z \in S$ and $c_i^\top z = q_i$, then we say that the ith inequality $c_i^\top z \le q_i$ in (9.19) is *tight* for z (that is, holds as an equality). A *face* F of S is obtained by requiring some of the n inequalities in (9.19) to be tight (as a result, F may be the empty set). That is, $F = \{z \in S \mid c_i^\top z = q_i \text{ for } i \in T\}$ for some set $T \subseteq \{1, \ldots, n\}$. There are 2^n ways of choosing T (some of which may define the same face F), so the number of faces is finite. If the face consists of a single point, this is called a *vertex* of the polyhedron. A bounded polyhedron is called a *polytope*. It can be shown that every polytope is the convex hull of its vertices, which agrees with our earlier definition of a polytope as the convex hull of a finite set of points in $\mathbb{R}^d$.

The mixed-strategy sets X and Y (6.6) are polytopes, because they are clearly bounded, and defined by a number of linear inequalities and one linear equation (any equation $v^\top z = \alpha$ can be written as two inequalities $v^\top z \le \alpha$ and $-v^\top z \le -\alpha$).

As a central concept of this section, we consider the following *best-response polyhedra* $\bar{P}$ for player I and $\bar{Q}$ for player II, for an $m \times n$ bimatrix game (A, B):

$$\begin{aligned} \bar{P} &= \{\, (x,v) \in X \times \mathbb{R} \mid B^\top x \le \mathbf{1}v \,\}, \\ \bar{Q} &= \{\, (y,u) \in Y \times \mathbb{R} \mid Ay \le \mathbf{1}u \,\}. \end{aligned} \tag{9.20}$$

For the game (9.1), the inequalities $B^\top x \le \mathbf{1}v$ that (in addition to $x \in X$) define $\bar{P}$ are

$$\begin{aligned} 2x_1 \quad\;\; + 4x_3 &\le v, \\ x_2 + \;\; x_3 &\le v, \end{aligned} \tag{9.21}$$

and the inequalities $Ay \le \mathbf{1}u$ that (in addition to $y \in Y$) define $\bar{Q}$ are

$$\begin{aligned} 2y_1 \qquad\;\; &\le u, \\ y_1 + 2y_2 &\le u, \\ 3y_2 &\le u. \end{aligned} \tag{9.22}$$

They are displayed in the upper envelope diagrams in Figures 9.2 and 9.1 where v and u are *at least* the payoff to player II and player I, respectively. (Because v and u have no upper bound, $\bar{P}$ and $\bar{Q}$ are polyhedra but not polytopes.) As soon as at least one of the inequalities in (9.21) and (9.22) is tight, v and u are *equal* to the best-response payoff because they cannot take a lower value. That is, for given $x \in X$ and $y \in Y$,

$$\begin{aligned} \min \{v \mid (x,v) \in \bar{P}\} &= \max \{\, (B^\top x)_j \mid 1 \le j \le n \,\}, \\ \min \{u \mid (y,u) \in \bar{Q}\} &= \max \{\, (Ay)_i \mid 1 \le i \le m \,\}, \end{aligned}$$

which is just the familiar definition of the upper envelope of expected payoffs. Assuming that at least one of the inequalities in $B^\top x \le \mathbf{1}v$ is tight (which defines v as the best-response payoff to x), the tight inequalities in $B^\top x \le \mathbf{1}v$ correspond to the pure best responses to x. Similarly, the tight inequalities in $Ay \le \mathbf{1}u$ (if at least one is tight) correspond to the pure best responses to y.

Hence, we can identify the *labels* that denote the pure strategies $1, \ldots, m + n$ of the players as in (9.2) with the tight inequalities in $\bar{P}$ and $\bar{Q}$. Let $(x,v) \in \bar{P}$ with v as best-response payoff to x, let $(y,u) \in \bar{Q}$ with u as best-response payoff to y, and $1 \le i \le m$ and $1 \le j \le n$. Then, in agreement with Definition 9.1,

$$\begin{aligned} x \text{ has label } i &\iff x_i = 0, \\ x \text{ has label } m + j &\iff (B^\top x)_j = v, \\ y \text{ has label } i &\iff (Ay)_i = u, \\ y \text{ has label } m + j &\iff y_j = 0. \end{aligned} \tag{9.23}$$

We can therefore rephrase Theorem 9.2 as follows.

Corollary 9.8. *Let $(x, v) \in \bar{P}$ and $(y, u) \in \bar{Q}$. Then (x, y) is an equilibrium of (A, B) with equilibrium payoffs u and v to players I and II if and only if (x, y) is completely labeled.*

In Definition 9.1, labels are defined in terms of the best-response regions $Y(i)$ and $X(j)$ in (9.3), and the zero faces of X and Y in (9.5). The best-response regions are polytopes, for example as in (9.14). While the best-response regions are convenient to draw in low dimension (in particular as subsets of a triangle), the definition as faces of the best-response polyhedra $\bar{P}$ and $\bar{Q}$ requires only a single tight inequality per label, by adding one more dimension for the payoff v or u to the other player.

We now simplify the best-response polyhedra further by eliminating these payoff variables, and replacing the mixed strategies x and y by arbitrary nonnegative vectors. Let

$$
\begin{aligned}
P &= \{\, x \in \mathbb{R}^m \mid \quad x \geq \mathbf{0}, \ B^\top x \leq \mathbf{1} \}, \\
Q &= \{\, y \in \mathbb{R}^n \mid A y \leq \mathbf{1}, \quad y \geq \mathbf{0} \,\}.
\end{aligned}
\tag{9.24}
$$

For the game (9.1), P is defined by the inequalities $x_1 \geq 0$, $x_2 \geq 0$, $x_3 \geq 0$, and

$$
\begin{aligned}
2x_1 \quad + 4x_3 &\leq 1, \\
x_2 + \ x_3 &\leq 1,
\end{aligned}
$$

so that P is the intersection of the "wedges" $x_3 \leq \frac{1}{4} - \frac{1}{2}x_1$ and $x_3 \leq 1 - x_2$ shown on the left in Figure 9.17. The constraints for Q are $y_1 \geq 0$, $y_2 \geq 0$, and

$$
\begin{aligned}
2y_1 \qquad &\leq 1, \\
y_1 + 2y_2 &\leq 1, \\
3y_2 &\leq 1,
\end{aligned}
\tag{9.25}
$$

which translate to $y_1 \leq \frac{1}{2}$, $y_2 \leq \frac{1}{2} - \frac{1}{2}y_1$, $y_2 \leq \frac{1}{3}$, and define a polygon in $\mathbb{R}^2$.

In general, we want P and Q to be polytopes, which is equivalent to $v > 0$ and $u > 0$ for any $(x, v) \in \bar{P}$ and $(y, u) \in \bar{Q}$, according to the following lemma.

Lemma 9.9. *Consider an $m \times n$ bimatrix game (A, B). Then P in (9.24) is a polytope if and only if the best-response payoff to any x in X is always positive, and Q in (9.24) is a polytope if and only if the best-response payoff to any y in Y is always positive.*

Proof. We prove the statement for Q; the proof for P is analogous. The best-response payoff to any mixed strategy y is the maximum entry of Ay, so this is not always positive if and only if $Ay \leq \mathbf{0}$ for some $y \in Y$. For such a y we have $y \geq \mathbf{0}$, $y \neq \mathbf{0}$, and $y\alpha \in Q$ for any $\alpha \geq 0$, which shows that Q is not bounded. Conversely, suppose the best-response payoff u to any y is always positive. Because

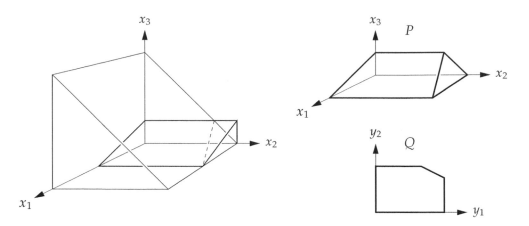

Figure 9.17 The polytopes P and Q for the example (9.12).

Y is compact and $\bar{Q}$ is closed, the minimum u' of $\{u \mid (y, u) \in \bar{Q}\}$ exists, $u' > 0$, and $u \ge u'$ for all (y, u) in $\bar{Q}$. Then the map

$$\bar{Q} \to Q - \{\mathbf{0}\}, \qquad (y, u) \mapsto y\tfrac{1}{u} \tag{9.26}$$

is a bijection with inverse $z \mapsto (z\tfrac{1}{\mathbf{1}^\top z}, \tfrac{1}{\mathbf{1}^\top z})$ for $z \in Q - \{\mathbf{0}\}$. Here, $\tfrac{1}{\mathbf{1}^\top z} \ge u'$ and thus $\mathbf{1}^\top z \le \tfrac{1}{u'}$, where $\mathbf{1}^\top z = \sum_{j=1}^{n} |z_j|$ because $z \ge \mathbf{0}$, which proves that Q is bounded and therefore a polytope.   □

As a sufficient condition that $v > 0$ and $u > 0$ hold for any (x, v) in $\bar{P}$ and (y, u) in $\bar{Q}$, we assume that

$$A \text{ and } B^\top \text{ are nonnegative and have no zero column} \tag{9.27}$$

(because then $B^\top x$ and Ay are nonnegative and nonzero for any $x \in X$ and $y \in Y$). We could simply assume $A > 0$ and $B > 0$, but it is useful to admit zero entries in a payoff matrix, like in our examples. Note that condition (9.27) is not necessary for positive best-response payoffs (which is still the case, for example, if the lower left entry of A in (9.1) is negative, as Figure 9.1 shows; see also Exercise 9.5). By adding a suitable positive constant to all payoffs of a player, which preserves the preferences of that player (and where the original payoff is recovered by simply substracting the constant), we can assume (9.27) without loss of generality.

With positive best-response payoffs, the polytope P (apart from its vertex $\mathbf{0}$) is obtained from $\bar{P}$ by dividing each inequality $\sum_{i=1}^{m} b_{ij} x_i \le v$ by v, which gives $\sum_{i=1}^{m} b_{ij}(x_i/v) \le 1$, and then treating x_i/v as a new variable that is again called x_i in P. Similarly, $Q - \{\mathbf{0}\}$ is obtained from $\bar{Q}$ by dividing each inequality in $Ay \le \mathbf{1}u$ by u. In effect, we have normalized the expected payoffs to be 1, and dropped the conditions $\mathbf{1}^\top x = 1$ and $\mathbf{1}^\top y = 1$. Conversely, nonzero vectors $x \in P$ and $y \in Q$ are multiplied by $v = \tfrac{1}{\mathbf{1}^\top x}$ and $u = \tfrac{1}{\mathbf{1}^\top y}$ to turn them into probability vectors. The scaling factors v and u are the expected payoffs to the other player.

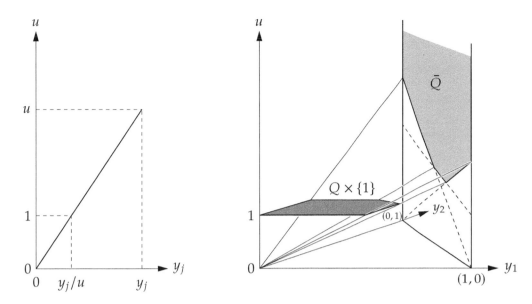

Figure 9.18 The projective transformation in (9.26), shown on the left for a single component y_j of y, on the right for $\bar{Q}$ as in Figure 9.1 and Q in (9.25).

Figure 9.18 shows a geometric interpretation of this division by the expected payoff, that is, the bijection (9.26), as a so-called "projective transformation". The left diagram shows the pair (y_j, u) as part of (y, u) in $\bar{Q}$ for any component y_j of y. The line that connects this pair to $(0, 0)$ contains the point $(y_j/u, 1)$. Hence, $(Q - \{\mathbf{0}\}) \times \{1\}$ is the intersection of the lines that connect any (y, u) in $\bar{Q}$ with $(\mathbf{0}, 0)$ in $\mathbb{R}^n \times \mathbb{R}$ with the "horizontal" hyperplane $\mathbb{R}^n \times \{1\}$. The vertex $\mathbf{0}$ of Q does not arise as such a projection, but corresponds to $\bar{Q}$ "at infinity".

The maps $(x, v) \mapsto x\frac{1}{v}$ from $\bar{P}$ to P and $(y, u) \mapsto y\frac{1}{u}$ from $\bar{Q}$ to Q preserve inequalities and when they are *tight*. Hence, we can label the inequalities in P and Q with the players' pure strategies $1, \ldots, m + n$, which is why we have written them in the order $x \geq \mathbf{0}$, $B^\top x \leq \mathbf{1}v$ and $Ay \leq \mathbf{1}v$, $y \geq \mathbf{0}$ in (9.24). Consequently, we obtain one more time the characterization of equilibria as completely labeled pairs of points, this time in $P \times Q - \{(\mathbf{0}, \mathbf{0})\}$.

Proposition 9.10. *Let (A, B) be an $m \times n$ bimatrix game so that (9.27) holds, and consider P and Q in (9.24) with their $m + n$ inequalities labeled $1, \ldots, m + n$, so that any $x \in P$ and $y \in Q$ have the labels of the tight inequalities. Then $(\bar{x}, \bar{y})$ is an equilibrium of (A, B) with equilibrium payoffs u and v to players I and II if and only if $(x, y) \in P \times Q - \{(\mathbf{0}, \mathbf{0})\}$ and (x, y) is completely labeled, $x = \bar{x}\frac{1}{v}$, $y = \bar{y}\frac{1}{u}$, $u = \frac{1}{\mathbf{1}^\top y}$, and $v = \frac{1}{\mathbf{1}^\top x}$.*

Proof. Condition (9.27) implies $u > 0$ and $v > 0$ for all $(\bar{x}, v) \in \bar{P}$ and $(\bar{y}, u) \in \bar{Q}$, so that the maps $(\bar{x}, v) \mapsto x = \bar{x}\frac{1}{v} \in P$ and $(\bar{y}, u) \mapsto y = \bar{y}\frac{1}{u} \in Q$ are well defined, and they preserve inequalities and when these inequalities are *tight*, so that the labels of $\bar{x}$ are the labels of x and the labels of $\bar{y}$ are the labels of y. Hence, if $(\bar{x}, \bar{y})$ is an equilibrium, then $(\bar{x}, \bar{y})$ and therefore (x, y) is completely labeled by Corollary 9.8.

Conversely, suppose $(x, y) \in P \times Q$ is completely labeled and $(x, y) \neq (\mathbf{0}, \mathbf{0})$. If $x = \mathbf{0}$ then $y \neq \mathbf{0}$, so that $y_j \neq 0$ for some $j \in \{1, \ldots, n\}$ and y does not have label $m + j$; this means that (x, y) is not completely labeled because x has only the labels $1, \ldots, m$, a contradiction. Hence, $x \neq \mathbf{0}$, and similarly $y \neq \mathbf{0}$. By the same reasoning, x has some label $m + j$, which by (9.23) means that at least one inequality in $B^\top x \leq \mathbf{1}$ is tight, so that $v = \frac{1}{\mathbf{1}^\top x}$ is the best-response payoff to $\bar{x} = xv$ for $(\bar{x}, v) \in \bar{P}$. Similarly, y in P has some label $i \in \{1, \ldots, m\}$, which means $(Ay)_i = 1$, so that $u = \frac{1}{\mathbf{1}^\top y}$ is the best-response payoff to $\bar{y} = yu$ for $(\bar{y}, u) \in \bar{Q}$. This proves the claim.   □

The last paragraph in the preceding proof shows that we do not need to worry about the possibilities $B^\top x < \mathbf{1}$ or $Ay < \mathbf{1}$ in P or Q (that is, no tight inequalities) when considering completely labeled pairs (x, y) in $P \times Q$. There is one obvious exception to this, namely the pair $(\mathbf{0}, \mathbf{0})$ in $P \times Q$, which takes the role of the artificial equilibrium for the Lemke–Howson algorithm in a natural way; it does not have to be added like in $\tilde{X} \times \tilde{Y}$. For our example (9.1), Figure 9.19 shows the Lemke–Howson paths on $P \times Q$ for missing label 2. They traverse the almost completely labeled *vertex pairs* of P and Q, which we identify by their sets of labels exactly as in (9.7) and (9.8).

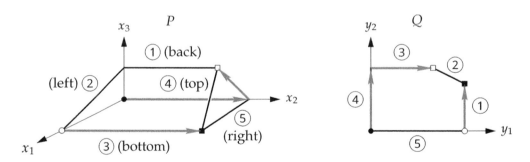

Figure 9.19 The Lemke–Howson paths from Figure 9.5 for missing label 2 on the polytopes P and Q.

9.7 Complementary Pivoting

We now describe how the Lemke–Howson algorithm works algebraically and in any dimension. As always, we consider an $m \times n$ bimatrix game (A, B). Let C be the $d \times d$ matrix defined by

$$d = m + n, \qquad C = \begin{pmatrix} 0 & A \\ B^\top & 0 \end{pmatrix}. \tag{9.28}$$

In agreement with (9.27) we assume, without loss of generality, that C is nonnegative and has no zero column. Then $S = P \times Q$ is a polytope, which can be written as

$$S = \{z \in \mathbb{R}^d \mid z \geq \mathbf{0},\ Cz \leq \mathbf{1}\} \tag{9.29}$$

where $z \in S = P \times Q$ means

$$z = (x, y), \quad z_i = x_i \ (1 \leq i \leq m), \quad z_{m+j} = y_j \ (1 \leq j \leq n). \tag{9.30}$$

We label both the d inequalities $z \geq \mathbf{0}$ and the d inequalities $Cz \leq \mathbf{1}$ in (9.29) with labels $1, \ldots, d$. As usual, any z in S gets the labels of the tight inequalities. The d labels correspond to the players' pure strategies as in (9.2) and (9.23). An equilibrium $(\bar{x}, \bar{y})$ of (A, B) then corresponds to a completely labeled point $z = (x, y)$ of $S - \{\mathbf{0}\}$ (that is, z has all labels $1, \ldots, d$), via the normalizations $\bar{x} = x\frac{1}{\mathbf{1}^\top x}$ and $\bar{y} = y\frac{1}{\mathbf{1}^\top y}$ in Proposition 9.10.

For an algebraic treatment, we write the inequalities $Cz \leq \mathbf{1}$ in (9.29) with nonnegative "slack variables" $w \in \mathbb{R}^d$ as the equality and nonnegativity constraints $Cz + w = \mathbf{1}, w \geq \mathbf{0}$. Then $z \in S$ has label $i \in \{1, \ldots, d\}$ if and only if $z_i = 0$ or $w_i = 0$, which can be written as $z_i w_i = 0$. If this holds for all $i = 1, \ldots, d$, then the vectors z and w are called *complementary*. Then the problem of finding a completely labeled point z of S means to find z and w in $\mathbb{R}^d$ so that

$$Cz + w = \mathbf{1}, \quad z \geq \mathbf{0}, \quad w \geq \mathbf{0}, \quad z_i w_i = 0 \ (1 \leq i \leq d). \tag{9.31}$$

In summary, if $z \in S$ then w is defined by $w = \mathbf{1} - Cz$, we have $z \geq \mathbf{0}$ and $w \geq \mathbf{0}$, and z is completely labeled if and only if z and w are complementary.

Recall that z has label i if $z_i = 0$ or $w_i = 0$. A *missing label* therefore means some $\ell \in \{1, \ldots, d\}$ so that $z_\ell > 0$ and $w_\ell > 0$ are allowed to hold. If with this exception all other conditions in (9.31) hold, then z and w are called *almost complementary*, which means that z is an *almost completely labeled* point of S, that is, z has all labels except possibly ℓ.

The Lemke–Howson algorithm starts with the trivial solution $z = \mathbf{0}, w = \mathbf{1}$ to (9.31), which defines the artificial equilibrium. For our example (9.1), the system $w = \mathbf{1} - Cz$ states

$$
\begin{aligned}
w_1 &= 1 & & & & -2z_4 \\
w_2 &= 1 & & & & -z_4 - 2z_5 \\
w_3 &= 1 & & & & -3z_5 \\
w_4 &= 1 - 2z_1 & & -4z_3 \\
w_5 &= 1 & & -z_2 - z_3 \, .
\end{aligned}
\tag{9.32}
$$

This is called a *dictionary*, which means that certain variables, here $w_1, \ldots, w_5$, are expressed in dependence of the others, here $z_1, \ldots, z_5$. The dependent variables on the left are called *basic variables* and the independent variables on the right are called *nonbasic variables*. The corresponding *basic solution* is obtained by setting all nonbasic variables to zero, so that the basic variables are equal to the constants in the dictionary. If these constants are all nonnegative, then the basic solution is called *feasible*, which is the case in (9.32) where the basic solution is $z = \mathbf{0}, w = \mathbf{1}$.

The algorithm proceeds by the operation of *pivoting* that exchanges a nonbasic variable with a basic variable in such a way that the new dictionary defines the same linear relationship between its variables, and so that the corresponding basic solution remains feasible. In geometric terms, each basic feasible solution defines a vertex of the polytope S, and pivoting means moving along an edge of S from one vertex to an adjacent vertex, as in Figure 9.19, as we will show.

A fixed missing label $\ell \in \{1, \ldots, d\}$ is chosen, here $\ell = 2$, which determines the *entering variable* z_ℓ. In the current basic solution, all nonbasic variables are set to zero, but now the entering variable, which in (9.32) is z_2, is allowed to assume positive values. The dictionary (9.32) describes how this changes the basic variables on the left, which is here only the variable w_5 because z_2 appears only in the last equation $w_5 = 1 - z_2 - z_3$. The constraint $w_5 \geq 0$ limits the increase of z_2, because w_5 becomes zero when $z_2 = 1$. This determines w_5 as the *leaving variable*.

The pivoting step is now to let z_2 *enter the basis* and w_5 *leave the basis*, which means rewriting the dictionary so that z_2 becomes basic and w_5 nonbasic. Here, this amounts to a simple exchange of w_5 with z_2, which defines the new dictionary

$$
\begin{aligned}
w_1 &= 1 & & & - 2z_4 \\
w_2 &= 1 & & & - z_4 - 2z_5 \\
w_3 &= 1 & & & - 3z_5 \\
w_4 &= 1 - 2z_1 & & - 4z_3 \\
z_2 &= 1 & - w_5 & - z_3 \, .
\end{aligned}
\tag{9.33}
$$

In the basic *solution* associated with this new dictionary, $z_2 = 1$ and $w_5 = 0$. In Figure 9.19 this is the endpoint of first arrow in P at the vertex $x = (0, 1, 0)$, where label 2 has been dropped and label 5 is picked up (because w_5 has become nonbasic). In Q, we still have $(y_1, y_2) = (z_4, z_5) = (0, 0)$.

The dictionary (9.33) has two nonbasic variables z_5 and w_5 with the same subscript 5, which is the duplicate label. The *complementary pivoting* rule now states that because w_5 has just left the basis, its complement z_5 is chosen as the next entering variable.

That is, starting from the basic solution associated with (9.33), z_5 is allowed to become positive, with all other nonbasic variables staying at zero. Restricted to this increase of z_5, the affected variables on the left-hand side are $w_2 = 1 - 2z_5$ and $w_3 = 1 - 3z_5$, both of which should remain nonnegative. The stronger of these constraints is $w_3 = 1 - 3z_5 \geq 0$, where w_3 becomes zero when $z_5 = \frac{1}{3}$ (while w_2 is still positive). Hence, w_3 is the leaving variable. The pivoting step means now to rewrite the third equation (which currently has w_3 on the left) as $z_5 = \frac{1}{3} - \frac{1}{3}w_3$ and to *substitute* this expression for z_5 where it appears elsewhere on the right. The only substitution is here $w_2 = 1 - z_4 - 2z_5 = 1 - z_4 - 2(\frac{1}{3} - \frac{1}{3}w_3)$, so that the resulting new dictionary is

$$
\begin{aligned}
w_1 &= 1 && - 2z_4 \\
w_2 &= \tfrac{1}{3} && - z_4 + \tfrac{2}{3}w_3 \\
z_5 &= \tfrac{1}{3} && - \tfrac{1}{3}w_3 \\
w_4 &= 1 - 2z_1 && - 4z_3 \\
z_2 &= 1 && - w_5 - z_3 \, .
\end{aligned}
\tag{9.34}
$$

In the preceding pivoting step, z_5 entered and w_3 left the basis. In Figure 9.19, label 5 was dropped and label 3 picked up in Q, with the new basic solution at the vertex $y = (0, \frac{1}{3})$ of Q.

By the complementary pivoting rule, the next entering variable in (9.34) is z_3 as the complement of w_3. The increase of z_3 creates the two constraints $w_4 = 1 - 4z_3 \geq 0$ and $z_2 = 1 - z_3 \geq 0$, where the former is stronger, with w_4 becoming zero when $z_4 = \frac{1}{4}$. Hence, the leaving variable is w_4, so that we rewrite the fourth equation as $z_3 = \frac{1}{4} - \frac{1}{2}z_1 - \frac{1}{4}w_4$ and substitute this expression for z_3:

$$
\begin{aligned}
w_1 &= 1 && - 2z_4 \\
w_2 &= \tfrac{1}{3} && - z_4 + \tfrac{2}{3}w_3 \\
z_5 &= \tfrac{1}{3} && - \tfrac{1}{3}w_3 \\
z_3 &= \tfrac{1}{4} - \tfrac{1}{2}z_1 && - \tfrac{1}{4}w_4 \\
z_2 &= \tfrac{3}{4} + \tfrac{1}{2}z_1 - w_5 && + \tfrac{1}{4}w_4 \, .
\end{aligned}
\tag{9.35}
$$

In the preceding pivoting step, z_3 entered and w_4 left the basis. This means that in P, label 3 was dropped and label 4 is picked up, and x moves to the new vertex $(0, \frac{3}{4}, \frac{1}{4})$ of P, which is already the white square in P in Figure 9.19.

As the complement of w_4 that has left the basis, the new entering variable in (9.35) is z_4. Of the two constraints $w_1 = 1 - 2z_4 \geq 0$ and $w_2 = \frac{1}{3} - z_4 \geq 0$, the second is stronger, where w_2 becomes zero when $z_4 = \frac{1}{3}$. We rewrite the second equation as $z_4 = \frac{1}{3} - w_2 + \frac{2}{3}w_3$ and substitute this expression for z_4:

$$
\begin{aligned}
w_1 &= \tfrac{1}{3} && + 2w_2 - \tfrac{4}{3}w_3 \\
z_4 &= \tfrac{1}{3} && - w_2 + \tfrac{2}{3}w_3 \\
z_5 &= \tfrac{1}{3} && - \tfrac{1}{3}w_3 \\
z_3 &= \tfrac{1}{4} - \tfrac{1}{2}z_1 && - \tfrac{1}{4}w_4 \\
z_2 &= \tfrac{3}{4} + \tfrac{1}{2}z_1 - w_5 && + \tfrac{1}{4}w_4 \, .
\end{aligned}
\tag{9.36}
$$

Because z_4 entered and w_2 left the basis, this means that in Q label 4 was dropped and label 2 picked up, with the new vertex $y = (\frac{1}{3}, \frac{1}{3})$ of Q, the white square in Q in Figure 9.19.

The algorithm *terminates* because the leaving variable has the missing label $\ell = 2$ as a subscript, so that there is now only one basic variable with that subscript and not two as in all the intermediate steps. The basic solution associated with the final dictionary (9.36) solves the system (9.31). Here, this solution is $z = (x, y)$ with $x = (0, \frac{3}{4}, \frac{1}{4})$ and $y = (\frac{1}{3}, \frac{1}{3})$. It defines the equilibrium $(\bar{x}, \bar{y})$ with $\bar{x} = x$ and payoff 1 to player II, and $\bar{y} = (\frac{1}{2}, \frac{1}{2})$ with payoff $\frac{3}{2}$ to player I.

In the rest of this section we describe complementary pivoting in general. For the moment, we assume that the polytope S in (9.29) is nondegenerate, which means that no point in S has more than d tight inequalities; this can be seen to be equivalent to the nondegeneracy of the game (A, B).

With the $d \times d$ identity matrix I, we write the equations in (9.31) as $Cz + Iw = \mathbf{1}$. This system has as variables the $2d$ components of z and w, the matrix $[C\ I]$ of coefficients, and a right-hand side $\mathbf{1}$. Any d linearly independent columns of the matrix $[C\ I]$ define an invertible matrix D, called the *basis matrix*. The corresponding *basic columns* have associated variables z_i or w_i called basic variables; the other columns and variables are called nonbasic.

Multiplying the system $Cz + Iw = \mathbf{1}$ with the inverse of the basis matrix D defines the equivalent system

$$D^{-1}Cz + D^{-1}w = D^{-1}\mathbf{1}. \tag{9.37}$$

This new system is a dictionary as in (9.32)–(9.36) above, except that we have kept both basic and nonbasic variables on the left-hand side of the equation. The reason is that in the system (9.37) the columns of the basic variables define the identity matrix, so that (9.37) shows how the basic variables depend on the nonbasic variables. When the nonbasic variables are set to zero, the basic variables are equal to $D^{-1}\mathbf{1}$, which (together with the zero nonbasic variables) defines the *basic solution* that corresponds to the system (9.37). This basic solution z, w is called *feasible* if it is nonnegative, that is, $D^{-1}\mathbf{1} \geq \mathbf{0}$. By nondegeneracy, this means $D^{-1}\mathbf{1} > \mathbf{0}$, that is, all basic variables have positive values.

Pivoting means to choose a nonbasic variable to enter the basis and a basic variable to leave the basis. Suppose the entering variable is z_j (the process is exactly the same for some w_j as entering variable). We assume that the current basic solution is feasible, and want to maintain a feasible solution while increasing z_j from zero. This increase of z_j changes the basic variables whose current value is equal to the vector $D^{-1}\mathbf{1}$, which we denote by q. Suppose the vector v is the nonbasic column in (9.37) that corresponds to z_j. With the increase of z_j, the vector of basic variables is given by $q - vz_j$ and should stay nonnegative, which means that $q_i - v_i z_j \geq 0$ for all rows $i = 1, \ldots, d$. Because $q \geq \mathbf{0}$ holds already, we only need to consider rows i so that $v_i > 0$. If v had no positive entry, we could increase z_j indefinitely, which would define an unbounded "ray" in S, but

S is bounded, so this cannot happen (complementary pivoting is also applied to general matrices C, and then this "ray termination" of the algorithm can happen).

Hence, the entering column v has at least one positive component v_i, and the increase of z_j is limited by the condition $q - vz_j \geq \mathbf{0}$, that is, $q_i \geq v_i z_j$ for all i such that $v_i > 0$, or equivalently

$$z_j \leq \min\{\frac{q_i}{v_i} \mid i = 1, \ldots, d, \ v_i > 0\}. \qquad (9.38)$$

Exactly when z_j is *equal* to the minimum in (9.38), one of the inequalities $q_i - v_i z_j \geq 0$ is tight. Such a row i determines the *leaving variable*. Finding the minimum in (9.38) is also called the *minimum ratio test*.

Nondegeneracy implies that the minimum in (9.38) and hence the leaving variable is *unique*. The reason is that if the minimum is taken for two (or more) rows, then only one of the corresponding two basic variables can be chosen to leave the basis, but the other will stay basic and after the pivoting step will have value zero, which means that the new basis is degenerate; we give an example in the next section.

The column of the system (9.37) for the entering variable z_j, and the row of the leaving variable, determine the *pivot element* v_i which by (9.38) is positive. The *pivoting* operation replaces the leaving basic variable by the entering variable, which means to replace the column of D for the leaving variable with the entering column. Rather than computing from scratch the inverse D^{-1} of the new basic matrix, it suffices to manipulate the system (9.37) with row operations that turn the entering column into the unit vector e_i. In these row operations, the pivot row is divided by the pivot element v_i, and suitable multiples of the pivot row are subtracted from the other rows. In our examples (9.32)–(9.35) above, these row operations reflect the algebraic substitutions to get from one dictionary to the next.

Pivoting, as described so far, is a general method of traversing the edges of a polyhedron to get from one vertex to another, where each vertex corresponds to a basic feasible solution. In linear programming, which is the maximization of a linear function over a polyhedron, pivoting is the main operation of the very important simplex algorithm.

What is special about *complementary* pivoting is the initialization and the choice of the next entering variable, with the goal to solve the system (9.31), that is, to find a complementary solution z, w. An initial complementary solution is $z = \mathbf{0}$ and $w = \mathbf{1}$. The freedom to move along a path is provided by relaxing the complementarity condition to allow for a single subscript ℓ (for the missing label) so that both z_ℓ and w_ℓ can take positive values and thus both be basic variables.

At the initial complementary solution $z = \mathbf{0}$ and $w = \mathbf{1}$, the first entering variable is z_ℓ. Once z_ℓ has entered the basis, a variable w_i has become nonbasic, which is now nonbasic alongside z_i. The complementary pivoting rule is to take the complement of the leaving variable as the next entering variable; the complement of w_i is z_i and vice versa (the same variable may enter and leave the basis multiple times, so it is also possible that some z_i leaves the basis and afterwards w_i enters the basis). The missing label ℓ provides a unique start of the algorithm, and subsequently the unique leaving variable and the complementary pivoting rule imply a unique continuation at each step. Hence, the algorithm progresses in a unique manner, and no almost complementary basis can be revisited. The algorithm terminates when either w_ℓ or z_ℓ leaves the basis, and then reaches a solution to (9.31).

Geometrically, this describes a path of edges of the polytope S. The points on that path have all labels in $\{1, \ldots, d\} - \{\ell\}$, and the end nodes of the path are completely labeled. With C as in (9.28), this is the LH path for missing label ℓ on $S = P \times Q$, as shown for our example in Figure 9.19.

9.8 Degeneracy Resolution

The Lemke–Howson algorithm needs, and has, an extension when the game is degenerate and the leaving variable is not always unique. As an example, consider the modification of our main example (9.1)

$$A = \begin{pmatrix} 2 & 0 \\ 1 & 2 \\ 0 & 3 \end{pmatrix}, \qquad B = \begin{pmatrix} 2 & 0 \\ 1 & 1 \\ 4 & 1 \end{pmatrix}. \tag{9.39}$$

where the second row of B has both columns as best responses, and the game is therefore degenerate. The last two rows of the initial dictionary (9.32) then need to be changed to

$$
\begin{aligned}
w_4 &= 1 - 2z_1 - z_2 - 4z_3 \\
w_5 &= 1 \qquad\quad - z_2 - z_3 .
\end{aligned}
\tag{9.40}
$$

With the missing label $\ell = 2$, the first entering variable z_2 means both w_4 and w_5 become simultaneously zero when $z_2 = 1$. Suppose that, as before, we choose w_5 as the leaving variable. Then (9.40) is rewritten as

$$
\begin{aligned}
w_4 &= 0 - 2z_1 + w_5 - 3z_3 \\
z_2 &= 1 \qquad\quad - w_5 - z_3 ,
\end{aligned}
\tag{9.41}
$$

which replaces the last two rows in (9.33). The next entering variable is z_5, with w_3 leaving, which produces the dictionary (9.34) except for the replacement of the last two rows by (9.41). The next entering variable is z_3 in (9.41), where the leaving variable is necessarily w_4 because z_3 cannot be increased at all while keeping w_4 nonnegative. The new (partial) dictionary obtained from (9.41) is

$$
\begin{aligned}
z_3 &= 0 - \tfrac{2}{3}z_1 + \tfrac{1}{3}w_5 - \tfrac{1}{3}w_4 \\
z_2 &= 1 + \tfrac{2}{3}z_1 - \tfrac{4}{3}w_5 + \tfrac{1}{3}w_4 ,
\end{aligned}
\tag{9.42}
$$

which represents the same basic feasible solution where $z_3 = 0$ and $w_4 = 0$ and hence the same vertex $(0, 1, 0)$ of P as the dictionary (9.41), except that the basis has changed. The only difference is that in (9.41) w_4 is basic and z_3 nonbasic, and vice versa in (9.42).

The equations (9.42) replace the last two rows in (9.35), where the new entering variable is z_4 as the complement of the variable w_4 that has left the basis. This leads to the same final pivoting step as in (9.35) with w_2 leaving, and the final dictionary composed of the first three rows of (9.36) and the two equations in (9.42). The resulting equilibrium $(\bar{x}, \bar{y})$ of the game (9.39) is $\bar{x} = (0, 1, 0)$ and $\bar{y} = (\frac{1}{2}, \frac{1}{2})$.

Here, the algorithm has still succeeded in finding an equilibrium of the game. However, we cannot guarantee any more that it terminates in general. Because of more than one possible leaving variable, the castle in Figure 9.6 now has rooms with more than two doors, and so it is well possible to reenter such a room through its third door, which could result in an endless loop.

This problem occurs when there are degenerate bases where the corresponding basic feasible solution has basic variables with value zero. The solution to this problem is to make sure, by a *perturbation* of the right-hand side $\mathbf{1}$ of the system $Cz + Iw = \mathbf{1}$, that basic variables in a basic feasible solution are always positive. The perturbation can be made vanishingly small so that it is in fact fictitious, and only induces a certain deterministic rule for the leaving variable.

Consider a new parameter $\varepsilon > 0$, which will be taken to be arbitrarily small. The modified system is $Cz + Iw = \mathbf{1} + (\varepsilon, \varepsilon^2, \ldots, \varepsilon^d)^\top$ and consequently the system (9.37) for a basis matrix D is

$$D^{-1}Cz + D^{-1}w = D^{-1}\mathbf{1} + D^{-1}(\varepsilon, \varepsilon^2, \ldots, \varepsilon^d)^\top. \qquad (9.43)$$

If $\varepsilon = 0$ this is just the original system (9.37), where the corresponding basic solution is feasible if $D^{-1}\mathbf{1} \geq 0$. We now want to strengthen this condition so that the basic solution is also positive when ε takes sufficiently small positive values. This holds trivially when $D^{-1}\mathbf{1} > 0$, but for a degenerate basis this vector may have zero components. Denote the ith row of the matrix $[D^{-1}\mathbf{1} \; D^{-1}]$ by $r_i = (q_{i0}, q_{i1}, \ldots, q_{id})$. Then the right-hand side in (9.43) is positive for all sufficiently small positive ε if and only if the *first nonzero entry* of r_i is positive for all $i = 1, \ldots, d$, because the ith row in (9.43) is $q_{i0} + q_{i1}\varepsilon + q_{i1}\varepsilon^2 + \cdots + q_{id}\varepsilon^d$, and the first nonzero summand, say $q_{ij}\varepsilon^j$, outweighs for small ε all the remaining summands which contain higher powers of ε. Note that r_i cannot be the all-zero row vector because its last d components are a row of D^{-1}. The matrix $[D^{-1}\mathbf{1} \; D^{-1}]$ is called *lexico-positive* if the first nonzero component of each of its rows r_i is positive. This is clearly the case initially when the basic variables are w with $D = I$, and will be maintained as a condition by the algorithm; it means that the perturbed system (9.43) is never degenerate for sufficiently small positive ε.

Consequently, the leaving variable in any pivoting step is always unique, and determined by the following extension of the minimum ratio test (9.38). Assume the entering variable is z_j (which could also be, completely equivalently, w_j) with entering column v, which has at least one positive component v_i. In the perturbed system (9.43), the constraints for the increase of z_j are now

$$z_j \le \frac{q_{i0}}{v_i} + \frac{q_{i1}}{v_i}\varepsilon + \frac{q_{i2}}{v_i}\varepsilon^2 + \cdots + \frac{q_{id}}{v_i}\varepsilon^d \tag{9.44}$$

for all rows i where $v_i > 0$. The coefficients of $1, \varepsilon, \varepsilon^2, \dots, \varepsilon^d$ in (9.44) are given by the row vector $\frac{1}{v_i} r_i$. We consider the *lexicographically smallest* of these row vectors, that is, the one with the smallest first component where any of these components differ; for example, among the rows

$$(0, 0, 2, -1), \qquad (0, 0, 1, 1000), \qquad (0, 3, -100, -8),$$

the second row is the lexicographically smallest. The lexicographically smallest row is unique, because if two rows were identical, say $\frac{1}{v_i} r_i$ and $\frac{1}{v_k} r_k$, this would imply two linearly dependent rows of D^{-1}. Choosing the lexico-smallest row $\frac{1}{v_i} r_i$ in (9.44) is called the *lexico-minimum ratio test*. Because this row is lexico-positive, this means that z_j, when it enters the basis in the perturbed system, assumes a (possibly very small) positive value in the new basic solution. Furthermore, after the pivoting step, all other basic variables stay positive in the perturbed system, and the new basis is again lexico-positive.

In short, the lexico-minimum ratio test determines the next leaving variable uniquely, and the perturbed system is always nondegenerate. What is nice is that the lexico-minimum ratio test involves only a comparison of $\frac{1}{v_i} r_i$ for the rows r_i of the matrix $[D^{-1}\mathbf{1}\ \ D^{-1}]$ where $v_i > 0$, and there is no actual perturbation at all, which is merely mimicked as if ε was positive but vanishingly small. Moreover, all the information about the rows r_i is already provided by the original dictionary (9.37) where the matrix D^{-1} is given by the columns for the variables w, so no additional information needs to be stored.

In summary, complementary pivoting, extended by lexico-minimum ratio test, proves the existence of an equilibrium in a bimatrix game, even if that game is degenerate. In this way, the Lemke–Howson algorithm provides an alternative and constructive equilibrium existence proof, independently of Brouwer's fixed point theorem.

The Lemke–Howson algorithm finds one equilibrium of a bimatrix game (A, B) but not necessarily all equilibria. The polytopes P and Q are also useful for finding *all equilibria*, with a different algorithm. Basically, such an algorithm computes all nonzero *vertices* x of P and y of Q, and outputs those pairs (x, y) that are completely labeled. If the game is degenerate, then some vertices x may have more than m labels, or some vertices y may have more than n labels. Then fewer labels are required for an equilibrium in the other polytope (as in the example in Figure 9.12), which may define a higher-dimensional *face* of P or Q (or both), all whose elements are equilibrium strategies. That face has a finite description either by the set of its vertices (as already computed in this process), or by its set of tight inequalities. For details of such algorithms see Avis, Rosenberg, Savani, and von Stengel (2010).

9.9 Further Reading

The subdivisions of the players' mixed-strategy simplices into labeled best-response regions, and the best-response diagrams like in Figure 9.16, are due to Shapley (1974). The example in Exercise 9.1 of a game where not all equilibria are found by the Lemke–Howson algorithm when started from the artificial equilibrium is shown in Shapley (1974, Figure 3) and is due to Robert Wilson. A rigorous definition of the LH graph that we described following Lemma 9.3 is given in von Stengel (2002).

The LH paths have an intrinsic *direction*, defined in terms of the determinants of the payoff matrices for the equilibrium supports, and the end nodes of the paths have an opposite sign called the *index* of the equilibrium. This was first proved by Shapley (1974); for a streamlined presentation see von Stengel (2021). The direction of the path accounts for the "D" in the complexity class "PPAD" defined by Papadimitriou (1994) that contains the computational problem of finding an equilibrium of a bimatrix game. This problem is in fact complete for PPAD (Chen, Deng, and Teng, 2009).

In their original publication, Lemke and Howson (1964) considered arbitrary nonnegative vectors instead of mixed strategies, like we do in (9.24) to define the polytopes P and Q. However, the payoff matrices A and B are replaced by strictly positive *cost matrices* $A' = \mathbf{1}\alpha\mathbf{1}^\top - A$ and $B' = \mathbf{1}\alpha\mathbf{1}^\top - B$ for some sufficiently large $\alpha \in \mathbb{R}$, to define the polyhedra

$$P' = \{\, x \in \mathbb{R}^m \mid \quad x \geq \mathbf{0},\ (B')^\top x \geq \mathbf{1}\},$$
$$Q' = \{\, y \in \mathbb{R}^n \mid A'y \geq \mathbf{1}, \quad\quad y \geq \mathbf{0}\}\,.$$

These are clearly unbounded, but do not have the artificial equilibrium $(\mathbf{0}, \mathbf{0})$ as a complementary solution. Instead, the algorithm is started from the "ray" of almost completely labeled pairs $(x, y) \in P' \times Q'$ with missing label ℓ given by $x = e_\ell \alpha$ so that $\alpha \in \mathbb{R}$ and $e_\ell \alpha \in P'$ (if $1 \leq \ell \leq m$, otherwise $y = e_{\ell-m}\alpha$) and the pure best response to ℓ as the corresponding vertex y in Q' (respectively, x in P').

The example (9.10) of strategic equivalence of two payoff matrices without a column shift is similar to Example 2.1 of Liu (1996). Moulin and Vial (1978) showed that strong strategic equivalence, defined in (9.11), only admits scaling and column shifts.

The polyhedra $\bar{P}$ and $\bar{Q}$ in (9.20) have been first described by Mangasarian (1964). As mentioned at the end of Section 9.8, these polyhedra, or the polytopes P and Q, can be used to find all equilibria of a bimatrix game, even for degenerate games; see Avis, Rosenberg, Savani, and von Stengel (2010). The projective transformation in Figure 9.18 has been described in von Stengel (2002), which is a general survey on computing equilibria in two-player games. For an introduction to polyhedra and polytopes see the first chapters of Ziegler (1995).

For a general $d \times d$ matrix C and a vector $q \in \mathbb{R}^d$ instead of $\mathbf{1}$, the conditions (9.31) define a so-called *linear complementarity problem*. If the right-hand side q has negative components, then $w = q$ and $z = \mathbf{0}$ is not a suitable initial solution. By extending this system with an extra column (usually the vector $-\mathbf{1}$) and extra variable z_0 to

$$Cz + w - \mathbf{1}z_0 = q, \qquad z \geq \mathbf{0}, \qquad w \geq \mathbf{0}, \qquad z_i w_i = 0 \quad (1 \leq i \leq d), \qquad (9.45)$$

then this system has a "primary ray" of solutions where $z = \mathbf{0}$ and $w = q + \mathbf{1}z_0$ for sufficiently large z_0. Minimizing z_0 in (9.45) so that $z = \mathbf{0}$ and $w \geq \mathbf{0}$ then allows starting the complementary pivoting algorithm, invented by Lemke (1965). A large body of work is concerned with whether Lemke's algorithm terminates with a solution to (9.45) where $z_0 = 0$, depending on properties of the matrix C (usually denoted by $-M$); see Cottle, Pang, and Stone (1992).

Pivoting is a general method to move from one basic feasible solution of a system of linear equations with nonnegative variables to another. It is the central step of the *simplex algorithm* of Dantzig (1963) for linear programming. Its explanation in terms of dictionaries is due to Chvátal (1983), who also explains (page 34) the lexicographic method for degeneracy resolution. Complementary pivoting is only special in terms of the choice of the entering variable as the complement of the variable that has just left the basis.

As shown in this chapter, the equilibrium structure of two-player games is best understood in terms of labeled polyhedra and polytopes. This insight can be used to apply known constructions of polytopes to questions about equilibria of bimatrix games, in particular the "dual cyclic polytopes" which have the largest possible number of vertices for a given dimension and number of inequalities (Ziegler, 1995, p. 10f). In von Stengel (1999), these polytopes are used to construct nondegenerate $n \times n$ games with a larger number of mixed equilibria than the coordination game (shown for $n = 3$ in Figure 9.7). These polytopes are also used in Savani and von Stengel (2006, 2016) to construct games with exponentially long Lemke–Howson paths for any missing label.

9.10   Exercises for Chapter 9

Exercise 9.1. Consider the symmetric 3×3 game $(A, A^\top)$ with A as in (9.12).

(a) Draw the best response diagram for this game.   [*Hint*: *You do not have to compute anything, but can use one of the pictures in Figure* 9.14 *with a suitable change of labels.*]

(b) Identify all equilibria of this game.

(c) Write down the Lemke–Howson path that starts at $(\mathbf{0}, \mathbf{0})$ for *each* missing label $1, 2, 3, 4, 5, 6$, as in (9.7), by suitably annotating the nodes in $\tilde{X}$ and $\tilde{Y}$. Which equilibrium is found for each missing label?   [*Hint*: *You can save half of the work by using the symmetry of the game.*]

Exercise 9.2. Consider the degenerate 2×3 game in Figure 6.12. Construct the best response diagram (note that X is only a line segment) and identify the set of all equilibria with the help of the labels.

Exercise 9.3. Consider the following 3×3 game.

(a) Find all pure strategy equilibria of this game.

(b) The picture below shows, for this game, the mixed-strategy sets X and Y for players I and II, but *not* which set is X and which is Y. It also shows their subdivisions into best-response regions, but without labels, which you are asked to provide. Explain which triangle is X and which is Y. [*Hint*: *consider the pure best responses first.*] For the three corners marked by a, b, e and p, s, u, respectively, indicate which pure strategy $(1, 0, 0)$, $(0, 1, 0)$, or $(0, 0, 1)$ they represent, and mark the triangles with the correct labels $\textcircled{1}, \ldots, \textcircled{6}$. Is this game degenerate?

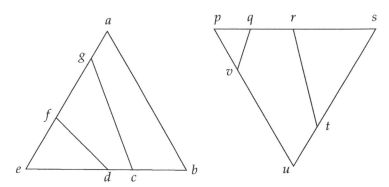

(c) Based on these labels, identify all equilibria of the game in mixed strategies (using the indicated nodes a, b, c, d, e, f, g and p, q, r, s, t, u, v).

(d) Show all Lemke–Howson paths for missing label 1, and how they connect the equilibria of the game.

Exercise 9.4. Consider a nondegenerate bimatrix game and an arbitrary pure-strategy equilibrium of the game (assuming one exists). Explain why the Lemke–Howson path started at $(\mathbf{0}, \mathbf{0})$ always finds this equilibrium for a suitable missing label. Suppose the game has at least two pure equilibria. Explain how the Lemke–Howson algorithm can be used to find a third (not necessarily pure) equilibrium.

Exercise 9.5. Give an example of a bimatrix game (A, B) where every column of A has a positive entry but where Q in (9.24) is not a polytope.

Exercise 9.6. Consider a nondegenerate bimatrix game that has a pure-strategy equilibrium. Show that there is a strategically equivalent game so that in this pure equilibrium, both players get a strictly higher payoff than for any other pair of pure strategies. Explain why the "dilemma" in the Prisoner's Dilemma has "gone away" when considering this equivalent game. Also, give an example that shows that nondegeneracy is needed to get strictly higher payoffs.

10

Game Trees with Imperfect Information

Although it appears late in this book, this is a core chapter about non-cooperative games. It studies extensive games, which are game trees with imperfect information. Typically, players do not always have full access to all the information which is relevant to their choices. Extensive games with *imperfect information* model exactly what information is available to the players when they make a move. The precise modeling and evaluation of strategic information is one of the strengths of game theory.

The central concept of *information sets* is introduced in Section 10.2 by means of a detailed example. The subsequent sections give the general definition of extensive games, and the definition of strategies and reduced strategies, which generalize these concepts as they have been introduced for game trees with perfect information. Essentially, all one has to do is to replace "decision node" by "information set".

The concept of *perfect recall*, treated in Section 10.6, means that a player always remembers what he knew or did earlier. This is a condition on the structure of the information sets of the player, not about how to interpret the game. In a game with perfect recall, the important Theorem 10.4 of Kuhn (1953) states that a player can replace any mixed strategy by a *behavior* strategy, which is the topic of Secton 10.8. In a behavior strategy, the player randomizes "locally" between his moves at each information set. In contrast, a mixed strategy first selects "globally" a pure strategy at the beginning of a play of the game, which the player then uses as a plan of action throughout play. Behavior strategies are much less complex than mixed strategies, as we will explain. In Section 10.9, we will illustrate optimal mixed and behavior strategies in a game-theoretic variant of the "Monty Hall" problem (with an opposite conclusion to the standard statistical version of this problem).

The final Section 10.10 treats *subgames* and subgame-perfect equilibria (SPE).

10.1   Prerequisites and Learning Outcomes

This chapter relies on Chapter 4 and Chapter 6.

After studying this chapter, you should be able to:

- explain extensive games and the concept of information sets;

- describe pure and reduced pure strategies for extensive games, and the corresponding strategic form and reduced strategic form of the game;

- define the concept of perfect recall, and see if this holds for a given game;

- explain the difference between behavior and mixed strategies, and state Kuhn's Theorem 10.4;

- construct extensive games from descriptions of games with imperfect information, like the game in Exercise 10.4;

- find all equilibria of simple extensive games, and represent the mixed strategies in these equilibria as behavior strategies;

- explain the concept of subgames and *subgame-perfect equilibria* (SPE), and how it generalizes and differs from backward induction. Understand that choosing single moves at a time as in backward induction *cannot* be applied when the game tree has imperfect information.

10.2   Information Sets

Consider the situation faced by a large software firm (player I) after a small startup firm (player II) has announced the deployment of a key new technology. The large firm has a large research and development operation. It is generally known that they have researchers working on a wide variety of innovations. However, only the large firm knows for sure whether or not they have made any progress on a product similar to the startup's new technology. The startup firm believes that there is a 50 percent chance that the large firm has developed the basis for a strong competing product. For brevity, when the large firm has the ability to produce a strong competing product, we will say that the large firm is "strong", and "weak" otherwise.

The large firm, after the announcement by the small firm, has two choices. It can either announce that it too will release a competing product, or it can cede the market for this product. The large firm will certainly condition its choice upon its private knowledge whether it is strong or weak. If the large firm has announced a product, then the startup firm has two choices: It can either negotiate a buyout to the large firm, or it can remain independent and launch its product. The startup firm does not have access to the large firm's private information on the status of its research. However, it does observe whether or not the large firm announces its

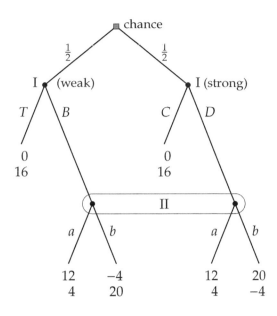

Figure 10.1 Game with imperfect information for player II, indicated by the information set that contains two decision nodes with the same moves a and b. As always, at a leaf of the tree, the top payoff is to player I, the bottom payoff to player II.

own product, and may attempt to infer from that announcement whether the large firm is strong or weak.

When the large firm is weak, the startup prefers to stay in the market over selling out to the large firm. When the large firm is strong, the opposite is true, and the startup is better off by selling out.

Figure 10.1 shows a game tree with imperfect information (also called an "extensive game") that models this situation. From the perspective of the startup firm, it is random whether the large firm is weak or strong. This is modeled by an initial chance move, here assumed to have probability $\frac{1}{2}$ for both possibilities.

When the large firm is weak, it can choose to cede the market to the startup, here written as move T, with immediate payoffs 0 and 16 to players I and II. It can also announce a competing product, modeled by move B, in the hope that the startup firm, player II, will sell out by choosing move a, with payoffs 12 and 4 to the two players. However, if player II then decides instead to stay in the market (move b), then it will even profit from the increased publicity and get payoff 20, with a negative payoff -4 to player I.

In contrast, when player I is strong, then it can again cede the market to player II (move C, with the same payoff pair $0, 16$ as before), or instead announce its own product (move D). It will quickly become apparent that player I should never choose C, but we keep it as a possible move. After move D of the "strong" player I, the payoffs to the two players are $12, 4$ if the startup sells out (move a) and $20, -4$ if the startup stays in (move b).

As an additional feature compared to a game tree with perfect information, some nodes of the players are enclosed by ovals that define *information sets*. An information set is a set of decision nodes, subject to certain restrictions that we will shortly describe in generality. The interpretation is that a player cannot distinguish among the nodes in an information set, given his knowledge at the time he makes the move. Because his knowledge at all nodes in an information set is the same, he makes the same choice at each node in that set. Here, the startup firm, player II, must choose between move a (sell out) and move b (stay in the market). These are the two choices at player II's information set, which has two nodes according to the different histories of play, which player II cannot distinguish.

Because player II is not informed about her position in a play of the game, *backward induction can no longer be applied.* If player II knew whether player I is weak or strong, then it would be better for her to choose b when player I is weak, and to choose a when player I is strong.

Player I also has information sets, namely one for each of his decision nodes. Each information set contains only a single node, which shows that the player is perfectly informed about the state of play. We do not draw these singleton information sets in the game tree; they are implicit when a decision node is shown without being surrounded by an oval.

One "optimal move" that can be seen directly from the game tree is move D when player I is strong. Choosing C gives player I payoff zero, whereas D gives him either 12 or 20, depending on whether player II chooses a or b, which are both larger than zero, so player I should choose D at any rate. Then the game reduces to a decision between move T and B for player I, and between a and b for player II.

The game tree shows that no pair of these moves for the two players is optimal against the other. For player I, if player II chooses b, then move T is better (because 0 is a higher payoff than -4). If player I chooses T, then the left node of the information set of player II is not reached. In that case, when player II has the choice between a and b, she knows with certainty that she is at the right node of her information set. Hence, she gets the higher payoff 4 when choosing a (rather than -4 when choosing b). In turn, if player II chooses a, then even a weak player I will choose B over T with its higher payoff 12 compared to 0. If player I chooses B, then player II will be with equal probability at either the left or the right node in her information set. Player II then has to use expected payoffs when comparing a and b. Clearly, move a will give her payoff 4, whereas move b will give her the expected payoff $\frac{1}{2} \cdot 20 + \frac{1}{2} \cdot (-4) = 8$. That expected payoff exceeds the payoff for move a. That is, against B, player II is better off choosing b. We have now come full circle: The sequence of best-response moves for the players is $b \to T \to a \to B \to b$, so there is no deterministic behavior of the players that defines an equilibrium.

	II : a	II : b
TC	16 / 0	16 / 0
TD	10 / 6	6 / 10
BC	10 / 6	18 / −2
BD	4 / 12	8 / 8

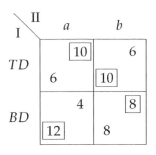

	II : a	II : b
TD	[10] / 6	6 / [10]
BD	4 / [12]	[8] / 8

Figure 10.2 Left: Strategic form of the game in Figure 10.1. Right: The 2×2 game that results after eliminating the strictly dominated strategies TC and BC, with best-response payoffs.

The game in Figure 10.1 can be analyzed in terms of the moves alone (we will give such an analysis at the end of this section), because for each player, only two moves matter. However, in general it is useful to consider the *strategic form* of the game that has already been defined for game trees with perfect information. This involves systematically considering the *strategies* of the players. The equilibrium concept is also defined in terms of strategies and not in terms of moves.

The left table in Figure 10.2 gives the strategic form of the game in Figure 10.1. As in the case of game trees with perfect information, a strategy of a player defines a move for each of the player's decision nodes. The difference, however, is that such a move is specified only *once* for all the decision nodes that belong to an information set, because by definition the choice of that move is the same no matter where the player is in the information set. In Figure 10.1, this means that player II has only two moves, which are the two moves a and b at her information set. Because player II only has one information set, her strategies are the same as her moves. Player I has two information sets, so a strategy for player I is a pair of moves, one move for the first and one move for the second information set. These pairs of moves are $TC, TD, BC,$ and BD.

The four strategies of player I correspond to the rows on the left in Figure 10.2, whereas the strategies a and b of player II correspond to the columns. Each of the eight cells in that table gives the expected payoffs as they are calculated from the payoffs at the leaves and the chance probabilities. For example, the strategy pair (BC, a) gives expected payoff $\frac{1}{2} \cdot 12 + \frac{1}{2} \cdot 0 = 6$ to player I and payoff $\frac{1}{2} \cdot 4 + \frac{1}{2} \cdot 16 = 10$ to player II. These expectations come from the possible chance moves, both of which have probability $\frac{1}{2}$ according to Figure 10.1. When the chance move goes to the left, the strategy pair (BC, a) means the play of the game follows move B and then move a and reaches the leaf with payoffs $12, 4$ to players I, II. When the chance move goes to the right, the strategy pair (BC, a) means that move C leads immediately to the leaf with payoffs $0, 16$.

The 4×2 game on the left in Figure 10.2 has two strictly dominated strategies, namely TC, which is strictly dominated by TD, and BC, which is strictly dominated by BD. This is in agreement with the earlier observation that for player I, move C is always worse than move D, no matter what player II does. However, one has to be careful to conclude that a "worse move" always leads to a *strictly* dominated strategy (the dominated strategy, for example TC instead of TD, would be obtained by replacing a given move, like D as part of the strategy TD, by the "worse move", here C). Namely, even if the payoffs from the "worse move" (here C) are strictly worse than those resulting from the given move (D), one has to make sure that the decision point with the moves in question (C and D) is *always reached with positive probability*. Only then the expected payoffs from the two strategies will always differ. This is here the case because the decision point of player I with moves C and D is reached by a chance move. If, in contrast, the chance move was replaced by a move of a third player, for example, then if that third player does not move to the right, the choice between C and D would *not affect the payoff* to player I. So in this case, we could only conclude that TD *weakly* dominates TC. In short, it is safer to look at the strategic form of the game, even though a careful look at the game tree may eliminate some strategies right away.

Elimination of the strictly dominated strategies TC and BC gives the 2×2 game on the right in Figure 10.2. There, each player has two strategies, which correspond to the two moves discussed earlier. The best-response payoffs indicated by boxes show that this game has no pure-strategy equilibrium.

The game has an equilibrium in mixed strategies, which are determined in the familiar way. Player I randomizes with equal probability $\frac{1}{2}$ between TD and BD so that the expected payoff to player II is 7 for both a and b. Player II chooses a with probability $\frac{1}{4}$ and b with probability $\frac{3}{4}$ so that the expected payoff to player I is $\frac{36}{4} = 9$ for both T and B.

The game in Figure 10.1 is a competition between two firms that is more familiar when viewed as a simple case of the card game *Poker*. We can think of player I being dealt with equal probability a weak or a strong "hand" of cards. Player I knows whether his hand is weak or strong (in a real Poker game, player I knows his hand but not necessarily whether it is better or worse than that of player II). Player I can then either "fold" (moves T and C), or "raise" (moves B and D). In turn, player II can either "pass" (move a) or "meet". The "fold" and "pass" moves entail a fixed outcome where the players receive payoffs that are independent of the hands they hold. Only the combinations of moves "raise" and "meet", corresponding to the strategy pair (BD, b), lead to outcomes where the players' hands matter, as can be seen from Figure 10.1 (in the strategic form, different strategy pairs give different payoffs because of the chance move).

When player I has a weak hand and decides to "raise" (move B), this amounts to *bluffing*, which means pretending that he has a strong hand, because player II does not know this. The purpose of bluffing, however, is not only to make player II "pass". Indeed, player II would not "pass" (move a) if player I bluffed all the time, as analyzed before. One main effect of bluffing is to leave player II *uncertain* as to whether player I has a strong or weak hand. The best possible outcome for player I is if player II meets a strong hand of player I (moves D and b), because this gives player I maximum payoff. If player I raises only if he has a strong hand, then player II would rather pass, which is not in the interest of player I. Above, we have reasoned that no deterministic combination of moves leads to an equilibrium. As described, this is made very vivid in terms of the Poker game interpretation.

To conclude the discussion of this example, we can also deduce in this case directly from the game tree the probabilities for the moves T and B, and for a and b, that are necessary in order to reach an equilibrium. Even without constructing the strategic form, we have seen that no deterministic choice of these moves gives an equilibrium. Hence, both players will randomize, and therefore must be indifferent between their choices. For player I, this indifference is very easy to determine: If he has to decide between T and B, then he gets the certain payoff 0 with T, so indifference results only when the expected payoff for B is also 0. If player II chooses a and b with probabilities $1 - q$ and q, respectively, that expected payoff to player I is $12(1 - q) - 4q$, so the equation $0 = 12(1 - q) - 4q = 12 - 16q$ gives $q = \frac{3}{4}$.

In turn, player II has to be indifferent between her moves a and b in order to randomize on her part. Here the situation is somewhat more complicated, because the expected payoff to player II depends on the probabilities of being either at the left or at the right node in her information set. Assuming, as we do, that player I chooses D with certainty, the probability of being at the right node is $\frac{1}{2}$, whereas the probability of being at the left node is $\frac{1}{2} \cdot p$, where p is the probability of player I choosing B. Because there is a nonzero probability $1 - p$ for T, a play of the game ends with probability $\frac{1}{2} \cdot (1 - p)$ at the leftmost leaf of the tree, and then the information set of player II will not be reached at all. Technically, what has to be computed is the *conditional* probability that player II will be at the left node versus being at the right node. Before going into the details, we note that from the description so far, the probability for the left node is p times the probability for the right node. Only these relative weights matter. Indifference of player II between a and b then gives the following equation, with the weighted payoffs for a and b on the left and right, respectively:

$$p \cdot 4 + 1 \cdot 4 = p \cdot 20 + 1 \cdot (-4), \tag{10.1}$$

which simplifies to $4p + 4 = 20p - 4$ or $8 = 16p$, that is, $p = \frac{1}{2}$.

The conditional probabilities for player II being at the left and right node are computed as follows: The left node has absolute probability $\frac{1}{2} \cdot p$ of being reached, the right node absolute probability $\frac{1}{2}$. The total probability of being at any of the two nodes (which are disjoint events) is the sum of these two probabilities, namely $\frac{p+1}{2}$. The conditional probability is then the absolute probability divided by the total probability, that is, for the left node, $\frac{p/2}{(p+1)/2}$ or $\frac{p}{p+1}$, and for the right node, $\frac{1/2}{(p+1)/2}$ or $\frac{1}{p+1}$. Using these conditional probabilities instead of the "weights" in (10.1) means dividing both sides of equation (10.1) by $p+1$. This does not change the solution to this equation, so what we have done is perfectly in order (as long as the total probability of reaching the information set is nonzero).

One more time, we interpret the equilibrium probabilities $p = \frac{1}{2}$ and $q = \frac{3}{4}$ in terms of the game tree: Player I, by choosing B with probability $\frac{1}{2}$, and D with certainty, achieves that player II is half as likely to be at the left node of her information set as at her right node. Therefore, the conditional probabilities for left versus right node are $\frac{1}{3}$ and $\frac{2}{3}$, respectively ($\frac{1}{4}$ and $\frac{1}{2}$ divided by the total probability $\frac{3}{4}$ of reaching the information set). Exactly in that case, player II is indifferent between a and b: The *conditional expected payoffs* to player II are then (clearly) 4 for move a, and $\frac{1}{3} \cdot 20 + \frac{2}{3} \cdot (-4) = \frac{12}{3} = 4$ for move b. The *overall* expected payoff to player II has to take the absolute probabilities into account: With probability $\frac{1}{4}$ (left chance move, player I choosing T), she gets payoff 16, and with probability $\frac{3}{4}$ the mentioned conditional payoff 4, overall $\frac{1}{4} \cdot 16 + \frac{3}{4} \cdot 4 = 4 + 3 = 7$. For player I, we have observed the indifference between T and B, so when the chance move goes to the left, player I gets payoff zero. When the chance move goes to the right, player I when choosing D will get $(1-q) \cdot 12 + q \cdot 20 = \frac{1}{4} \cdot 12 + \frac{3}{4} \cdot 20 = 18$. The overall payoff to player I is $\frac{1}{2} \cdot 0 + \frac{1}{2} \cdot 18 = 9$.

10.3   Extensive Games

In this section, we give the general definition of an *extensive game*, also known as a game in extensive form, or as a game tree with imperfect information.

Similar to the game trees with perfect information considered in Chapter 4, the basic structure of an extensive game is a *tree*. As before, it has three types of nodes: Terminal nodes, called *leaves*, with a payoff for each player; non-terminal nodes which are *chance* nodes (typically drawn as squares), where the next node is chosen randomly according to a given probability that is specified in the tree; or non-terminal nodes that are *decision* nodes, which belong to a particular player. The outgoing edges from each decision node are labeled with the names of choices or *moves* which uniquely identify the respective edges.

The decision nodes are partitioned into *information sets*. "Partitioned" means that every decision node belongs to exactly one information set, and that no information set is empty. When drawing the extensive game, information sets are shown as ovals around the nodes that they contain (except when the information set contains only one node); some people draw information sets by joining the nodes in the set with a single dashed line instead.

The *interpretation* of an information set is that during a play of the game, a player is only told that he is at *some* node in an information set, but not at which particular node. Consequently, the information sets have to fulfill certain conditions: In an information set, all decision nodes belong to the same player; all decision nodes in the set have the same number of outgoing edges; and the set of moves (given by the labels on the outgoing edges) is the same for each node in the information set. In Figure 10.1, for example, the information set of player II has two moves, *a* and *b*. When the player has reached an information set, he chooses a move, which is by definition the *same* move no matter where he is in the information set.

These conditions are trivially fulfilled if every information set is a singleton (contains only one node; recall that we usually do not draw these singleton information sets). If all information sets are singletons, then the game has *perfect information* and the information sets can as well be omitted. This gives a game tree with perfect information as considered earlier. Only for information sets that contain two or more nodes do these conditions matter.

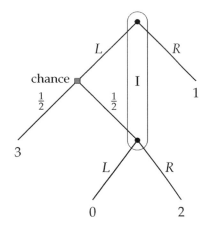

Figure 10.3 Game with an information set where two nodes share a path, which is usually not allowed in an extensive game.

In addition to the above conditions, we will impose additional constraints on the structure of the information sets. The first of these is that no two nodes in an information set share a path, as in the example in Figure 10.3. The single-player game in that figure has two decision nodes of player I, which belong to a single information set that has the two moves *L* and *R*. The only other non-terminal node is a chance move. Every condition about extensive games is met. However, the two decision nodes of player I share a path, because the first node is the root, from which there is a path to the second node. Information sets with this feature have a very problematic interpretation, because we require that a *move*, for example *L*, at such an information set is the same no matter where the player is. However, in Figure 10.3, what would it mean for player I to make move L? It certainly means going left from the root, but if afterwards the chance move goes to the right so that player I can move again, does he automatically have to make move L again?

The problem with the information set in Figure 10.3 is that, by the interpretation of information sets, player I must have *forgotten* if he made a move at all when called to choose between L and R, and it is not even clear whether he is allowed to choose again or not. An information set can represent that a player forgets what he knew or did earlier. In that case, the player is said to have *imperfect recall*. However, players with imperfect recall do not conform to our idea of "rational" players. Moreover, it is not clear how often a player can move if one node in an information set is the descendant of another. We will therefore assume that all players have *perfect recall*, which is defined, and discussed in detail, in Section 10.6 below. In particular, we will see that perfect recall implies that no two nodes in an information set are connected by a path in the tree (see Proposition 10.2 below), so that a situation like in Figure 10.3 cannot occur.

To summarize the definition of game trees with imperfect information: The only new concept is that of an information set, which models a player's *lack* of information about where he is in the game.

10.4   Strategies for Extensive Games and the Strategic Form

The definition of a *strategy* in an extensive game generalizes the definition of a strategy for a game tree with perfect information: A (pure) strategy of a player specifies a move for each information set of the player.

A formal definition of strategies requires some notation to refer to the information sets of a player, and to the moves available at each information set. Suppose player i is a player in the extensive game (like player I or player II in a two-player game). Then H_i denotes the set of information sets of player i. A particular information set of player i is typically called h. We denote the set of *choices* or moves at the information set h by C_h for $h \in H_i$. For example, if h is the left information set of player I in Figure 10.1, then $C_h = \{T, B\}$.

The set Σ_i of strategies of player i is then

$$\Sigma_i = \underset{h \in H_i}{\times} C_h , \tag{10.2}$$

that is, the Cartesian product of the choice sets C_h of player i. An element of Σ_i is a tuple (vector) of moves, one move c from C_h for each information set h of player i.

If N is the set of players, like $N = \{I, II\}$ in the notation we use for a two-player game, then the set of pure *strategy profiles* is $\times_{i \in N} \Sigma_i$, so a strategy profile specifies one pure strategy for each player.

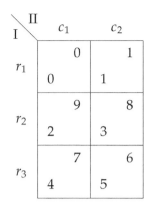

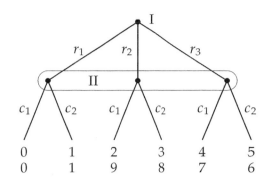

Figure 10.4 A 3×2 game and an extensive game that has this game as its strategic form.

The *strategic form* of the extensive game is given by the set N of players, the set of strategies Σ_i for each player i, and the expected payoff that each player gets for each strategy profile. If the game has no chance moves, then the payoff for each strategy profile is given deterministically by the leaf of the game tree that is reached when the players use their strategies. If the game has chance moves, then more than one leaf may be reached with a given strategy profile, and then expected payoffs have to be calculated, as in the example of Figure 10.2.

With the help of information sets, it is possible to model *simultaneous play* in an extensive game. This is not possible with a game tree with perfect information where the sequential moves mean that the player who moves second is informed about the move of the first player.

In particular, given a game in strategic form, there is an extensive game that has exactly this strategic form. For example, consider an $m \times n$ game. It does not matter which player moves first, so we can assume this is player I, who has m moves, say $r_1, r_2, \ldots, r_m$ for the m rows of the game, at the root of the game tree. Each of these moves leads to a decision node of player II. All these nodes belong to a single information set. At this information set, player II has n moves $c_1, c_2, \ldots, c_n$ that correspond to the n columns of the game. Each move of player II leads to a leaf of the game tree. Because m possible moves of player I are succeeded by n moves of player II, the game tree has mn leaves. If the given game has payoffs a_{ij} and b_{ij} to players I and II when they play row i and column j, respectively, then these two payoffs are given at the leaf of the game tree that is reached by move r_i followed by move c_j. Because each player has only one information set, his strategies coincide with his moves. The strategic form of the constructed extensive game is then the original $m \times n$ game. Its rows are marked with the strategies $r_1, \ldots, r_m$ and its columns with $c_1, \ldots, c_n$ because these are the names of the moves of the two players in the extensive game. Figure 10.4 gives an example.

⇒ Exercise 10.1 asks you to repeat this argument but also with player II moving
first and player I moving second.

10.5   Reduced Strategies

Consider the extensive game in Figure 10.5. Player II has three information sets, call
them h, h', and h'', with $C_h = \{l, r\}$, $C_{h'} = \{a, b\}$, and $C_{h''} = \{c, d\}$. Consequently,
she has $2 \times 2 \times 2 = 8$ pure strategies. Each pure strategy is a triple of moves, an
element of $\Sigma_{II} = C_h \times C_{h'} \times C_{h''}$, for example lac (we write the triple of moves by
just writing the three moves next to each other for brevity). If player I chooses T,
then this strategy lac of player II will lead to the leftmost leaf of the game tree with
the payoff pair $(1, 5)$, via the moves l, T, and then a. Clearly, if player I plays T
and player II plays the strategy lad, then this will lead to the same leaf. In fact,
irrespective of the strategy chosen by player I, the strategies lac and lad of player II
always have the same effect. This can be seen directly from the game tree, because
when player II makes move l, she will *never* reach the information set h'' where
she has to decide between c and d.

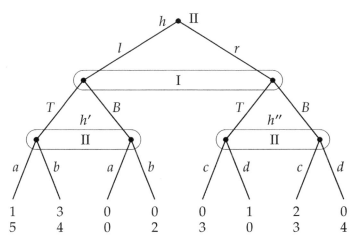

Figure 10.5   Extensive game with one information set for player I and three information sets h, h', h'' for player II. Note that h is a singleton and not drawn explicitly.

This leads to the concept of a *reduced strategy*, which we have already met for a
game tree with perfect information. A reduced strategy of player i specifies a move
for each information set of player i, *except* for those information sets of player i
that are *unreachable* because of an *own* earlier move (that is, because of a move
of player i). When writing down a reduced strategy, we list the moves for each
information set (in a fixed order of the information sets), where an unspecified
move is denoted by an extra symbol, the star "$*$", which stands for any move at
that unreachable information set.

In Figure 10.5, the information set h'' of player II is unreachable after her own earlier move l, and her information set h' is unreachable after her own earlier move r. Her reduced strategies are therefore $la*$, $lb*$, $r*c$, $r*d$. Note that the star stands for any unspecified move that would normally be at that place when considering a full (unreduced) strategy. So player II's reduced strategy $la*$ can be thought of representing both her pure strategies lac and lad.

The *reduced strategic form* of an extensive game is the same as its strategic form, except that for each player, only reduced strategies are considered and not pure strategies. By construction, the resulting expected payoffs for each player are well defined, because the unspecified moves never influence which leaves of the game tree are reached during play. Figure 10.6 gives the reduced strategic form of the game in Figure 10.5.

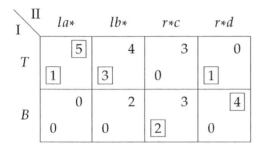

Figure 10.6 Reduced strategic form of the game in Figure 10.5. In the strategies of player II, a star * indicates an unspecified move at an information set that is unreachable due to an own earlier move. As usual, the numbers in boxes are best-response payoffs.

10.6 Perfect Recall

What do the information sets of player II in Figure 10.5 describe? At the information set h, player II knows for certain that she is at the root of the game. At the information set h', she does not know the move of player I, but she knows that she made the earlier move l at the root. At the information set h'', she again does not know the move of player I, but she knows that she made the move r earlier at the root.

For comparison, consider the game in Figure 10.7, which has the same tree and the same payoffs as the game in Figure 10.5, except that the information sets h' and h'' have been united to form a new set h''' which has four nodes. Moreover, in order to conform with the definition of an extensive game, the moves at h''' have to be the same at all nodes; they are here called a and b, which replace the earlier moves c and d at h''. When player II is to make a move at h''', she knows nothing about the state of play except that she has made some move at the beginning, followed by either move T or B of player I. In particular, the interpretation of h''' implies that player II *forgot* her earlier move l or r that she made at the root. We say that player II has *imperfect recall* in the game in Figure 10.7.

Imperfect recall creates difficulties when interpreting the game, and contradicts our idea of "rational" players. All the extensive games that we will analyze have *perfect recall*, which we now proceed to define formally. For example, Figure 10.5 shows a game with perfect recall. There, player II has two information sets h' and h'' that have more than one node, so these are information sets where player II has some lack of information. We can test whether this lack of information involves her earlier moves. Namely, each node in an information set, say in h', is reached by a unique path from the root of the game tree. Along that path, we look at the moves that the players make, which is move l of player II, followed by move T of player I for the left node of h', or followed by move B for the right node of h'. (In general, this may also involve chance moves.) Among these moves, we *ignore* all moves made by the other players (which is here only player I) or by chance. Consequently, we see that each node in h' is preceded by move l of player II. That is, each node in the information set has the *same sequence of own earlier moves*. This is the definition of perfect recall.

Definition 10.1. Player i in an extensive game has *perfect recall* if for every information set h of player i, all nodes in h are preceded by the same sequence of moves of player i. The extensive game is said to have perfect recall if all its players have perfect recall.   □

To repeat, the crucial condition about the nodes in an information set h of player i is that they are preceded by the same *own* earlier moves. We have to disregard the moves of the other players because there must be something that player i is lacking information about if player i does not have perfect information.

> ⇒ Make sure you clearly understand the distinction between perfect information and perfect recall. How are the respective concepts defined?

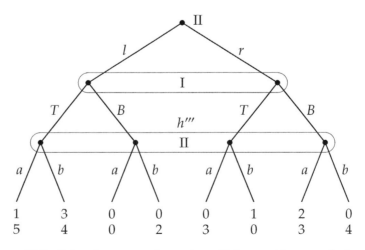

Figure 10.7 Extensive game where player II does not have perfect recall.

We can verify that all players have perfect recall in Figure 10.5. We have already verified the condition of Definition 10.1 for h'. The other information sets of player II are h and h''. For h, the condition holds trivially because h has only a single node (so we can make the easy observation that in a game with perfect information, that is, singleton information sets, all players have perfect recall). For h'', both nodes in h'' are preceded by the sequence of own moves that consists of the single move r. This shows that player II has perfect recall. Player I has only one information set, and for both nodes in that set the sequence of own earlier moves is *empty*, because player I has not made a move before. It is useful to consider the empty sequence, often denoted by $\emptyset$ like the empty set, as a sequence of moves, so that Definition 10.1 can be applied.

In Figure 10.7, player II does *not* have perfect recall because, for example, the leftmost node of her information set h''' is preceded by her own earlier move l, whereas the rightmost node is preceded by the move r. So the sequences of own earlier moves are not the same for all nodes in h'''.

One way to understand Definition 10.1 is via the interpretation of information sets. They represent a player's lack of knowledge about the state of play, that is, at which of the nodes in the information set the player is at that moment. When the sequence of own earlier moves is the *same* for all these nodes, then the player can indeed not get additional information about where he is in the information set from knowing what moves he already played (as should normally be the case). If, on the other hand, some nodes in the information set are preceded by *different* own earlier moves, then the player can only fail to know where he is in the information set when he has *forgotten* these moves. For that reason, we say he has imperfect recall. So a player with perfect recall cannot have such information sets, because they would not accurately reflect his knowledge about the state of play.

However, Definition 10.1 is not about how to *interpret* the game. Perfect recall is a *structural property of the information sets*. This property is easily verified by considering for each information set the nodes therein and checking that they are reached by the same sequence of earlier own moves, which can be done without interpreting the game at all.

Figure 10.8 shows a game with only one player, who has an information set (with moves L and R) that contains two nodes. Both nodes are preceded by an own earlier move named "B", so it seems that player I has perfect recall in this game. However, this would obviously not be the case if the two singleton information sets had moves that are named differently, for example T and B at the left information set and C and D at the right information set (as in Figure 10.1). Indeed, player I does *not* have perfect recall in the game in Figure 10.8 because moves at *different* information sets are always considered *distinct*. A formal way of saying this is that the choice sets C_h and $C_{h'}$ are disjoint whenever $h \neq h'$. This can be assumed without loss of generality, either by re-naming the moves suitably, or by assuming that in addition to the name of the move that is indicated in the game tree, part of the "identity" of a move $c \in C_h$ is the information set h where that move is made.

In the game in Figure 10.8, player I has imperfect recall, because he forgot his *earlier knowledge*, namely the outcome of the chance move. In contrast, player II in Figure 10.7 has not forgotten any earlier knowledge (all she knew earlier was that she makes the first move in the game, which she still knows when she reaches the information set h'''), but she has forgotten the move that she made. For that reason, it is sometimes said that a player has perfect recall if she never forgets what she "knew or did earlier". Definition 10.1 is a precise way of stating this.

> $\Rightarrow$ Exercises 10.2 and 10.3 test your understanding of perfect recall.

Perfect recall implies that a situation like in Figure 10.3 cannot occur.

Proposition 10.2. *If a player has perfect recall, then no two nodes in an information set of that player are connected by a path.*

Proof. Suppose otherwise, that is, h is an information set of a player who has perfect recall, so that there are nodes in h where one node precedes the other on a path in the game tree. If there are several such nodes, let u be the earliest such node (that is, there is no other node in h from which there is a path to u), and let v be a node in h so that there is a path from u to v. (An example is Figure 10.3 with u as the root of the game tree and v as the second node in player I's information set.) On the path from u to v, the player makes a move c at the decision node u. This move c precedes v. However, there is no other node in h where move c precedes u, by the choice of u, so u is not preceded by c. Consequently, the *sets* of the player's own moves that precede v and u are already different, so the sequences of own

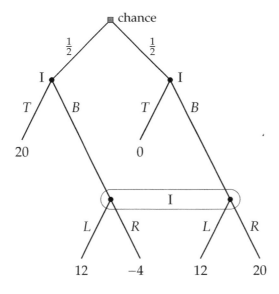

Figure 10.8 Single-player game where player I does not have perfect recall. Moves at different information sets are always considered as distinct, even if they have the same name (for example, here B).

earlier moves for the two nodes certainly differ as well. So the player does not have perfect recall, which is a contradiction.   □

The examples in Figures 10.3 and 10.8 show that a game with a *single* player probably does not have perfect recall if it has information sets that contain two or more nodes.

> ⇒ Construct a one-player game with perfect recall where not all information sets are singletons. Why do you need a chance move?

10.7 Behavior Strategies

Our next topic concerns ways of randomizing in an extensive game. Unlike games with perfect information, which always have a equilibrium in pure strategies that can be found by backward induction, games with imperfect information may require randomized strategies, as the example in Figure 10.1 has demonstrated.

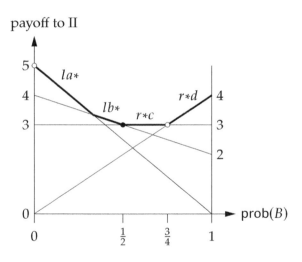

Figure 10.9 Upper envelope of expected payoffs to player II for the game in Figure 10.6. The white and black dots mark equilibrium strategies of player I.

We find all equilibria of the game in Figure 10.5 that has the reduced strategic form shown in Figure 10.6. The only equilibrium in pure strategies is the strategy pair $(T, la*)$. In order to find further mixed equilibria, if any, we use the upper envelope method. Figure 10.9 shows the expected payoffs to player II as a function of the probability that player I chooses B. As this figure shows, the only pairs of strategies between which player II can be indifferent are $la*$ and $lb*$ when $\text{prob}(B) = \frac{1}{3}$; or $lb*$ and $r*c$ when $\text{prob}(B) = \frac{1}{2}$; or $r*c$ and $r*d$ when $\text{prob}(B) = \frac{3}{4}$. The first of these cannot be played in an equilibrium because against both $la*$ and $lb*$, the best response of player I is T, so that player I cannot be made indifferent between T and B, which is necessary for player I wanting to mix between T and B in equilibrium. In the other two cases, however, the pure best responses of player I differ, and it is possible to obtain equilibria. These equilibria are $((\frac{1}{2},\frac{1}{2}),(0,\frac{2}{5},\frac{3}{5},0))$ and $((\frac{1}{4},\frac{3}{4}),(0,0,\frac{1}{3},\frac{2}{3}))$.

What do these mixed strategies mean in terms of the extensive game? For player I, the mixed strategy $(\frac{1}{2},\frac{1}{2})$ means that he chooses T and B with equal probability. For player II, the mixed strategy $(0,\frac{2}{5},\frac{3}{5},0)$ amounts to choosing $lb*$ with probability $\frac{2}{5}$, and $r*c$ with probability $\frac{3}{5}$. The interpretation of such a mixed strategy is that each player chooses *at the beginning of a play of the game* a reduced pure strategy, which is a plan of action of what do at each reachable information set. The player then sticks to this plan of action throughout the play. In particular, if player II has chosen to play $lb*$, then she first makes the move l with certainty, and then move b at her second information set h'. Her reduced pure strategy $r*c$ is a similar plan of action.

There is a different way of playing such a mixed strategy, which is called a *behavior strategy*. A behavior strategy is a randomized choice of a *move* for each information set. That is, rather than using a lottery to determine a (reduced) pure strategy at the beginning of playing the game, the player uses a "local" lottery whenever she reaches an information set. In our example, player II's mixed strategy $(0,\frac{2}{5},\frac{3}{5},0)$ is then played as follows: At her first information set h, she plays moves l and r with probabilities $\frac{2}{5}$ and $\frac{3}{5}$, respectively. When she reaches her information set h' (which is only possible when she has made the earlier move l), she chooses move a with probability zero and move b with probability one; this "behavior" at h' is deterministic. Similarly, when she reaches her information set h'' (having made move r earlier), then she chooses move c with certainty.

Definition 10.3. In an extensive game, a *behavior strategy* β of player i is defined by a probability distribution on the set of moves C_h for each information set h of player i. It is given by a probability $\beta(c)$ for each $c \in C_h$, that is, a number $\beta(c) \geq 0$ for each $c \in C_h$ so that $\sum_{c \in C_h} \beta(c) = 1$. In a *reduced* behavior strategy, these numbers are left unspecified if h is unreachable because all earlier own moves by player i that allow her to reach h have probability zero under β. □

The concept of a reduced behavior strategy is illustrated by the other mixed equilibrium of the game in Figure 10.5, with the mixed strategy $(0,0,\frac{1}{3},\frac{2}{3})$ of player II. This is a randomization between the two pure strategies $r*c$ and $r*d$. It can also be played as a (reduced) behavior strategy. Namely, at her first information set h, player II chooses l and r with probability 0 and 1, respectively. Then the information set h' is unreachable because it is only reachable when player II makes move l with positive probability, which is not the case here. Consequently, the behavior probabilities for making moves a and b are not specified, as in a reduced behavior strategy. (To obtain an "unreduced" behavior strategy, one would specify some arbitrary probabilities for a and b.) At her information set h'', player II chooses c and d with probability $\frac{1}{3}$ and $\frac{2}{3}$, respectively.

A behavior strategy can be played by "delaying" the player's random choice until she has reached an information set, when she makes the next move. However, nothing prevents the player making these random choices in advance for each information set, which will then give her a pure strategy that she plays throughout the game. Then the behavior strategy is used like a mixed strategy. In other words, *behavior strategies can be considered as special mixed strategies.*

A behavior strategy β of player i defines a mixed strategy μ by defining a suitable probability $\mu(\pi)$ for every pure strategy π of player i, which is found as follows. The pure strategy π defines a move $\pi(h)$ for each information set $h \in H_i$. When player i uses the behavior strategy β, that move has a certain probability $\beta(\pi(h))$. Because these random moves at the information sets h of player i are made independently, the probability that *all* the moves specified by π are made is the product of these probabilities. That is, the mixed strategy probability is

$$\mu(\pi) = \prod_{h \in H_i} \beta(\pi(h)). \tag{10.3}$$

We have already used equation (10.3) when we interpreted a mixed strategy, for example $\mu = (0, 0, \frac{1}{3}, \frac{2}{3})$ in Figure 10.5, as a behavior strategy β. We explain how this works even when, like here, the mixed strategy is defined with probabilities for reduced pure strategies. The behavior strategy β was given by $\beta(l) = 0, \beta(r) = 1$, $\beta(c) = \frac{1}{3}$, and $\beta(d) = \frac{2}{3}$, so that $\frac{2}{3} = \mu(r*d) = \beta(r) \cdot \beta(*) \cdot \beta(d)$. This is expressed as a product of *three* behavior probabilities, one for each information set, because the pure strategies are given by a move for each information set; note that in (10.3), the product is taken over all information sets h of player i. Because we consider reduced pure strategies, the term $\beta(*)$ represents the behavior probability for making *any* move at h' (in this case, as part of the reduced pure strategy $r*d$), which is unspecified. Clearly, that probability is equal to 1. That is, we take $\beta(*) = 1$ whenever (10.3) is applied to a reduced pure strategy π.

The convention $\beta(*) = 1$ is easily seen to agree with the use of (10.3) for unreduced pure strategies. For example, consider the unreduced strategic form where $r*d$ stands for the two pure strategies rad and rbd. Each of these has a certain probability under both μ and β (when extended to the unreduced strategic form), where μ is the representation of β as a mixed strategy. Whatever these probabilities are (obtained by splitting the probability for $r*d$ into probabilities for rad and rbd), we certainly have

$$\mu(rad) + \mu(rbd) = \beta(r)\beta(a)\beta(d) + \beta(r)\beta(b)\beta(d) = \beta(r) \cdot (\beta(a) + \beta(b)) \cdot \beta(d).$$

In this equation, the left-hand side is the proper interpretation of the mixed strategy probability $\mu(r*d)$, and the right-hand side of the probability $\beta(r) \cdot \beta(*) \cdot \beta(d)$, in agreement with $\beta(*) = \beta(a) + \beta(b) = 1$.

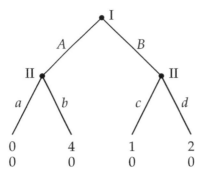

Figure 10.10 Game with four pure strategies of player II. Here, not all mixed strategies of player II are representable as behavior strategies.

While every behavior strategy can be considered as a mixed strategy, the converse does not hold. Figure 10.10 shows a game where player II has the four pure strategies ac, ad, bc, bd, which are also her reduced pure strategies. Consider the mixed strategy $\mu = (\frac{1}{2}, 0, 0, \frac{1}{2})$ of player II, where she chooses the pure strategy ac, that is, moves a and c, with probability $\frac{1}{2}$, otherwise the pure strategy bd. Clearly, every move is made with positive probability overall (in fact, probability $\frac{1}{2}$ for each single move). Then if (10.3) was true, every move combination would have to have positive probability as well! However, this is not the case, because the strategy ad, for example, has probability zero under μ. The reason is that when player II uses μ, the random choice between the moves a and b is *correlated* with the random choice between c and d. That is, knowing that move a has been made gives some information about whether move c or d has been made (here, the correlation is maximal: move a implies move c). If these random choices were made *independently*, as required in a behavior strategy, then the random choice between a and b would not affect the random choice between c and d.

In this example, the "behavior" of player II at the left information set in Figure 10.10 described by μ is to choose a with probability $\frac{1}{2}$ and b with probability $\frac{1}{2}$, and ditto for moves c and d. This defines a behavior strategy β of player II. This behavior strategy defines the mixed strategy $(\frac{1}{4}, \frac{1}{4}, \frac{1}{4}, \frac{1}{4})$ because it chooses each pure strategy with equal probability. This mixed strategy is clearly different from μ.

A second example for Figure 10.10 is the mixed strategy $\mu' = (\frac{1}{6}, \frac{1}{3}, \frac{1}{4}, \frac{1}{4})$. The corresponding behavior strategy β' has move probabilities $\beta'(a) = \beta'(b) = \frac{1}{2}$, because $\beta'(a) = \mu'(ac) + \mu'(ad) = \frac{1}{6} + \frac{1}{3}$. Similarly, $\beta'(c) = \mu'(ac) + \mu'(bc) = \frac{1}{6} + \frac{1}{4} = \frac{5}{12}$ and $\beta'(d) = \mu'(ad) + \mu'(bd) = \frac{1}{3} + \frac{1}{4} = \frac{7}{12}$. Again, making these moves independently with β' results in different probabilities for the pure strategies, for example in $\beta'(a) \cdot \beta'(c) = \frac{5}{24}$ for ac, which is different from $\mu'(ac)$.

What we have done in this example is to take a mixed strategy μ of a player and "observe" at each information set of that player the "behavior" resulting from μ. This behavior defines a certain behavior strategy β. In a similar way, β' is obtained from μ'. The behavior strategy may differ from the original mixed strategy because with the behavior strategy, random choices at different information sets are no longer correlated. However, what does *not* change are the probabilities of reaching the nodes of game tree, given some (pure or mixed) strategy of the other players. We say that the strategies μ and β are *realization equivalent*; similarly, μ' and β' are realization equivalent. This realization equivalence holds in any extensive game with perfect recall, and is the topic of the next section.

10.8 Kuhn's Theorem: Behavior Strategies Suffice

Information sets were introduced by Kuhn (1953). His article also defined perfect recall and behavior strategies, and proved an important connection between the two concepts. This connection is the central property of extensive games and is often called "Kuhn's theorem". It says that a *player who has perfect recall can always replace a mixed strategy by a realization equivalent behavior strategy.*

Why should we even care about whether a player can use a behavior instead of a mixed strategy? The reason is that a behavior strategy is much *simpler to describe* than a mixed strategy, significantly so for larger games. As an example, consider a simplified Poker game where each player is dealt only a single card, with 13 different possibilities for the rank of that card. A game tree for that game may start with a chance move with 13 possibilities for the card dealt to player I, who learns the rank of his card and can then either "fold" or "raise". Irrespective of the rest of the game, this defines already $2^{13} = 8192$ move combinations for player I. Assuming these are all the pure strategies of player I, a mixed strategy for player I would have to specify 8191 probabilities (which determine the remaining probability because these probabilities sum to one). In contrast, a behavior strategy would specify one probability (of folding, say) for each possible card, which are 13 different numbers. So a behavior strategy is much less complex to describe than a mixed strategy.

This example also gives an intuition *why* a behavior strategy suffices. The moves at the different decision nodes of player I can be randomized independently because they concern disjoint parts of the game. There is no need to correlate the moves for different cards because in any one game, only one card is drawn.

In general, a player may make moves *in sequence* and a priori there may be some gain in correlating an earlier move with a subsequent move. This is where the condition of perfect recall comes in. Namely, the knowledge about any earlier move is already captured by the information sets because perfect recall says that the player knows all his earlier moves. So there is no need to "condition" a later move on an earlier move because all earlier moves are known.

For illustration, consider the game in Figure 10.7 which does not have perfect recall. It is easily seen that this game has the same strategic form, shown in

Figure 10.6, as the game in Figure 10.5, except that the pure strategies of player II are la, lb, ra, rb instead of the reduced strategies $la*, lb*, r*c, r*d$. Here, the mixed strategy $(0, \frac{2}{5}, \frac{3}{5}, 0)$ of player II, which selects the pure strategy lb with probability $\frac{2}{5}$, and ra with probability $\frac{3}{5}$, *cannot* be expressed as a behavior strategy. Namely, this behavior strategy would have to choose moves l and r with probability $\frac{2}{5}$ and $\frac{3}{5}$, respectively, but the probabilities for moves a and b, which would be $\frac{3}{5}$ and $\frac{2}{5}$, are *not* independent of the earlier move (because otherwise the pure strategy la, for example, would not have probability zero). So this mixed strategy, which is part of an equilibrium, cannot be expressed as a behavior strategy. Using a mixed strategy gives player II more possibilities of playing in the game.

In contrast, the mixed strategy $(0, \frac{2}{5}, \frac{3}{5}, 0)$ of player II in the perfect-recall game in Figure 10.5 chooses $lb*$ and $r*c$ with probability $\frac{2}{5}$ and $\frac{3}{5}$, respectively. As shown earlier, this is in effect a behavior strategy, because correlating move l with move b, and move r with move c, is done via the information sets h' and h'', which are separate in Figure 10.5 but merged in Figure 10.7. That is, player II *knows* that she made move l when she has to make her move b at h', and she knows that she made move r when she has to make her move c at h''.

Before stating Kuhn's theorem, we have to define what it means to say that two randomized strategies μ and ρ of a player are *realization equivalent*. This is the case if for any fixed strategies of the other players, every node of the game tree is reached ("realized") with the same probability when the player uses μ as when he uses ρ. Two realization equivalent strategies μ and ρ always give every player the same expected payoff because the leaves of the game tree are reached with the same probability no matter whether μ or ρ is used.

When are two strategies μ and ρ realization equivalent? Suppose these are strategies of player i. Consider a particular node, say u, of the game tree. Equivalence of μ and ρ means that, given that the other players play in a certain way, the probability that u is reached is the same when player i uses μ as when he uses ρ. In order to reach u, the play of the game has to make a certain sequence of moves, including chance moves, that lead to u. This move sequence is unique because we have a tree. In that sequence, some moves are made by chance or by players other than player i. These moves have certain given probabilities because the chance move probabilities and the strategies adopted by the other players are fixed. We can assume that the overall probability for these moves is not zero, by considering fixed strategies of the other players which allow reaching u (otherwise u is trivially reached with the same zero probability with μ as with ρ). So what matters in order to compare μ and ρ are the moves by player i on the path to u. If the probability for playing the *sequence of moves of player i* on the path to u is the *same* under μ and ρ, and if this holds for every node u, then μ and ρ are realization equivalent. This is the key ingredient of the proof of Kuhn's theorem.

Theorem 10.4 (Kuhn, 1953). *If player i in an extensive game has perfect recall, then for any mixed strategy μ of player i there is a realization equivalent behavior strategy β of player i.*

Proof. We illustrate the proof with the extensive game in Figure 10.11, with player II as player i. The payoffs are omitted; they are not relevant because only the probabilities of reaching the nodes of the game tree matter.

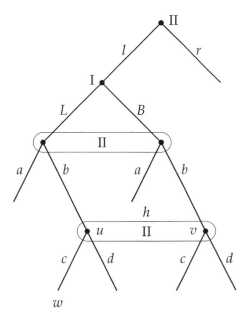

Figure 10.11 Game that illustrates sequences of moves of player II, and the proof of Kuhn's Theorem 10.4. Payoffs are omitted.

Given μ, we want to construct a realization equivalent behavior strategy β. Essentially, the probability $\beta(c)$ for any move c at an information set h of player i will be the "observed" probability that player i makes move c when she uses μ. In order to obtain this probability, we consider *sequences* of moves of player i. Such a sequence is obtained by considering any node x of the game tree and taking the moves of player i on the path from the root to x; the resulting sequence shall be denoted by $\sigma^i(x)$. For example, if x is the leaf w of the game tree in Figure 10.11 and $i = \text{II}$, then that sequence is $\sigma^{\text{II}}(w) = lbc$, that is, move l followed by b followed by c. If all players are considered, then node w is reached by the sequence of moves $lLbc$ which includes move L of player I, but that move is ignored in the definition of $\sigma^{\text{II}}(w)$; for player I, the sequence of moves defined by w is just the move L, that is, $\sigma^{\text{I}}(w) = L$.

According to Definition 10.1, the condition of perfect recall states that the sequence of own earlier moves is the same for all nodes in an information set. That is,

$$h \in H_i \quad \Longrightarrow \quad \forall u, v \in h \quad \sigma^i(u) = \sigma^i(v).$$

An example is given in Figure 10.11 by the information set h where player II has the two moves c and d, which has two nodes u and v with $\sigma^i(u) = \sigma^i(v) = lb$. Consequently, the sequence of moves *leading to* h can be defined uniquely as $\sigma_h = \sigma^i(u)$ for any $u \in h$, where i is identified as the player who moves at h, that is, $h \in H_i$. Furthermore, we can extend the sequence σ_h that leads to h by any move c at h, which we write as the longer sequence $\sigma_h c$, to be read as "σ_h followed by c". In the example, the sequence that leads to h is $\sigma_h = lb$, and it can be extended either by move c or by move d, giving the longer sequences lbc and lbd. Any sequence of moves of player i is either the empty sequence $\emptyset$ (for example $\sigma^i(r) = \emptyset$ if r is the root of the tree), or it has a last move c made at some information set h of player i, and then it can be written as $\sigma_h c$. Hence, these sequences $\sigma_h c$ for $h \in H_i$, $c \in C_h$ and $\emptyset$ are all possible sequences of moves of player i.

We now consider the *probability* $\mu[\sigma]$ that such a sequence σ of player i is played when player i uses μ. This probability $\mu[\sigma]$ is simply the combined probability under μ for all the pure strategies that *agree* with σ in the sense that they prescribe all the moves in σ. For example, if $\sigma = lb$ in Figure 10.11, then the pure strategies of player i that agree with σ are the strategies that prescribe the moves l and b, which are the pure strategies lbc and lbd. So

$$\mu[lb] = \mu(lbc) + \mu(lbd), \tag{10.4}$$

where $\mu(lbc)$ and $\mu(lbd)$ are the mixed strategy probabilities for the pure strategies lbc and lbd.

The longer the sequence σ is, the smaller is the set of pure strategies that agree with σ. Conversely, shorter sequences have more strategies that agree with them. The extreme case is the empty sequence $\emptyset$, because every pure strategy agrees with that sequence. Consequently, we have

$$\mu[\emptyset] = 1. \tag{10.5}$$

We now look at a different interpretation of equation (10.4). In this special example, lbc and lbd are also sequences of moves, namely $lbc = \sigma_h c$ and $lbd = \sigma_h d$ with $\sigma_h = lb$. So the combined probability $\mu[lbc]$ of all pure strategies that agree with lbc is simply the mixed strategy probability $\mu(lbc)$, and similarly $\mu[lbd] = \mu(lbd)$. Clearly, $\mu[lb] = \mu[lbc] + \mu[lbd]$. In general, considering any information set h of player i, we have the following important equation:

$$\mu[\sigma_h] = \sum_{c \in C_h} \mu[\sigma_h c]. \tag{10.6}$$

This equation holds because because the sequence σ_h that leads to h is extended by some move c at h, giving the longer sequence $\sigma_h c$. When considering the pure strategies that agree with these extended sequences, we obtain exactly the pure strategies that agree with σ_h.

We are now in a position to define the behavior strategy β that is realization equivalent to μ. Namely, given an information set h of player i so that $\mu[\sigma_h] > 0$, let

$$\beta(c) = \frac{\mu[\sigma_h c]}{\mu[\sigma_h]}. \tag{10.7}$$

Because of (10.6), this defines a probability distribution on the set C_h of moves at h, as required by a behavior strategy. In (10.7), the condition of perfect recall is used because otherwise the sequence σ_h that leads to the information set h would not be uniquely defined.

What about the case $\mu[\sigma_h] = 0$? This means that no pure strategy that agrees with σ_h has positive probability under μ. In other words, h is unreachable when player i uses μ. We will show soon that $\mu[\sigma] = \beta[\sigma]$ for all sequences σ of player i. Hence, h is also unreachable when the player uses β, so that $\beta(c)$ for $c \in C_h$ does not need to be defined, that is, β is a *reduced* behavior strategy according to Definition 10.3. Alternatively, we can define some behavior at h (although it will never matter because player i will not move so as to reach h) in an arbitrary manner, for example by giving each move c at h equal probability, setting $\beta(c) = 1/|C_h|$.

We claim that β, as defined by (10.7), is realization equivalent to μ. We can define $\beta[\sigma]$ for a sequence σ of player i, which is the probability that all the moves in σ are made when the player uses β. Clearly, this is simply the product of the probabilities for the moves in σ. That is, if $\sigma = c_1 c_2 \cdots c_n$, then $\beta[\sigma] = \beta(c_1)\beta(c_2) \cdots \beta(c_n)$. When $n = 0$, then σ is the empty sequence $\emptyset$, in which case we set $\beta[\emptyset] = 1$. So we get by (10.7)

$$\beta[c_1 c_2 \cdots c_n] = \beta(c_1)\beta(c_2) \cdots \beta(c_n) = \tfrac{\mu[c_1]}{\mu[\emptyset]} \tfrac{\mu[c_1 c_2]}{\mu[c_1]} \cdots \tfrac{\mu[c_1 c_2 \cdots c_{n-1} c_n]}{\mu[c_1 c_2 \cdots c_{n-1}]} = \mu[c_1 c_2 \cdots c_n]$$

because all the intermediate terms cancel out, and by (10.5). (This is best understood by considering an example like $\sigma = lbc = c_1 c_2 c_3$ in Figure 10.11.) Because $\beta[\sigma] = \mu[\sigma]$ for all sequences σ of player i, the strategies β and μ are realization equivalent, as explained before Theorem 10.4; the reason is that any node u of the game tree is reached by a unique sequence of moves of *all* players and of chance. The moves of the other players have fixed probabilities, so all that matters is the sequence $\sigma^i(u)$ of moves of player i, which is some sequence σ, which has the same probability no matter whether player i uses β or μ. $\qquad\square$

It is instructive to reproduce the preceding proof for the examples after Definition 10.3 or Figure 10.10 above, which gives also some practice in finding behavior strategies in small games.

10.9 Behavior Strategies in the Monty Hall Problem

We illustrate the concept of behavior strategies for a game-theoretic, non-standard version of the *Monty Hall problem*. The original version has been posed as a problem in probability in the context of an early television game show with the showmaster named "Monty Hall". A contestant can win a prize by choosing one of three closed doors. Behind one of these doors is the prize (a car), and behind each of the other two doors is something the contestant does not want (a goat). Once the contestant has selected one of the doors, Monty Hall opens one of the remaining doors, which reveals a goat, and offers the contestant the option to *switch* to the other closed door. The question is, should she switch doors? It seems that the contestant is now facing equal chances between two options, the originally chosen door and the other unopened door, so that there is no point in switching. However, the correct solution is that the original probability $\frac{1}{3}$ that the contestant has chosen the correct door has not changed by Monty Hall's action of revealing a goat behind another door, because Monty Hall knows where the prize is and can always choose such a door. That probability stays at $\frac{1}{3}$, so that the probability that the other unopened door has the prize is $\frac{2}{3}$, and therefore the contestant should indeed switch.

This analysis runs counter to the spontaneous reaction by many people who rather see the opened door with the goat as *confirmation* that they had chosen the correct door in the first place (this is incorrect). Moreover, they are suspicious that Monty Hall only tries to lure them away from their correct choice. That suspicion is correct if Monty Hall would only *sometimes* open another door with a goat behind it and offer the switch to the remaining door. Rather obviously, if Monty Hall only offered the switch when the contestant has chosen the correct door, then switching would mean that she loses the prize, and therefore she should never switch. In short, the original "solution" to the Monty Hall problem is only valid when Monty Hall *always* offers the switch.

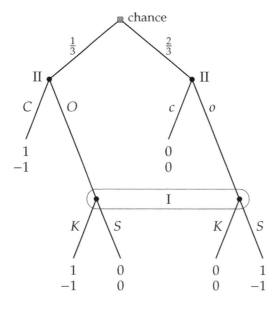

Figure 10.12 The (modified) Monty Hall problem as a game between player I as the contestant who has chosen the correct door with probability $\frac{1}{3}$ (left chance move), and the television show host Monty Hall as player II who can keep the doors closed (moves C and c), or can open another non-prize door (moves O and o). When a door is opened, player I can keep her original choice (K) or switch to the remaining closed door (S). Payoff 1 means that player I gets the prize, paid by player II, payoff 0 means no prize.

Figure 10.12 shows a new game between the contestant as player I and Monty Hall as a player II who has choices after the contestant has chosen one of the three doors. We assume that the prize is hidden behind each door with equal probability. Monty Hall knows whether or not the contestant has chosen the correct door, shown by his two decision nodes with the two moves C and O at the left node and c and o at the right node, which he can choose independently. The moves C and c mean that Monty keeps the doors closed, moves O and o that he opens another door with a goat behind it. In that case the contestant (who does not know if she has chosen the correct door) can keep to her choice (K) or switch to the other unopened door (S), and then gets the prize or not. We assume that the hosting television station prefers not to the pay the prize, so that this is zero-sum game.

This game has the same structure as the introductory game in Figure 10.1 of this chapter, although with reversed roles of the players, and different payoffs and chance probabilities. It has the following strategic form:

$$
\begin{array}{c|c|c|c|c|}
 & Cc & Co & Oc & Oo \\
\hline
K & \begin{smallmatrix}&-\frac{1}{3}\\ \frac{1}{3}&\end{smallmatrix} & \begin{smallmatrix}&-\frac{1}{3}\\ \frac{1}{3}&\end{smallmatrix} & \begin{smallmatrix}&-\frac{1}{3}\\ \frac{1}{3}&\end{smallmatrix} & \begin{smallmatrix}&-\frac{1}{3}\\ \frac{1}{3}&\end{smallmatrix} \\
\hline
S & \begin{smallmatrix}&-\frac{1}{3}\\ \frac{1}{3}&\end{smallmatrix} & \begin{smallmatrix}&\boxed{-1}\\ \boxed{1}&\end{smallmatrix} & \begin{smallmatrix}&\boxed{0}\\ 0&\end{smallmatrix} & \begin{smallmatrix}&-\frac{2}{3}\\ \frac{2}{3}&\end{smallmatrix} \\
\hline
\end{array}
\qquad (10.8)
$$

Although the game is zero-sum, we show the expected payoffs to both players because the best responses are easier to see. If the contestant chooses K, which is the top row in (10.8), then her expected payoff is always $\frac{1}{3}$, and all pure strategies of player II are best responses. This is therefore a degenerate game. Payoff $\frac{1}{3}$ also results when Monty Hall never opens a door (strategy Cc), where player I does not even have the opportunity to switch doors. Hence, K and Cc are mutual best responses with the pure equilibrium (K, Cc), and $\frac{1}{3}$ as the value of the game.

We now study the set of *all* equilibrium strategies of the players. A second pure equilibrium is (K, Oc), where Monty Hall only opens another door if the contestant has chosen the correct door in the first place (what we called "trying to lure her to switch"). In this equilibrium, the unique best response to Oc is K. Hence, player I's unique max-min strategy (and strategy in every equilibrium) is K, by Proposition 8.3 about zero-sum games.

On the other hand, there are many equilibrium strategies of player II, because all of his pure strategies are best responses to K, and even the strategies Co and Oo can potentially have positive probability in a mixed equilibrium. One such mixed strategy is $y = (0, 0, \frac{1}{2}, \frac{1}{2})$, that is, to play Oc and Oo with probability $\frac{1}{2}$ each. Then the top row K has expected payoff $-\frac{1}{3}$ to player II (it always has), and the bottom row S has also expected payoff $-\frac{1}{3}$, so y is also a max-min strategy (in terms of his own payoffs) of player II. A second max-min strategy is $y' = (0, \frac{1}{3}, \frac{2}{3}, 0)$ where the same applies. With y', Monty Hall plays with probability $\frac{1}{3}$ the strategy Co, with the seemingly strange offer to the contestant to switch *only* if she initially chose the wrong door, so that switching would give her the prize for sure; however, y' also chooses with probability $\frac{2}{3}$ the "opposite" strategy Oc. The strategy Oo, chosen with probability $\frac{1}{2}$ by y, is the strategy of always opening another door with a goat behind it, as in the original Monty Hall problem, and payoff $\frac{2}{3}$ when player I chooses S. Nevertheless, in the game the contestant should never switch, because doing so with any positive probability is not a max-min strategy (with a worse payoff than $\frac{1}{3}$ against Oc).

It can be shown that all optimal mixed strategies of player II in this game are convex combinations of the described max-min strategies Cc, Oc, y, and y'. Except for y', these are all behavior strategies. As pure strategies, Cc and Co are trivially behavior strategies. With the strategy y, the behavior of player II is to always choose O at the left decision node and to randomize between c and o with probability $\frac{1}{2}$ at the right decision node. The strategy y' is not a behavior strategy, as its distribution on the move pairs Cc, Co, Oc, Oo in the following table shows:

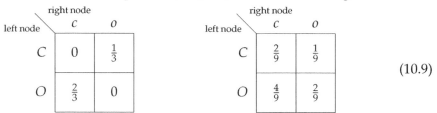

$$(10.9)$$

The right table in (10.9) shows the behavior strategy β' that, by Kuhn's theorem, is realization equivalent to y'. It randomizes between C and O with probabilities $\frac{1}{3}$ and $\frac{2}{3}$, and between c and o with probabilities $\frac{2}{3}$ and $\frac{1}{3}$. Translated back to probabilities for the pure strategies Cc, Co, Oc, Oo, this behavior strategy β' is the mixed strategy $(\frac{2}{9}, \frac{1}{9}, \frac{4}{9}, \frac{2}{9})$, which gives expected payoff $-\frac{1}{3}$ to player II in row S.

Figure 10.13 gives a geometric view of these max-min strategies of player II. The left picture shows the mixed-strategy simplex, and the right picture the square that represents the pairs of behavior strategy probabilities. If p is the probability that he plays O (vertical coordinate of the square, going downwards) and q is the probability that he plays o (horizontal coordinate), then his four pure strategies Cc, Co, Oc, Oo are played with probabilities $(1-p)(1-q)$, $(1-p)q$, $p(1-q)$, and pq, respectively. For $p, q \in \{\frac{1}{4}, \frac{1}{2}, \frac{3}{4}\}$, they define a grid of lines which is shown in both pictures. However, as a subset of the mixed-strategy simplex, this grid is "twisted" and *not convex* because a convex combination of behavior strategies is a general mixed strategy and usually not a behavior strategy. The square is convex, and so is the set of max-min behavior strategies as a subset of that square. The behavior strategy β' belongs to that set, and is a convex combination of Cc and y.

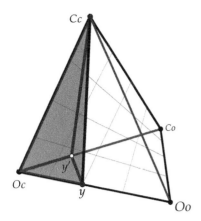 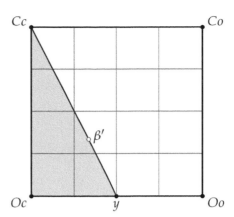

Figure 10.13   Left: Simplex of mixed strategies (always for player II) for the game (10.8), and in gray its subset of max-min strategies with vertices Cc, Oc, y, y'. The "twisted grid" shows the behavior strategies for move probabilities that are multiples of $\frac{1}{4}$. Right: The square of behavior strategies for the extensive game in Figure 10.12, with the behavior strategy β' that is realization equivalent to y', and the same grid. The shaded triangle is the set of max-min behavior strategies.

The pure strategies are the four corners of the square (in the same positions as the cells in (10.9)), and are also the vertices of the mixed-strategy simplex.

10.10   Subgames and Subgame-Perfect Equilibria

In an extensive game with perfect information as studied in Chapter 4, a subgame is any subtree of the game, given by a node of the tree as the root of the subtree and all its descendants. This definition has to be refined when considering games with imperfect information, where every player has to *know* that they have entered the subtree.

Definition 10.5. In an extensive game, a *subgame* is any subtree of the game tree so that every information set is either a subset of or disjoint from the nodes of the subtree.   □

In other words, if any information set h contains some node of the subtree, then h must be completely contained in the subtree. This means that the player to move at h knows that he or she is in the subgame. Information sets that have no nodes in the subtree do not matter because they are disjoint from the subtree.

In Figure 10.14, the subtree with player I's decision node as root defines a subgame of the game. On the other hand, a node in player II's second information set is not the root of a subgame, as shown in the figure for the node that follows move B of player I, because this information set is neither contained in nor disjoint from the subtree with that root.

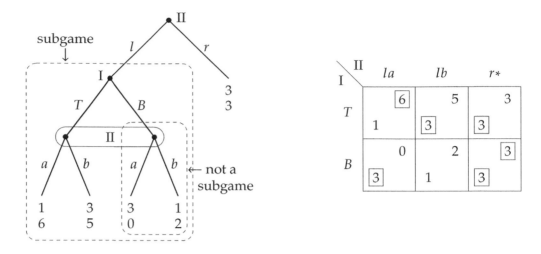

Figure 10.14 Extensive game, with its subgame, and the reduced strategic form of the game. Only one equilibrium of the game is subgame-perfect, see Figure 10.15.

A trivial subgame is either the game itself, or any leaf of the game tree. Some games have no other subgames, for example the games in Figure 10.1 or 10.5.

If every player has perfect recall, then any mixed strategy can be replaced by a realization equivalent behavior strategy by Kuhn's Theorem 10.4. The strategies in any equilibrium can therefore be assumed to be behavior strategies. A behavior strategy can easily be restricted to a subgame, by ignoring the behavior probabilities for the information sets that are disjoint from the subgame. This restriction of a strategy would be much more difficult to define for a profile of mixed strategies.

Definition 10.6. In an extensive game with perfect recall, a *subgame-perfect equilibrium* (SPE) is a profile of behavior strategies that defines an equilibrium for every subgame of the game.   □

In Figure 10.14, the reduced strategic form of the extensive game is shown on the right. Figure 10.15 shows the non-trivial subgame that starts with the move of player I, and its strategic form. This subgame has a unique mixed equilibrium, where player I chooses T, B with probabilities $\frac{2}{3}, \frac{1}{3}$, and player II chooses a, b with probabilities $\frac{1}{2}, \frac{1}{2}$. The resulting payoffs to players I and II are 2 and 4, which are substituted for the subgame to obtain a smaller game where player II chooses between l and r.

This smaller game is shown on the right in Figure 10.15. Because this small game has perfect information, an SPE is found by backward induction, which here means that player II chooses l with certainty. In the overall resulting behavior strategies, player I mixes T, B with probabilities $\frac{2}{3}, \frac{1}{3}$, and player II chooses l for

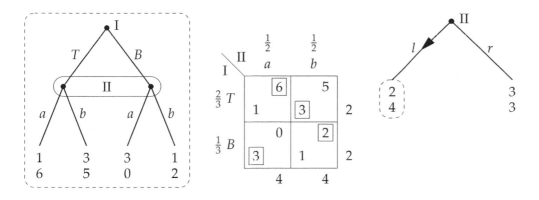

Figure 10.15 The subgame of the game in Figure 10.14, and its unique mixed equilibrium with payoffs 2 and 4. In the full game, the subgame is replaced by these payoffs, as shown on the right, so that the optimal move of player II at the root is l. This defines the SPE.

sure and mixes a, b with probabilities $\frac{1}{2}, \frac{1}{2}$. This profile of behavior strategies is an SPE.

The step of computing an equilibrium of a subgame and substituting the resulting equilibrium payoffs for the subgame generalizes the backward induction step in games with perfect information. In games with perfect information, this backward induction step amounts to finding an optimal move of a single player in a very simple subgame where each move leads directly to a payoff. The optimal move defines an equilibrium in that subgame. In a game with imperfect information, the subgame may involve more than one player, and the equilibrium is usually more difficult to find.

The described SPE of the extensive game in Figure 10.10 is the mixed-strategy equilibrium $((\frac{2}{3}, \frac{1}{3}), (\frac{1}{2}, \frac{1}{2}, 0))$ of the reduced strategic form, which also defines a pair of behavior strategies. Another equilibrium is the pair (B, r_*) of reduced pure strategies. The strategy r_* is only a reduced behavior strategy for player II, because it does not specify the probabilities for the moves a and b at her second information set (which is unreachable when she makes move r). However, this equilibrium is not subgame-perfect, because we have already determined the SPE, which in this game is unique.

In general, just as for games with perfect information, an equilibrium of an extensive game can only be called subgame-perfect if each player's behavior (or pure) strategy is not reduced, but is fully specified and defines a probability distribution (or deterministic move) for each information set of the player.

We described for the example in Figure 10.14 how to find an SPE, which generalizes the backward induction process. We describe this generally in the proof of the following theorem.

Theorem 10.7. *Any extensive game with perfect recall has an SPE in behavior strategies.*

Proof. If the game tree consists of a single node, which is simultaneously the root of the game tree and its only leaf with a payoff to each player, then the theorem holds trivially because every player has only one strategy (doing nothing) and this defines an equilibrium of the game, which is also trivially subgame-perfect.

Hence, assume that the game tree does not just consist of a single node. A SPE of the game is then found inductively as as follows. First, find a subgame of the game with root u that is not just a leaf and that has itself no further nontrivial subgames. Such a subgame exists, because if this does not hold for the root of the entire game, then consider a nontrivial subgame with root u instead, and continue in that manner until a subgame that has no further subgames is found.

Second, consider an equilibrium of this subgame, which exists by Nash's Theorem 6.5, and which can be represented using behavior strategies by Kuhn's Theorem 10.4. This is an SPE of the subgame because the subgame has itself no subgames. In the entire game, replace the root u of the subgame with a leaf whose payoffs are the expected payoffs in the equilibrium of the subgame. The resulting game tree then has fewer nodes, so that this process can be repeated, until u is the root of the original game. The resulting behavior strategies are then combined to define a profile of behavior strategies for the original game that by construction is subgame-perfect.   □

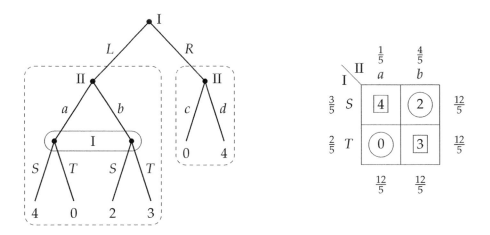

Figure 10.16 Left: Zero-sum game, with payoffs to player I, and its nontrivial subgames. Right: Strategic form of the left subgame, and its optimal mixed strategies.

We illustrate the use of subgames for the example in Figure 10.16, which is a zero-sum game; payoffs are to player I. The game has two nontrivial subgames, each of which starts with a move of player II. In the right subgame, in an SPE player II as the minimizer clearly chooses move c with payoff 0 to player II. The left subgame is a 2×2 game with the strategic form shown on the right in Figure 10.16. This game has a unique mixed equilibrium, found with the "difference trick" as $((\frac{3}{5}, \frac{2}{5}), (\frac{1}{5}, \frac{4}{5}))$ and payoff $\frac{12}{5} = 2.4$ to player I, and -2.4 to player II. It describes the optimal behavior of the players in the left subgame, and is part of a behavior strategy for the whole game. At the root of the game tree, player I will choose L, as this gives him the higher payoff 2.4, which is the unique value of the game because the game is zero-sum.

It is also possible to find all equilibria of this game, including those that are not subgame-perfect, by considering the game tree and without creating the strategic form. Because the game is zero-sum, its payoff is the max-min payoff to each player in any equilibrium, and therefore 2.4 for player I and -2.4 for player II. If R is played with positive probability, then the minimizing player II has to play c because otherwise her expected payoff would be lower. With player II choosing c, this would lower player I's payoff compared to playing L and then the max-min strategy that mixes S and T as described above. Hence, L is played for certain in any equilibrium, with player I's behavior strategy as described before. However, this means that the right decision node of player II is never reached, and player II can mix between c and d as long as the expected payoff to player I does not exceed 2.4, that is, as long as $\text{prob}(d) \le 0.6$. This behavior at the right node, plus the behavior for mixing a, b with probabilities $\frac{1}{5}, \frac{4}{5}$ as described above, is the set of behavior strategies of player II in an arbitrary equilibrium of the game.

10.11 Further Reading

The example of the competition between software firms in Section 10.2 is adapted from Turocy and von Stengel (2002).

Information sets were introduced by Harold Kuhn (1953). Realization equivalent strategies are just called "equivalent" by Kuhn. His Definition 17 states that player i has perfect recall if for every pure strategy π of player i and every information set h of player i that is reachable ("relevant") when playing π, all nodes in h are reachable ("possible") when playing π. This is equivalent to our Definition 10.1 of perfect recall, which we have adapted from Selten (1975, p. 27).

The concept of a reduced behavior strategy is not standard, but we have introduced it here because it corresponds to a behavior strategy in the same way as a reduced pure strategy corresponds to a pure strategy.

Other textbook treatments of extensive games with imperfect information are Osborne and Rubinstein (1994, Chapter 11) and Gibbons (1992, Section 2.4).

Our proof of Kuhn's Theorem 10.4 is based on sequences of moves. These sequences play a central role in the *sequence form* of von Stengel (1996). This is a strategic description that instead of pure strategies considers sequences σ, which are played with probabilities $\mu[\sigma]$ that fulfill the equations (10.6) for all $h \in H_i$ and (10.5). Whereas the strategic form may be exponentially larger than the extensive game, the sequence form has the same size. A zero-sum game, for example, can then be solved with the help of a linear program of the same size as the game tree.

The Monty Hall problem was posed in Selvin (1975). Its behavioral assumptions are discussed and tested in Friedman (1998). Even though our game-theoretic analysis suggests that "not switching doors" is very justifiable when Monty Hall does not always offer a switch, this is still considered a "choice anomaly".

10.12   Exercises for Chapter 10

Exercise 10.1. Explain the two ways, with either player I or player II moving first, of representing an $m \times n$ bimatrix game as an extensive game that has the given game as its strategic form. Giving an example may be helpful. In both cases, how many decision nodes do the players have? What is the number of terminal nodes of the game tree? Why do we need information sets here?

Exercise 10.2. Which of the following extensive games have perfect recall and if not, why not? For each extensive game with perfect recall, find all its equilibria in mixed (including pure) strategies.

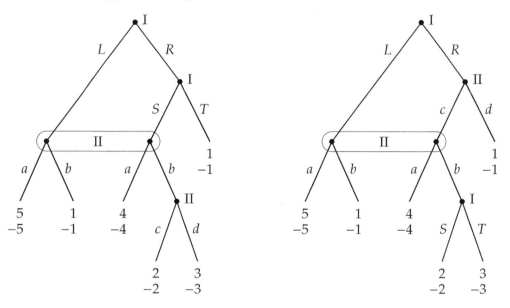

Exercise 10.3. Which of the following extensive games have perfect recall and if not, why not? For each extensive game with perfect recall, find all its equilibria in mixed (including pure) strategies.

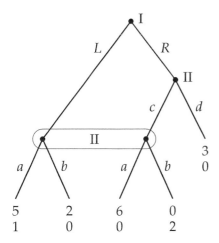 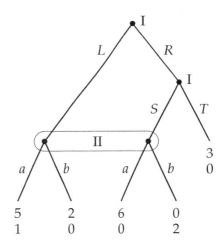

Exercise 10.4. Consider the following zero-sum game, a simplified version of Poker adapted from Kuhn (1950). A deck has three cards (of rank High, Middle, and Low), and each player is dealt a card. All deals are equally likely, and of course the players get different cards. A player does not know the card dealt to the other player. After seeing his hand, player I has the option to Raise (R) or Fold (F). When he folds, he loses one unit to player II. When he raises, player II has the option to meet (m) or pass (p). When player II chooses "pass", she has to pay one unit to player I. When player II chooses "meet", the higher card wins, and the player with the lower card has to pay two units to the winning player.

(a) Draw a game in extensive form that models this game, with information sets, and payoffs to player I at the leaves.

(b) Simplify this game by assuming that at an information set where a player's move is always at least as good as the other move, no matter what the other player's move or the chance move were or will be, then the player will choose that move. Draw the simplified extensive form game.

(c) What does the simplification in (b) have to do with weakly dominated strategies? Why is the simplification legitimate here?

(d) Give the strategic form of the simplified game from (b) and find an equilibrium of that game. What are the behavior probabilities of the players in that equilibrium? What is the unique payoff to the players in equilibrium?

Exercise 10.5. For an extensive game with perfect recall, show that the information set that contains the root (starting node) of any *subgame* is always a singleton (that is, contains only that node). [*Hint: Use Proposition* 10.2.]

Exercise 10.6. Find all equilibria in (pure or) mixed strategies of this extensive game. Which of them are subgame-perfect?

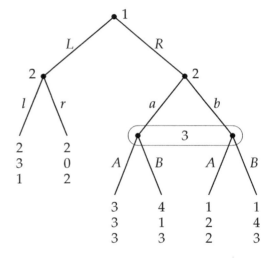

Exercise 10.7. Consider this three-player game in extensive form. At a terminal node, the top, middle, and bottom payoffs are to player 1, 2, 3, respectively. Find all *subgame-perfect equilibria* (SPE) in mixed strategies.

Exercise 10.8. Consider this three-player game in extensive form. Find all *subgame-perfect equilibria* (SPE) in mixed strategies.

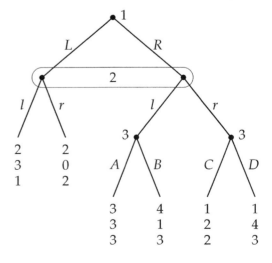

11

Bargaining

This chapter presents models of bargaining. Bargaining occurs in real life between the buyer and seller of a house, for example. This is a game-theoretic problem because it is useful to think about the situation of the other player. Both players have something to gain by reaching an agreement.

A basic bargaining model is that of "splitting a pie" in a way that is acceptable to both players, but where within that range a player can only gain at the expense of the other. The question is how to split the pie fairly. In his undergraduate thesis, later published in Nash (1950), John Nash proposed a system of "axioms", or natural conditions, for arriving at a fair "bargaining solution". Nash showed that these axioms always entail a unique solution, which is also easy to calculate. Section 11.2 shows how a bargaining set arises from a bimatrix game, here a variant of the Prisoner's Dilemma, where the players can benefit from cooperation. General bargaining sets and corresponding axioms are presented in Sections 11.3 and 11.4, with a geometric characterization of the bargaining solution in Section 11.5.

Nash's bargaining model is not a non-cooperative game in strategic or extensive form that we have considered so far. Rather, it belongs to *cooperative game theory*, where players are assumed to be able to *agree* in some way and then adhere to their agreement. The model describes which agreements are possible and how the players benefit from them, but does not describe individual actions that lead to this agreement, nor an enforcement mechanism. The typical approach of cooperative game theory is to propose *solutions* to the game (ideally a unique solution, such as a specific agreement) that have to fulfill certain *axioms*. The Nash bargaining solution is of this kind.

The later sections of this chapter go back to a non-cooperative bargaining model, due to Rubinstein (1982). This model uses a game tree with perfect information where the players make alternating offers that the other player can accept or reject. The analysis of this game uses the study of commitment games from Section 4.7. The most basic form, described in Section 11.7, is the "Ultimatum game" that consists of a single take-it-or-leave-it offer which can only be accepted or rejected.

With alternating offers over several rounds, described in Sections 11.8 and 11.9, there is a chance at each stage that the game does not proceed to the next round so that players wind up with nothing because they could not agree. The effect of this is a "discount factor" which quantifies to what extent an earlier agreement is better that a later one. By thinking ahead, the optimal strategies of the players define sequences of alternating offers, with agreement in the first round. This can be done with a fixed number of rounds, or with "stationary strategies" for an infinite number of bargaining rounds where, in essence, the game repeats every two rounds when it comes back to the same player to make an offer (see Section 11.10). The nice conclusion of this alternating-offers bargaining model, shown in Section 11.11, is that a discount factor that represents very patient players leads to the *same* outcome as Nash's axiomatic bargaining solution. Nash's "cooperative" solution concept is thereby justified with a model of self-interested behavior.

11.1   Prerequisites and Learning Outcomes

Prerequisites for this chapter are: two-player games in strategic form (Chapter 3), the concept that utility functions admit positive-affine transformations (Section 5.6), concave utility functions (Section 5.9), commitment games (Section 4.7), and max-min strategies (Section 8.2).

After studying this chapter, you should be able to:

- explain the concepts of bargaining sets, bargaining axioms, Nash bargaining solution, threat point, and Nash product;
- draw bargaining sets derived from bimatrix games, and for bargaining over a "unit pie" when players have concave utility functions;
- explain, and solve specific examples of, the model of bargaining over multiple rounds of demands and counter-demands, and of stationary SPE for infinitely many rounds.

11.2   Bargaining Sets

Consider the variant of the Prisoner's Dilemma game on the left in Figure 11.1. Player I's strategy B dominates T, and r dominates l, so that the unique equilibrium of the game is (B, r) with payoff 1 to both players. The outcome (T, l) would give payoffs 2 and 3 to players I and II, which is clearly better.

Suppose now that the players can talk to each other and have agreed on a certain course of action that they intend to follow. We do not model how the players can talk, agree, or enforce their agreement, but instead try to give guidelines about which agreement they should reach.

The first step is to model what players can achieve by an agreement, by describing a "bargaining set" of possible payoff pairs that they can bargain over. Suppose that the players can agree to play any of the cells of the payoff table in Figure 11.1. The resulting utility pairs (u, v) are points in a two-dimensional diagram with horizontal coordinate u (utility to player I) and vertical coordinate v (utility to player II). These points are shown on the right in Figure 11.1, marked with the respective strategy pairs. For example, T, l marks the utility pair $(2, 3)$.

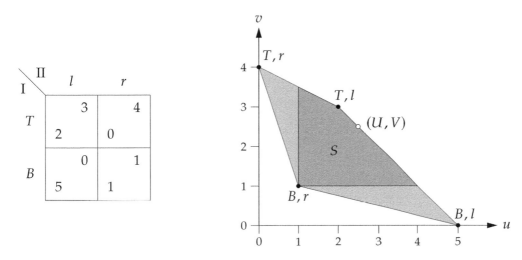

Figure 11.1   Left: 2×2 game similar to the Prisoner's Dilemma. Right: corresponding bargaining set S (dark gray), and its bargaining solution (U, V).

The *convex hull* (set of convex combinations) of these utility pairs is shown in light and dark gray in Figure 11.1. Any such convex combination is obtained by a joint lottery that the players have decided to play as a possible result of their agreement. For example, the utility pair $(1.5, 2)$ would be the expected utility if the players have decided to toss a fair coin, and, depending on the outcome, to either play (T, l) or else (B, r); we assume that this coin-toss and subsequent joint action is part of the players' agreement and they will follow through with it.

Making the bargaining set convex with the help of lotteries is not as strange as it sounds. For example, if two people find something that both would like to have but that cannot be divided, tossing a coin to decide who gets it is often accepted as a fair solution.

Not every point in the convex hull is "reasonable" because we always assume that the players have the option *not to agree* to it in advance. For example, player I would not agree that both players choose with certainty the strategy pair (T, r), where player I gets payoff zero. The reason is that without any agreement, player I can use his *max-min* strategy B that guarantees him a payoff of 1 irrespective of what the other player does. As shown in Chapter 8 on zero-sum games, a max-min strategy may be a mixed strategy; in the Prisoner's Dilemma game in Figure 11.1 it is a pure strategy. We ask that every agreed outcome of the bargaining situation should give each player at least his max-min payoff (which is, as usual, defined in terms of that player's own payoffs).

The *bargaining set* S derived from a two-player game in strategic form is thus defined as the convex hull of utility pairs (as specified in the game), with the additional constraint that $u \geq u_0$ and $v \geq v_0$ for all (u, v) in S, where u_0 is the max-min payoff to player I and v_0 is the max-min payoff to player II. With $u_0 = v_0 = 1$, the resulting set S is shown in dark gray in Figure 11.1.

In this example, the "Nash bargaining solution" (which is not to be confused with "Nash equilibrium", a very different concept) will give the utility pair $(U, V) = (2.5, 2.5)$, as we will explain (the fact that $U = V$ is a coincidence). This expected payoff pair is a convex combination of the utility pairs $(2, 3)$ and $(5, 0)$. It is achieved by a joint lottery that chooses (T, l) with probability $\frac{5}{6}$ and (B, l) with probability $\frac{1}{6}$. In effect, player II always plays l and player I chooses between T and B with probabilities $\frac{5}{6}$ and $\frac{1}{6}$. However, this is not a "mixed strategy" but rather a lottery performed by an independent device, with an outcome that both players have in advance decided to accept.

11.3   Bargaining Axioms

We consider bargaining situations modeled as bargaining sets, and try to obtain a "solution" for any such situation. The conditions about this model are called *axioms* and concern both the bargaining set and the solution one wants to obtain.

Definition 11.1 (Axioms for bargaining set). A *bargaining set* S with *threat point* (u_0, v_0) is a nonempty subset of $\mathbb{R}^2$ so that

(a) $u \geq u_0, v \geq v_0$ for all $(u, v) \in S$;

(b) S is compact;

(c) S is convex.   □

Each pair (u, v) in a bargaining set S represents a utility u for player I and a utility v for player II, which the players can achieve by reaching a specific agreement. At least one agreement is possible because S is nonempty. The threat point (u_0, v_0) is the pair of utilities that each player can (in expectation) guarantee for himself. Axiom (b) in Definition 11.1 states that S is bounded and closed, with the aim of defining a "bargaining solution" by some sort of maximization process. The convexity axiom (c) means that players have, if necessary, access to joint lotteries as part of their agreement.

In Section 11.2, we have constructed a bargaining set S from a bimatrix game (A, B). We show that this set fulfills the axioms in Definition 11.1. This is clear for (b) and (c). By construction, (a) holds, but we have to show that S is not empty. The threat point (u_0, v_0) is obtained with u_0 as the max-min payoff to player I (in terms of his payoffs) and v_0 as the max-min payoff to player II (in terms of her payoffs). Thus, if player I plays his max-min strategy $\hat{x}$ and player II plays her max-min strategy $\hat{y}$, then the payoff pair $(u, v) = (\hat{x}^\top A \hat{y}, \hat{x}^\top B \hat{y})$ that results in the game is an element of S because $u \geq u_0$ and $v \geq v_0$, so S is not empty.

However, in the construction of a bargaining set from a bimatrix game, the threat point (u_0, v_0) need not itself be an element of S, as shown in Exercise 11.2(ii). For this reason, the threat point (u_0, v_0) is explicitly stated together with the bargaining set. If the threat point belongs to S and (a) holds, then there can be no other threat point. The axioms in Definition 11.2 below would be easier to state if the threat point was included in S, but then we would have to take additional steps to obtain such a bargaining set from a bimatrix game. In our examples and drawings we will typically consider sets S that include the threat point.

From now on, we simply assume that the bargaining set S fulfills these axioms. We will later consider bargaining sets for other situations, which are specified directly and not from a bimatrix game.

In the following definition, the axioms (d)–(h) continue (a)–(c) from Definition 11.1. We discuss them afterwards.

Definition 11.2 (Axioms for bargaining solution). For a given bargaining set S with threat point (u_0, v_0), a *bargaining solution* $N(S)$ is a pair (U, V) so that

(d) $(U, V) \in S$;

(e) the solution (U, V) is *Pareto-optimal*, that is, for all $(u, v) \in S$, if $u \geq U$ and $v \geq V$, then $(u, v) = (U, V)$;

(f) it is *invariant under utility scaling*, that is, if $a, c > 0$ and $b, d \in \mathbb{R}$ and S' is the bargaining set $\{(au + b, cv + d) \mid (u, v) \in S\}$ with threat point $(au_0 + b, cv_0 + d)$, then $N(S') = (aU + b, cV + d)$;

(g) it *preserves symmetry*, that is, if $u_0 = v_0$ and $(u, v) \in S$ implies $(v, u) \in S$, then $U = V$;

(h) it is *independent of irrelevant alternatives*: If S, T are bargaining sets with the same threat point and $S \subset T$, then either $N(T) \notin S$ or $N(T) = N(S)$. ☐

The notation $N(S)$ is short for "Nash bargaining solution". The solution is a pair (U, V) of utilities for the two players that belongs to the bargaining set S by axiom (d).

The Pareto-optimality stated in (e) means that it should not be possible to improve the solution for one player without hurting the other player. Graphically, the solution is therefore on the upper-right (or "north-east") border of the bargaining set, which is also called the *Pareto-frontier* of the set.

Invariance under utility scaling as stated in (f) means the following. The utility function of a player can be changed by changing its "scale" and "origin" without changing its meaning, just like a temperature scale (see Chapter 5). Such a change for player I means that instead of u we consider $au + b$ where a and b are fixed real numbers and $a > 0$. Similarly, for player II we can replace v by $cv + d$ for fixed c, d and $c > 0$. When drawing the bargaining set S, this means that (by adding b) the first coordinate of any point in that set is moved by the amount b (to the right or left depending on whether b is positive or negative), and stretched by the factor a (if $a > 1$, and shrunk if $a < 1$). In (f), the set S' and the new threat point are obtained from S by applying such a "change of scale" for both players.

Because this scaling does not change the meaning of the utility values (they still represent the same preference), the bargaining solution should not be affected either, as stated in axiom (f). In a sense, this means the bargaining solution should not depend on a direct comparison of the players' utilities. For example, if you want to split a bar of chocolate with a friend you cannot claim more because you like chocolate "ten times more than she does" (which you could represent by changing your utility from u to $10u$), nor can she claim that you should get less because you already have 5 bars of chocolate in your drawer (represented by a change from u to $u + 5$).

Preservation of symmetry as stated in (g) means that a symmetric bargaining set S, where for all (u, v) we have $(u, v) \in S \Leftrightarrow (v, u) \in S$, and with a threat point of the form (u_0, u_0), should also have a bargaining solution of the form $N(S) = (U, U)$.

The independence of irrelevant alternatives stated in (h) is the most complicated axiom. As illustrated in Figure 11.2, it means that if a bargaining set S is enlarged to a larger set T that has additional points (u, v), but without changing the threat

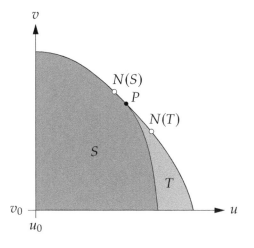

Figure 11.2 Bargaining sets S (dark gray) and T (dark and light gray). Independence of irrelevant alternatives states that if S is a subset of T, then $N(T)$ is either not in S, like here, or equal to $N(S)$, but cannot be another point in S, like P.

point, then either T has a bargaining solution $N(T)$ that is one of these new points, or else it is the same solution as the solution $N(S)$. It is not allowed that the solution for T moves to a different point, like P in Figure 11.2, that still belongs to S.

The stated axioms (a)–(h) for a bargaining set and a possible bargaining solution are reasonable requirements that one can defend as general principles.

11.4 The Nash Bargaining Solution

This section shows that the bargaining axioms (a)–(h) in Definitions 11.1 and 11.2 imply that a bargaining solution exists, which is, moreover, unique.

Theorem 11.3. *Under the Nash bargaining axioms, every bargaining set S that contains a point (u,v) with $u > u_0$ and $v > v_0$ has a unique solution $N(S) = (U, V)$, which is obtained as the unique point (u,v) that maximizes the product (also called the "Nash product") $(u - u_0)(v - v_0)$ for $(u,v) \in S$.*

Proof. First, consider the set $S' = \{(u - u_0, v - v_0) \mid (u,v) \in S\}$, which is the set S translated so that the threat point (u_0, v_0) is the origin $(0,0)$, as shown in Figure 11.3(a) and (b). Maximizing the Nash product then amounts to maximizing uv for $(u,v) \in S'$. Let (U, V) be a utility pair (u,v) where this happens, where by assumption $UV > 0$; Figure 11.3(b) shows the hyperbola $uv = c$ for the maximum number c so that the hyperbola still intersects S'. Next, we re-scale the utilities so that $(U, V) = (1, 1)$, by replacing S with the set $S'' = \{(\frac{1}{U}u, \frac{1}{V}v) \mid (u,v) \in S'\}$, shown in Figure 11.3(c). To simplify notation, we rename S'' to S.

Next, consider the set $T = \{(u,v) \mid u \geq 0, v \geq 0, u + v \leq 2\}$, see Figure 11.3(d). For the bargaining set T, the solution is $N(T) = (1, 1)$, because T is a symmetric set, and $(1, 1)$ is the only symmetric point on the Pareto-frontier of T. We claim that $S \subseteq T$. Then $N(T) = (1, 1) \in S$, so that the independence of irrelevant alternatives implies $N(S) = N(T)$.

We prove $S \subseteq T$ using the convexity of S. Namely, suppose there was a point $(\bar{u}, \bar{v})$ in S which is not in T. Because $\bar{u} \geq 0, \bar{v} \geq 0$, this means $\bar{u} + \bar{v} > 2$. Then, as shown in Figure 11.3(d), even if the Nash product $\bar{u} \cdot \bar{v}$ is not larger than 1 (which is the Nash product UV), then the Nash product uv of a suitable convex combination $(u,v) = (1 - \varepsilon)(U, V) + \varepsilon(\bar{u}, \bar{v})$ for some small positive ε is larger than 1.

This is illustrated in Figure 11.3(d): The hyperbola $\{(u,v) \mid uv = 1\}$ touches the set T at the point $(1, 1)$. The tangent to the hyperbola is the line through points $(2, 0)$ and $(0, 2)$, which is the Pareto-frontier of T. Because the point $(\bar{u}, \bar{v})$ is to the right of that tangent, the line segment that connects $(1, 1)$ and $(\bar{u}, \bar{v})$ intersects the hyperbola and therefore contains points (u,v) to the right of the hyperbola, where $uv > 1$. This is a contradiction, because $(u,v) \in S$ since S is convex, but the maximum Nash product of a point in S is 1.

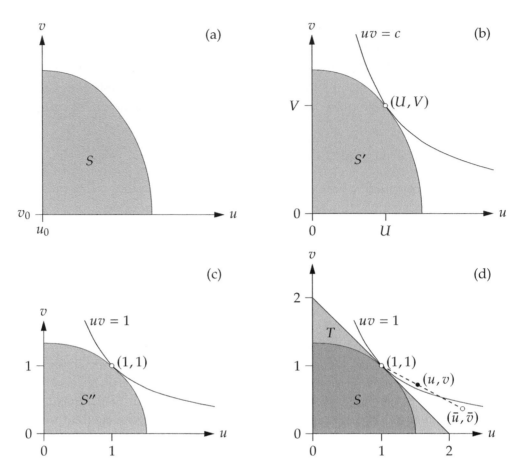

Figure 11.3 Proof steps for Theorem 11.3. The dashed line in (d) shows that the point $(\bar{u}, \bar{v})$, which is outside the triangle T (dark and light gray), cannot belong to S (dark gray), which shows that $S \subseteq T$.

To prove this formally, consider the Nash product for the convex combination (u, v) of $(U, V) = (1, 1)$ and $(\bar{u}, \bar{v})$:

$$
\begin{aligned}
uv &= [(1 - \varepsilon)U + \varepsilon\bar{u}][(1 - \varepsilon)V + \varepsilon\bar{v}] \\
&= [U + \varepsilon(\bar{u} - U)][V + \varepsilon(\bar{v} - V)] \\
&= UV + \varepsilon[\bar{u} + \bar{v} - (U + V) + \varepsilon(\bar{u} - U)(\bar{v} - V)]. \quad (11.1)
\end{aligned}
$$

Now, because $\bar{u} + \bar{v} > 2$, but $U + V = 2$, the term $\bar{u} + \bar{v} - (U + V)$ is positive, and $\varepsilon(\bar{u} - U)(\bar{v} - V)$ can be made smaller than that term in absolute value, for suitably small $\varepsilon > 0$, even if $(\bar{u} - U)(\bar{v} - V)$ is any negative number. Consequently, $uv > 1 = UV$ for sufficiently small positive ε, which shows that UV was not the maximum Nash product in S, which contradicts the construction of (U, V). This proves that, indeed, $S \subseteq T$, and hence $N(S) = N(T)$ as claimed.

In order to show that the point $(U, V) = (1, 1)$ is unique, we can again look at the hyperbola, and argue similarly as before. Namely, suppose that $(u, \frac{1}{u})$ is another point in S on the hyperbola with $u \neq 1$. Because S is convex, the line segment that connects $(1, 1)$ and $(u, \frac{1}{u})$ is also in S, but its interior points have a larger Nash product. For example, the mid-point $\frac{1}{2} \cdot (1, 1) + \frac{1}{2} \cdot (u, \frac{1}{u})$ has the Nash product $(1 + u + \frac{1}{u} + u\frac{1}{u})/4$, which is larger than 1 if and only $u + \frac{1}{u} > 2$, which holds because this is equivalent to $(u - 1)^2 > 0$, which holds because $u \neq 1$. Hence, again because S is convex, the Nash product has a unique maximum in S. $\qquad\square$

The assumption that S has a point (u, v) with $u > u_0$ and $v > v_0$ has the following purpose: Otherwise, for $(u, v) \in S$, all utilities u of player I or v of player II (or both) fulfill $u = u_0$ or $v = v_0$, respectively. Then the Nash product $(u - u_0)(v - v_0)$ for $(u, v) \in S$ would always be zero, so it has no unique maximum unless S is just the singleton $\{(u_0, v_0)\}$. Then the bargaining set S is a singleton or a line segment $\{u_0\} \times [v_0, V]$ or $[u_0, U] \times \{v_0\}$. In this rather trivial case, the Pareto-frontier consists of a single point (U, V), which defines the unique Nash solution $N(S)$.

> $\Rightarrow$ If the bargaining set is a line segment as discussed in the preceding paragraph, why is its Pareto-frontier a single point? Is this always so if the line segment is not parallel to one of the coordinate axes?

It can be shown relatively easily that the maximization of the Nash product as described in Theorem 11.3 gives a solution (U, V) that fulfills the axioms in Definitions 11.1 and 11.2.

> $\Rightarrow$ In Exercise 11.1, you are asked to verify this for axiom (f).

We describe how to find the bargaining solution (U, V) for the bargaining set S in Figure 11.1. The threat point (u_0, v_0) is $(1, 1)$. The Pareto-frontier consists of two line segments: a first line segment that joins $(1, 3.5)$ to $(2, 3)$, and a second line segment that joins $(2, 3)$ to $(4, 1)$, where the latter is part of a longer line segment that extends from $(2, 3)$ to $(5, 0)$ (the first line segment is also part of the longer segment that connects $(4, 0)$ to $(2, 3)$, but this will not be needed). The Nash product $(u - u_0)(v - v_0)$ for a point (u, v) on the Pareto-frontier is the size of the area of the rectangle between the threat point (u_0, v_0) and the point (u, v). In this case, the diagram shows that the second line segment creates the larger rectangle in that way. A point on the second line segment is of the form $(1 - p)(2, 3) + p(5, 0)$ for $0 \leq p \leq 1$, that is, $(2 + 3p, 3 - 3p)$. Because $(u_0, v_0) = (1, 1)$, the resulting Nash product is $(1 + 3p)(2 - 3p)$ or $2 + 3p - 9p^2$. The unique maximum over p of this expression is obtained for $p = \frac{1}{6}$ because this is the zero of its derivative $3 - 18p$, with negative second derivative, so this is a maximum. The point (U, V) is therefore given as $(2.5, 2.5)$. (In general, such a maximum can result for $p < 0$ or $p > 1$, which implies that the left or right endpoint of the line segment with $p = 0$

or $p = 1$, respectively, gives the largest value of the Nash product. Then one has to check the adjoining line segment for a possibly larger Nash product, which may occur at the corner point.) This value of p allows representing the Nash bargaining solution as a convex combination of payoff pairs that correspond to cells in the bimatrix game, namely via a lottery that chooses (T, l) with payoff pair $(2, 3)$ with probability $1 - p = \frac{5}{6}$ and (B, r) with payoff pair $(5, 0)$ with probability $p = \frac{1}{6}$. Both players have agreed in advance to accept the outcome of this lottery (even though with probability $\frac{1}{6}$ the pair $(5, 0)$ gives payoff zero to player II, which is less than at the threat point where she gets $v_0 = 1$; however, her expected payoff 2.5 for the lottery is higher than v_0).

11.5   Geometry of the Bargaining Solution

In this section, we give a geometric characterization of the bargaining solution. This characterization helps finding the bargaining solution quickly, and is needed later in Section 11.11. We first state an easy lemma.

Lemma 11.4. *Consider a triangle T with vertices $(0, 0)$, $(a, 0)$, $(0, b)$ with $a, b > 0$. Then the rectangle of maximal area contained in T with vertices $(0, 0)$, $(U, 0)$, (U, V) and $(0, V)$ is given by $U = a/2$ and $V = b/2$. Furthermore, if $S \subseteq T$ and $(U, V) \in S$, then (U, V) is also the point in S that maximizes the product UV.*

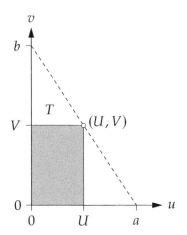

Figure 11.4   Illustration of Lemma 11.4. The product UV, which is the area of the gray rectangle inscribed in the triangle T, is maximal for $U = a/2$ and $V = b/2$.

Proof. The product UV is the area of the rectangle inscribed in T as shown in Figure 11.4. The maximum value of uv for $(u, v) \in T$ clearly results when (u, v) is on the line that joins $(b, 0)$ and $(a, 0)$, shown as a dashed line in Figure 11.4, that is, when $v = b - \frac{b}{a}u$. The derivative of the function $u \mapsto u \cdot (b - \frac{b}{a}u)$ is $b - 2\frac{b}{a}u$, which is zero when $u = a/2$; this is a maximum of the function because its second derivative is negative. Hence, UV is maximal for $U = a/2$ and $V = b/2$ as claimed.

If $S \subseteq T$, then the maximum of uv for $(u,v) \in S$ is clearly at most the maximum of uv for $(u,v) \in T$, and therefore attained for $(u,v) = (U,V)$ if $(U,V) \in S$. □

The following proposition gives a useful condition for finding the bargaining solution geometrically. The bargaining set is assumed to be convex as per axiom (c), to have threat point $(0,0)$, and to have a Pareto-frontier of the form $(u, f(u))$ for a decreasing function f where, for definiteness, $0 \le u \le 1$; any other right endpoint of the interval for u is also possible.

Proposition 11.5. *Suppose that the Pareto-frontier of a bargaining set S with threat point $(0,0)$ is given by $\{(u, f(u)) \mid 0 \le u \le 1\}$ for a decreasing and continuous function f with $f(0) > 0$ and $f(1) = 0$. Then the bargaining solution (U,V) is the unique point $(U, f(U))$ where the bargaining set has a tangent line with slope $-f(U)/U$. If f is differentiable, then this slope is the derivative $f'(U)$ of f at U, that is, $f'(U) = -f(U)/U$.*

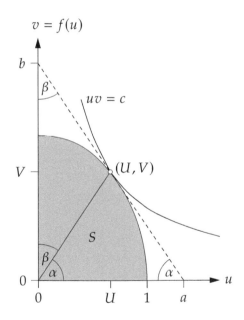

Figure 11.5 Illustration of Proposition 11.5 and its proof. The dashed line is the tangent to the hyperbola $\{(u,v) \mid uv = c = UV\}$ at the point (U,V), which is also a tangent to the bargaining set S at (U,V). The endpoints $(a,0)$ and $(0,b)$ of this tangent fulfill $a = 2U$ and $b = 2V$ with angles α and β as shown.

Proof. Let (U,V) be the bargaining solution for the bargaining set S. It is on the Pareto-frontier, so $V = f(U)$. Let $c = UV$ and consider the hyperbola $\{(u,v) \mid uv = c\}$, which intersects S in the unique point (U,V) by Theorem 11.3. Because (U,V) maximizes the Nash product, c is the maximal value of $c = uv$ for $(u,v) \in S$. The function $u \mapsto c/u$ is differentiable with derivative $-c/u^2$, which for $u = U$ is equal to $-V/U$. Consider the line through (U,V) with slope $-V/U$, which is the tangent to the hyperbola at (U,V), shown as a dashed line in Figure 11.5. This line intersects the u-axis at the point $(a,0) = (2U,0)$ and the v-axis at $(0,b) = (0,2V)$, and defines a triangle T with vertices $(0,0)$, $(a,0)$, and $(0,b)$. Figure 11.5 corresponds to Figure 11.3(b), and if we re-scale the axes by replacing u with u/U and v with v/V then we obtain Figure 11.3(d) where T is the triangle with vertices $(0,0)$, $(2,0)$, and $(0,2)$. With the help of (11.1) we showed $S \subseteq T$, where the assumption $(U,V) = (1,1)$ was not used, so this applies also to Figure 11.5, without re-scaling the axes. Hence, the dashed line is also a tangent to the set S, as claimed.

If the function f is not differentiable, then there may be more than one tangent to the set S at a point. In that case it remains to prove that the described construction of (U,V) is unique. Suppose that a tangent of S touches S at the point (U,V) and has the downward slope $-V/U$, and intersects the u-axis at $(a,0)$ and the v-axis at $(0,b)$ as in Figure 11.5 (where we now ignore the hyperbola). The slope $-V/U$ of the tangent implies that $a = 2U$ and $b = 2V$. Being a tangent to the set S means that S is a subset of the triangle T with vertices $(0,0)$, $(a,0)$, and $(0,b)$. By Lemma 11.4, UV is the maximum value of uv for $(u,v) \in S$. This shows that (U,V) is unique.

If the function f is differentiable, then the tangent to the set S is unique at every point, and we can argue directly as follows. The bargaining solution maximizes the Nash product $u \cdot f(u)$ on the Pareto-frontier. This requires the derivative with respect to u to be zero, that is, $f(u) + u \cdot f'(u) = 0$ or $f'(u) = -f(u)/u$, so U has to solve this equation, as claimed. $\square$

In Figure 11.5, the point (U,V) is the midpoint of the long side of the right-angled triangle with vertices $(0,0)$, $(a,0)$, and $(0,b)$. The line from $(0,0)$ to (U,V) divides that triangle into two isosceles triangles. The first triangle has apex (U,V) and equal (mirrored) angles α at its vertices $(0,0)$ and $(a,0)$. These angles correspond to the slopes V/U and $-V/U$, equal except for their sign, of the two lines that connect these vertices to (U,V). The second triangle has apex (U,V) and equal (mirrored) angles β at its vertices $(0,0)$ and $(0,b)$.

Figure 11.6 shows bargaining sets S with a Pareto-frontier given by line segments, as it arises when S is derived from a bimatrix game as in Figure 11.1 (which with the threat point shifted to $(0,0)$ defines the set shown in Figure 11.6(b)). Then the Nash bargaining solution is found easily with the help of Lemma 11.4: Find a tangent to the bargaining set S with endpoints $(a,0)$ and $(0,b)$ so that the midpoint (U,V) of that tangent belongs to S (Proposition 11.5 asserts that this tangent exists). Because the boundary of S consists of line segments, the tangent at an interior point of the line segment has the same slope as that line segment. Only finitely many slopes need to be checked. In Figure 11.6(a), the middle line segment contains the midpoint (U,V) of the corresponding tangent.

The case of a Pareto-frontier with only two line segments is particularly easy. In Figure 11.6(b), the right line segment contains the midpoint (U,V) of the corresponding tangent. That tangent and the line segment have the same right endpoint $(a,0)$, which implies $U = \frac{a}{2}$. Similarly, in Figure 11.6(c), the top line segment contains the midpoint (U,V) of the corresponding tangent. That tangent and the line segment have the same left endpoint $(0,b)$, which implies $V = \frac{b}{2}$. In

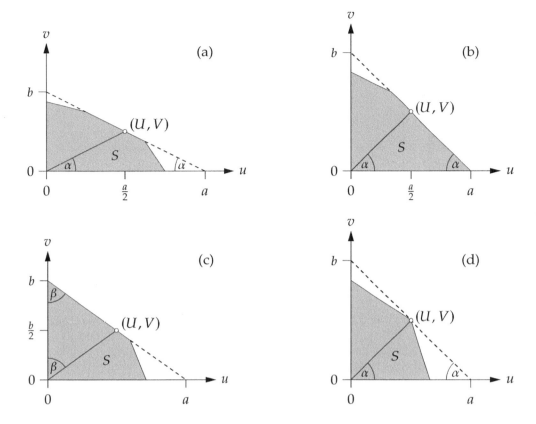

Figure 11.6   Various bargaining sets S with a piecewiese linear Pareto-frontier. By Proposition 11.5, the Nash bargaining solution (U,V) is the midpoint of the tangent to S (dashed line) intersected with the positive quadrant, with endpoints $(a,0)$ and $(0,b)$. In (d) the point (U,V) is a vertex of S, where there are many tangents to S through that point. One of these tangents has the same angle α with the u-axis as the upwards line from $(0,0)$ to (U,V).

Figure 11.6(d), neither line segment defines a tangent whose midpoint belongs to S, because the right line segment is too steep and the top line segment too shallow. Then the vertex (U,V) of the bargaining set S is the Nash bargaining solution; this can be verified by considering the dashed line through that vertex with slope $-V/U$, which is the negative of the slope V/U of the line from $(0,0)$ to (U,V), with angle α with the u-axis. That line should be a tangent to S, as it is here.

11.6   Splitting a Unit Pie

In the remainder of this chapter, we consider a bargaining situation where player I and player II have to agree to split a "pie" into an amount x for player I and y for player II. The total amount to be split is normalized to be one unit, so this is called "a unit pie". The possible splits (x, y) have to fulfill $x \geq 0$, $y \geq 0$, and $x + y \leq 1$. If the two players cannot agree, they both receive zero, which defines the threat point $(0, 0)$. Splits (x, y) so that $x + y < 1$ are not Pareto-optimal, but we admit such splits in order to obtain a convex set.

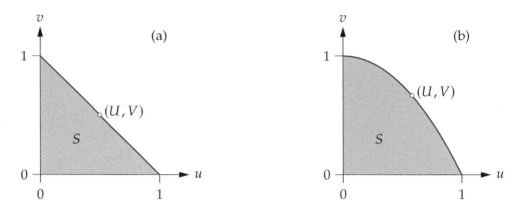

Figure 11.7 Bargaining over a unit pie, with bargaining set S as defined in (11.2). In (a), the utility functions are $u(x) = x$ and $v(y) = y$, with bargaining solution $(U, V) = (\frac{1}{2}, \frac{1}{2})$. In (b), they are $u(x) = \sqrt{x}$ and $v(y) = y$, with bargaining solution $(U, V) = (1/\sqrt{3}, \frac{2}{3}) \approx (0.577, 0.667)$.

If player I gets utility x and player II gets utility y, then the bargaining set is the triangle of all pairs (x, y) with $x \geq 0$, $y \geq 0$ and $x + y \leq 1$, shown in Figure 11.7(a). The split (x, y) is Pareto-optimal if $x + y = 1$, so the Pareto-frontier is given by the splits $(x, 1 - x)$, shown as the bold line in the diagram. Because the threat point is $(0, 0)$, the Nash product is $x(1 - x)$, which is maximized for $x = \frac{1}{2}$. The Nash bargaining solution $(U, V) = (\frac{1}{2}, \frac{1}{2})$ results also from the symmetry axiom in Definition 11.2(g).

More generally, we consider the split of a unit pie where the players have utility functions u and v that are not the identity function. We assume that the utility functions are $u : [0, 1] \to [0, 1]$ for player I and $v : [0, 1] \to [0, 1]$ for player II, which are normalized so that $u(0) = v(0) = 0$ and $u(1) = v(1) = 1$. These utility functions are assumed to be increasing, *concave* (see (5.46) and Figure 5.5), and continuous (it can be shown that concavity implies continuity except at endpoints of the domain of the function). The bargaining set of resulting utility pairs to the two players is given by

$$S = \{(u(x), v(y)) \mid x \geq 0, \ y \geq 0, \ x + y \leq 1\}. \tag{11.2}$$

The following proposition asserts that S fulfills the axioms in Definition 11.1.

Proposition 11.6. *Let $u, v : [0, 1] \to \mathbb{R}$ be increasing, concave, and continuous utility functions of player I and II, respectively, with $u(0) = v(0) = 0$ and $u(1) = v(1) = 1$.*

Then the bargaining set S defined (11.2) has threat point $(0,0)$ and is compact and convex. Moreover, S is downward closed, *that is,*

$$(\bar{u},\bar{v}) \in S, \quad 0 \le u' \le \bar{u}, \quad 0 \le v' \le \bar{v} \quad \Rightarrow \quad (u',v') \in S. \tag{11.3}$$

Proof. We first show (11.3). Let $0 \le u' \le \bar{u}$ and $0 \le v' \le \bar{v}$ with $(\bar{u},\bar{v}) \in S$, that is, $\bar{u} = u(\bar{x})$ and $\bar{v} = u(\bar{y})$ for some $\bar{x} \ge 0$ and $\bar{y} \ge 0$ with $\bar{x} + \bar{y} \le 1$. Because $u(0) = 0 \le u' \le u(\bar{x})$ and u is continuous, we have $u(x') = u'$ for some $x' \in [0,\bar{x}]$ by the intermediate value theorem, and similarly $v(y') = y'$ for some $y' \in [0,\bar{y}]$, which implies $x' + y' \le \bar{x} + \bar{y} \le 1$ and thus $(u',v') \in S$, which shows (11.3).

The set S is the image of a compact set under the continuous function $(x,y) \mapsto (u(x),v(y))$ and therefore is also compact. To show that S is convex, consider a convex combination $(u',v') = (1-\lambda)(u(x),v(y)) + \lambda(u(\tilde{x}),v(\tilde{y}))$ of two points in S, where $\lambda \in [0,1]$. Then $u' \le u((1-\lambda)x + \lambda\tilde{x}) = \bar{u}$ and $v' \le v((1-\lambda)y + \lambda\tilde{y}) = \bar{v}$ because u and v are concave functions, where clearly $(\bar{u},\bar{v}) \in S$. Moreover, $u' \ge 0$ and $v' \ge 0$. Hence, (11.3) implies $(u',v') \in S$, which shows that S is convex. □

For $u(x) = \sqrt{x}$ and $v(y) = y$, the set S is shown in Figure 11.7(b). The Pareto-frontier is given by the points $(u(x),v(1-x))$ for $0 \le x \le 1$. Because $u = \sqrt{x}$ and $v = 1 - x = 1 - u^2$, these are pairs $(u, 1-u^2)$, so v is easy to draw as the function $1 - u^2$ of u. In general, one may think of traversing the points $(u(x),v(1-x))$ on the Pareto-frontier when changing the parameter x, from the upper endpoint $(0,1)$ of the frontier when $x = 0$, to its right endpoint $(1,0)$ when $x = 1$.

The Nash bargaining solution (U,V) is found by maximizing the Nash product $u(x)v(1-x)$ on the Pareto-frontier. In Figure 11.7(b), the Nash product is $\sqrt{x}(1-x)$ or $x^{1/2} - x^{3/2}$ with derivative $\frac{1}{2}x^{-1/2} - \frac{3}{2}x^{1/2}$ or $\frac{1}{2}x^{-1/2}(1-3x)$, which is zero when $x = \frac{1}{3}$, with a negative second derivative, so this is a maximum. Hence, the pie is split into $\frac{1}{3}$ for player I and $\frac{2}{3}$ for player II. The corresponding pair (U,V) of utilities is $(1/\sqrt{3}, \frac{2}{3})$ or about $(0.577, 0.667)$. For a similar calculation with different utility functions see Exercise 11.4.

11.7 The Ultimatum Game

Assume that two players have to split a unit pie, as described in the previous section. In the following, we will develop the "alternating offers" bargaining game applied to this situation. This is a non-cooperative game that models how players may interact in a bargaining situation. The solution to this non-cooperative game, in its most elaborate version of "stationary strategies" (see Section 11.10), will approximate the Nash bargaining solution. The model justifies the Nash solution via a different approach, which is not axiomatic, but instead uses the concept of subgame-perfect equilibria in game trees.

The model is that the pie is split according to the suggestion of one of the players. Suppose player I suggests splitting the pie into x for player I and $1 - x$ for player II, so that the players receive utility $u(x)$ and $v(1 - x)$, respectively. The amount x proposed by player I is also called his *demand* because that is the share of the pie that he receives in the proposed split.

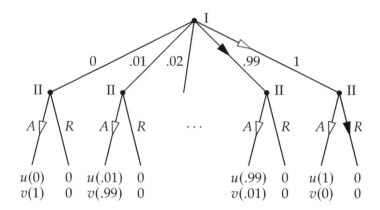

Figure 11.8 Discrete version of the Ultimatum game. The black and white arrows denote the two subgame-perfect equilibria, which are the same except for player II's response to the full demand $x = 1$, and player I's best response of $x = .99$ or $x = 1$.

In the most basic form, this model is called the *Ultimatum game*, where player II can either accept the split or reject it. If player II accepts, the payoff pair is $(u(x), v(1 - x))$, and if she rejects, it is $(0, 0)$, in which case neither player receives anything. Figure 11.8 shows a discretized version of this game where the possible demand x is a multiple of 0.01. Hence, player I has 101 possible actions for $x \in \{\frac{i}{100} \mid i = 0, 1, \ldots, 100\}$, and for each proposed x player II can either choose A (accept) or R (reject). Player II therefore has 2^{101} pure strategies, given by the different combinations of A and R for each possible x.

The game is shown in Figure 11.8, and we look for a subgame-perfect equilibrium (SPE) of the game, that is, because the game has perfect information, we apply backward induction. Whenever $1 - x > 0$, the utility $v(1 - x)$ to player II is positive, because otherwise

$$0 = v(1 - x) = v(x \cdot 0 + (1 - x) \cdot 1) < 1 - x = x \cdot v(0) + (1 - x) \cdot v(1),$$

which contradicts concavity of v (and is also seen graphically). Hence, the SPE condition implies that player II accepts any positive amount $1 - x$ offered by player I. These choices, determined by backward induction, are shown as arrows in Figure 11.8, and they are unique except when $x = 1$ where player I demands the whole pie. Then player II is indifferent between accepting and rejecting because she will get nothing in either case. Consequently, in terms of pure strategies, both $AA \cdots AA$ and $AA \cdots AR$ are SPE strategies of player II. Here, the first 100 A's denote acceptance for each $x = 0, 0.01, \ldots, 0.99$, and the last A or R is the choice when $x = 1$, shown by a white or black arrow, respectively, in Figure 11.8. Given this strategy of player II, player I's best response is to maximize his payoff, which he does as shown by the corresponding white or black arrow for player I. If player II accepts the demand $x = 1$, then player I demands $x = 1$, and the pair $(1, AA \cdots AA)$ is an SPE, and if player II rejects the demand $x = 1$, then player I demands $x = 0.99$ (because for $x = 1$ he would get nothing), and the resulting SPE is $(0.99, AA \cdots AR)$. These are two subgame-perfect equilibria in pure strategies. If $u(0.99) < 1$, they are also unique; if $u(0.99) = 1$ then the demand $x = .99$ is also a best response by player I to the strategy $AA \cdots AA$ of player II that accepts every demand.

In essence, player II receives virtually nothing in the Ultimatum game because of the backward induction assumption. This does not reflect people's behavior in real bargaining situations of this kind, as confirmed in laboratory experiments where people play the Ultimatum game. Typically, such an experiment is conducted where player I states, unknown to player II, his demand x, and player II declares simultaneously, unknown to player I, her own "reserve demand" z; the two players may even be matched anonymously from a group of many subjects that take the respective roles of players I and II. Whenever $1 - x \geq z$, that is, the offer to player II is at least as large as her reserve demand, the two players get x and $1 - x$, otherwise nothing. (So player II may declare a small reserve demand like $z = 0.01$ but still receive 0.4 because player I's demand is $x = 0.6$; both can be regarded as reasonable choices, but they are not mutual best responses.) In such experiments, player II often makes positive reserve demands like $z = 0.4$, and player I makes cautious demands like $x = 0.5$, which contradict the SPE assumption. A better description of these experiments may therefore be equilibria that are not subgame-perfect, with "threats" by player II to reject offers $1 - x$ that are too low. Nevertheless, we assume SPE because they allow definite conclusions, and because the model will be refined, and made more realistic, by allowing rejections to be followed by "counter-demands". (Exercise 11.3 asks you to find *all* equilibria of the discrete Ultimatum game.)

The described two SPE of the discrete Ultimatum game are found whenever the unit pie can be split into $M + 1$ possible splits $(x, 1 - x)$ for $x = \frac{i}{M}$, for $i = 0, 1, \ldots, M - 1, M$, like $M = 100$ above. Of her 2^{M+1} pure strategies, player II can play only two in an SPE, namely always choosing A whenever $x < 1$, and choosing either A or R in response to $x = 1$. The respective demand by player I is $x = 1$ or $x = 1 - \frac{1}{M}$. With a very fine discretization where M becomes very large, these two SPE are nearly identical, with player I demanding all (or nearly all) of the pie, and player II receiving almost nothing.

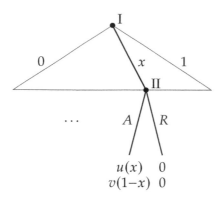

Figure 11.9 The Continuous Ultimatum game. Player I makes any demand $x \in [0,1]$, which player II either accepts (A) or rejects (R). Only one decision point of player II is shown.

In the *Continuous Ultimatum* game, the possible demand x by player I is any number in the interval $[0,1]$, which defines an infinite number of strategies for player I. A strategy of player II is then given by a function $f : [0,1] \rightarrow \{A, R\}$, where $f(x) = A$ means that player II accepts the demand x, and $f(x) = R$ means that player II rejects it. (Equivalently, a strategy of player II is an arbitrary subset of $[0,1]$ that consists exactly of those demands x that she accepts.) This infinite game is indicated in Figure 11.9. The infinitely many choices x are indicated by the triangle, whose baseline represents the interval $[0,1]$. Every point of that baseline has a separate decision point for player II where she can decide between A and R.

We want to show that the Continuous Ultimatum game has a *unique* SPE. This requires an additional assumption, namely that the utility function u of player I is *strictly* increasing. This is the case if $x' = 1$ in the following lemma.

Lemma 11.7. *Let $u : [0,1] \rightarrow \mathbb{R}$ be an increasing, concave, and continuous function with $u(0) = 0$ and $u(1) = 1$. Then u is strictly increasing on an interval $[0, x']$, where $u(x) = 1$ for all $x \in [x', 1]$.*

Proof. The set $I = \{x \in [0,1] \mid u(x) = 1\}$ is not empty (it contains 1) and closed because u is continuous. Let x' be the smallest element of I, with $u(x') = u(1) = 1$. Then $I = [x', 1]$ because $x' \le x \le 1$ implies $1 = u(x') \le u(x) \le u(1) = 1$. We have $x' > 0$ because $u(0) = 0 < 1$. To see that u is strictly increasing on the interval $[0, x']$, suppose there were $\bar{x}$ and $\hat{x}$ in that interval with $\bar{x} < \hat{x}$ but $u(\bar{x}) = u(\hat{x}) < u(x') = 1$. Then it is easy to see (a drawing will help) that $u(\hat{x})$ is below the line that connects $u(\bar{x})$ and $u(x')$, which contradicts the concavity of u. □

If $x' < 1$ in Lemma 11.7, then player I is perfectly satisfed by receiving only x', which is less than a full share of the pie. Then any demand x with $x' \le x < 1$ is accepted by player II because she receives $v(1 - x) > 0$, and x is a possible strategy of player I in an SPE. In that case, we do not get a unique SPE. We therefore require that u is strictly increasing, or equivalently, by Lemma 11.7, that $u(x) = 1$ implies $x = 1$. Hence, we assume that the utility function u of player I, and similarly the utility function v of player II, is strictly increasing.

The following proposition states that the *only* SPE of the Continuous Ultimatum game is where player I demands the whole pie, with $x = 1$, and where player II accepts any demand, including the demand $x = 1$ on the equilibrium path where she is indifferent between accepting and rejecting. The reasoning is more subtle than the SPE just being a "limit" of the discretized Ultimatum game.

Proposition 11.8. *Let* $u, v : [0, 1] \to \mathbb{R}$ *be strictly increasing, concave, and continuous utility functions of player I and II, respectively, with* $u(0) = v(0) = 0$ *and* $u(1) = v(1) = 1$. *Then the Continuous Ultimatum game has the unique SPE where the demand* x *of player I is* $x = 1$ *and player II always chooses A.*

Proof. The described strategies define an SPE. To show uniqueness, first observe that player II has to accept whenever $x < 1$ because then she receives a positive payoff $v(1 - x)$. Suppose there is an SPE where player I's demand is x so that $x < 1$. Because u is strictly increasing, player I would get more by demanding a little more, but still less than 1, for example $\frac{x+1}{2}$, which player II still accepts. Hence, $x = 1$ is the only strategy by player I in an SPE. On the equilibrium path, player II is indifferent, and can choose either A or R in response to $x = 1$, and in fact could randomize between A and R. Suppose that R is played with any positive probability $\varepsilon > 0$, giving player I expected payoff $(1 - \varepsilon)u(1) + \varepsilon u(0)$, which is $1 - \varepsilon$. But then player I's best response to that strategy of player II would not be $x = 1$. Instead, player I could improve his payoff, for example by demanding $1 - \frac{\varepsilon}{2}$ because then player II would accept for sure and player I receives $u(1 - \frac{\varepsilon}{2}) \geq 1 - \frac{\varepsilon}{2} > 1 - \varepsilon$. However, no demand less than 1 is possible in an SPE. Hence, if player II does accept the demand $x = 1$ with certainty, then this reaction of player II does not define an SPE, so the described SPE is the only SPE. $\square$

In summary, with demands that can be chosen continuously, the demand of a player in an SPE makes the other player *indifferent between accepting and rejecting*, which the other player nevertheless *accepts* with certainty. This will also be the guiding principle in SPE of more complicated games, which we study next.

11.8 Alternating Offers Over Two Rounds

Next, we extend the bargaining model so that once a player has rejected the demand of the other player, she can make a counter-demand that, in turn, the other player can accept or reject. In order to find an SPE of this game by backward induction, these alternating offers have to end after some time. The simplest case is bargaining over two rounds, discussed in this section. In the first round, player I makes a demand x, which player II can accept or reject. In the latter case, she can make a counter-demand y that represents the share of the pie she claims for herself, so the pie is split into $1 - y$ for player I and y for player II. If player I accepts that demand, he receives utility $u(1 - y)$ and player II receives utility $v(y)$. If player I rejects the final demand, both players receive zero.

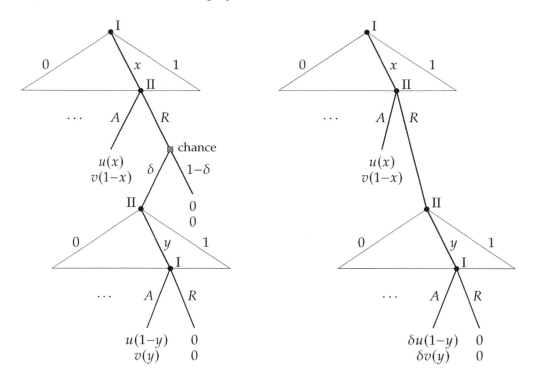

Figure 11.10 Two-round bargaining with demand x by player I in round one and counter-demand y by player II in round two. Left: A second round takes place with probability δ, otherwise termination with payoffs $(0,0)$. Right: Simplified, equivalent game with δ as a discount factor applied to the second-round utility values.

In this game, player II's demand in the final round is final. Hence, this subgame is the Ultimatum game with the roles of the players exchanged, where by Proposition 11.8 player II can demand the whole pie. Hence, there seems to be no incentive for her to accept anything offered to her in the first round. For that reason, the model has an additional feature shown in Figure 11.10. Namely, when player II rejects the first demand, there is a positive probability $1 - \delta$ that bargaining terminates with no agreement, where we assume $0 < \delta < 1$. This is the unsuccessful outcome, with payoff zero to both, given as the result of a chance move. The game proceeds to player II's counter-demand only with probability δ.

By computing expected values, the chance move can be eliminated from the game, which gives the game shown on the right in Figure 11.10. In that game, the pair $(u(1 - y), v(y))$ of utilities when agreement is reached in the second round is multiplied by δ, which is a reduction of the payoffs because $\delta < 1$. The usual interpretation of δ is that of a *discount factor*, which expresses that later payoffs are worth less than earlier payoffs.

A discount factor δ applied to future payoffs is a realistic assumption. In monetary terms, it can express lost interest payments (for money today as opposed to money in the future). More importantly, future payoffs are less secure because, for example, the buyer or seller in a deal may change her mind, or because of other unforeseen circumstances. This is modeled with the chance move in the left game in Figure 11.10. In addition, such a chance move justifies why both players should have the same discount factor. Alternatively, one may use different discount factors δ_I and δ_{II} for the two players, which give second-round payoffs $\delta_I u(1 - y)$ and $\delta_{II} v(y)$. If these discount factors are different, a higher discount factor represents a more patient player whose future payoff differs less from the present payoff. However, the resulting game would be more complicated to analyze, and would not lead to a direct connection with the Nash bargaining solution. We therefore use only a single discount factor, as represented by the model where δ is the probability that bargaining continues into the next round.

An SPE of the game is found as follows. We use backward induction and analyze the last round first. When player II makes her demand y to player I, he can only accept or reject. This is the Ultimatum game in Proposition 11.8 (with exchanged players). Hence, player II demands the whole pie with $y = 1$, and player I accepts any offer made to him, in particular this demand $y = 1$ (recall that player I's strategy is a function of the demand y that he faces), and player I will get nothing. Here, it does not matter if the last stage represents the subgame that starts with player II's demand y in the left or right game in Figure 11.10, because the payoffs in the latter are merely scaled with the positive scalar δ. These subgames in the two games in Figure 11.10 are strategically equivalent and have the same subgame-perfect equilibria. With this assumption, if player II rejects player I's demand x in the first round and chooses R, then player I gets payoff zero, and player II gets $\delta v(1)$, which is δ.

By backward induction, what should player I do in the first round? Player II's reaction is unique except when she is indifferent. This means that she will accept the proposed split if she is offered a utility $v(1 - x)$ (now undiscounted, because this is the first round) that exceeds the amount δ that she can expect to get by rejection; she is indifferent if $v(1 - x) = \delta$; and she will reject the split if $v(1 - x) < \delta$. By the same reasoning as for Proposition 11.8, the demand x by player I in the SPE is chosen so that $v(1 - x) = \delta$, and player II accepts this on the equilibrium path. (If she did not accept, player I would offer her a little bit more and make her accept. However, the only SPE is where player I makes player II indifferent.)

The resulting SPE is unique. Player I demands x so that $v(1 - x) = \delta$, which is accepted immediately by player II, so that the game terminates with player I receiving $u(x)$ and player II receiving $v(1 - x) = \delta$. The full equilibrium strategies are (with this x): Player II accepts any demand x' so that $0 \le x' \le x$, and rejects any $x' > x$. Any counter-demand by player II in the second round is $y = 1$, and player I accepts any amount in the second round. The strategies have to specify the actions of the players in the second stage in order to perform the backward induction analysis.

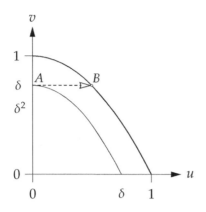

Figure 11.11 Graphical solution of the two-round bargaining game in Figure 11.10, for $\delta = \frac{3}{4}$. The inner curve is the set of discounted payoff pairs for the second round.

Player I's demand x in the first round is easily found graphically as shown in Figure 11.11. This shows, as the outer curve, the Pareto-frontier of the bargaining set, here for the utility functions $u(x) = \sqrt{x}$ and $v(y) = y$ used earlier. Recall that for $0 \leq x \leq 1$, one obtains any point $(u(x), v(1-x))$ on that Pareto-frontier. Multiplying any such pair of real numbers with the factor δ gives a "shrunk" version of the Pareto-frontier, shown as the curve with its lower endpoint $(\delta, 0)$ and upper endpoint $(0, \delta)$. This second curve is traversed when looking at the points $(\delta u(1-y), \delta v(y))$ when y is changed from 0 to 1, which are the discounted payoffs after agreement in the second round. In the subgame given by the second round, player I can only accept or reject, so this Ultimatum game has the upper endpoint $(0, \delta)$ of the curve, denoted by A in Figure 11.11, as its payoff pair. Player I's consideration in round one is to find a point B on the curve of first-round payoffs so that player II is indifferent between accepting and rejecting (where she will accept in the SPE). So this point B has the same vertical coordinate δ as A, that is, B is of the form $(u(x), \delta)$. As shown by the arrow, B is found by moving horizontally from point A until hitting the outer curve. In the example, x is the solution to the equation $v(1-x) = \delta$, which is simply $x = 1 - \delta$ because v is the identity function. The resulting utility $u(x)$ to player I is $\sqrt{1-\delta}$. Figure 11.11 shows this for $\delta = \frac{3}{4}$, where $u(x) = \frac{1}{2}$ and $v(1-x) = \frac{3}{4}$, that is, $B = (\frac{1}{2}, \frac{3}{4})$.

11.9   Alternating Offers Over Several Rounds

The bargaining game with three rounds of alternating offers is shown in Figure 11.12. Player I makes a demand x in the first round, which can be accepted or rejected by player II. If she rejects, she can make a counter-demand y, which in turn player I can accept or reject. If player I accepts, both players' payoffs are discounted,

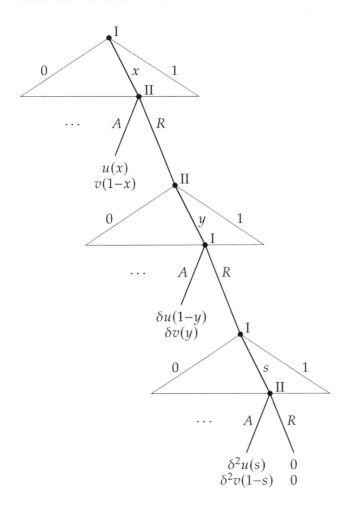

Figure   11.12
Bargaining over
three rounds.

giving the expected payoff pair $(\delta u(1 - y), \delta v(y))$. If player I rejects, then he
can make in the third and final round a counter-counter-demand s, which if
accepted gives the pair of twice-discounted payoffs $(\delta^2 u(s), \delta^2 v(1 - s))$. If player II
rejects the final demand s, both players get zero. These expected payoffs result
because any rejection independently incurs a risk that, with probability $1 - \delta$, the
game terminates with no agreement and payoff pair $(0,0)$, and continues with
probability δ into the next round. Figure 11.12 shows directly the discounted
payoffs and not the chance moves between the bargaining rounds, which would
have to be included in a detailed model like the left game in Figure 11.10.

The SPE of the three-round game is found by backward induction. The third
and final round is an Ultimatum game where player I demands $s = 1$, which is
accepted by player II. The set of twice discounted payoffs $(\delta^2 u(s), \delta^2 v(1 - s))$ is the
inner curve on the left in Figure 11.13. Player I's demand $s = 1$ gives the payoff pair
$(\delta^2, 0)$, shown as point A in the figure. In the previous round, player II maximizes
her payoff $\delta v(y)$ by making player I indifferent between accepting and rejecting.

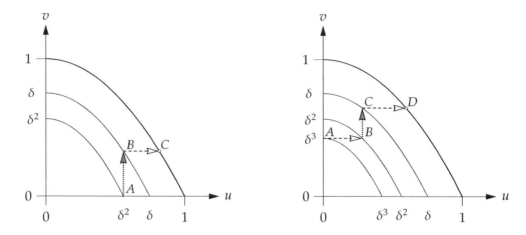

Figure 11.13 Graphical solution of the bargaining game over three rounds (left) and four rounds (right).

Hence, y is the solution to the equation $\delta u(1-y) = \delta^2$. This is point B in the figure, which is the point on the curve of second-round payoff pairs $(\delta u(1-y), \delta v(y))$ with the same horizontal coordinate as A. With y determined like that, player I's demand x in the first round is the solution to the equation $v(1-x) = \delta v(y)$ because this makes player II indifferent between accepting and rejecting in the first round. The corresponding point $C = (u(x), v(1-x))$ on the set of first-round payoff pairs has the same vertical coordinate as B. On the equilibrium path, player I demands x in the first round, which is immediately accepted by player II. The complete equilibrium strategies are as before, where a player accepts any demand up to and including the demand in the SPE, and otherwise rejects and makes the described counter- or counter-counter-demand in the next round.

In our example with utility functions $u(x) = \sqrt{x}$ and $v(y) = y$, the SPE of the game with three rounds is found as follows (according to the left picture in Figure 11.13). The vertical arrow from point A to point B gives the equation $\delta u(1-y) = \delta^2$ or $u(1-y) = \sqrt{1-y} = \delta$, that is, $y = 1 - \delta^2$. The horizontal arrow from B to C represents the equation $v(1-x) = \delta v(y)$, that is, $1-x = \delta(1-\delta^2) = \delta - \delta^3$, which has the solution $x = 1 - \delta + \delta^3$. This is player I's demand in the first round in the SPE; his payoff is $u(x) = \sqrt{1 - \delta + \delta^3}$. For $\delta = \frac{3}{4}$, as in Figure 11.13, we have $y = \frac{7}{16}$ and $x = \frac{43}{64}$ with $u(x) = \sqrt{43}/8$ and $v(1-x) = \frac{21}{64}$. These are the coordinates of point C, approximately $(0.82, 0.33)$.

The game with four bargaining rounds has the same structure as the game with three rounds, except that if player II rejects player I's demand s in round three, she can make a final demand t of her own in the fourth and final round, which player I can only accept or reject. If player I accepts, the players receive the three-times discounted payoffs $(\delta^3 u(1-t), \delta^3 v(t))$. The corresponding SPE is solved graphically as shown on the right in Figure 11.13. The last round is an Ultimatum game where player II can demand the whole pie with $t = 1$, which determines the payoff pair $(0, \delta^3)$, which is point A in the figure. In the previous round, the third round of the game, player I demands s so that player II is indifferent between accepting and rejecting, which gives point B, which is the pair of discounted payoffs $(\delta^2 u(s), \delta^2 v(1 - s))$ so that $\delta^2 v(1 - s) = \delta^3$. This determines s. In the round before that, player II makes player I indifferent between accepting and rejecting with the demand y so that $\delta u(1 - y) = \delta^2 u(s)$, shown as point C. This equation determines y. In the first round, player I demands x so as to make player II indifferent, according to $v(1 - x) = \delta v(y)$, which defines point $D = (u(x), v(1 - x))$. This is also the equilibrium outcome because that demand x is accepted by player II.

For $u(x) = \sqrt{x}$ and $v(y) = y$, the right picture in Figure 11.13 gives the following SPE. The arrow from A to B gives the equation $\delta^2 v(1-s) = \delta^3$ or $v(1-s) = \delta$, which determines player I's demand $s = 1 - \delta$ in round three. The arrow from B to C gives the equation $\delta u(1 - y) = \delta^2 u(s)$ or $u(1 - y) = \delta u(s)$, that is, $\sqrt{1 - y} = \delta\sqrt{1 - \delta}$ or equivalently $1 - y = \delta^2(1 - \delta) = \delta^2 - \delta^3$, which has the solution $y = 1 - \delta^2 + \delta^3$ that determines player II's demand y in the second round. Finally, the arrow from C to D gives the equation $v(1 - x) = \delta v(y)$ or $1 - x = \delta(1 - \delta^2 + \delta^3)$, which gives player I's demand x in the first round as $x = 1 - \delta + \delta^3 - \delta^4$.

Note that in the game of bargaining over four rounds, the arrow from A to B defines the same equation as in the two-round game in Figure 11.11; essentially, the fourth round in the four-round game describes the same situation, and gives the same picture, as the second round in the two-round game, except that all payoffs are shrunk by the factor δ^2.

The bargaining game with alternating offers can be defined for any number of rounds. In each odd-numbered round (the first, third, fifth round, etc.), player I makes a demand, which can be accepted or rejected by player II. After rejection, the game progresses with probability δ into the next round, and otherwise terminates with payoffs $(0, 0)$. Player II can make a demand in each even-numbered round. If the total number of rounds is even, then player II can make the last demand, which is the whole pie, which is advantageous for player II. If the total number of rounds is odd, then player I can make the last demand, and in turn demand the whole pie for himself.

The graphical solutions in Figures 11.11 and 11.13 can be extended to any number of rounds, with payoffs in the kth round given by $(\delta^{k-1} u(z), \delta^{k-1} v(1 - z))$ for some $z \in [0, 1]$. These payoff pairs define a curve that is the original Pareto-frontier of the bargaining set shrunk by the factor δ^{k-1}. The examples for a total number of two, three, or four rounds show that the resulting SPE defines a point $(u(x), v(1 - x))$. This is the pair of payoffs to the two players because they agree in round one.

324 Chapter 11. Bargaining

In general, if the total number of rounds is odd, player I is always better off than if the total number of rounds is even, which favours player II who can make the last demand. However, with an increasing total number of rounds, the resulting points $(u(x), v(1-x))$ move towards each other. We should expect that with a very large number of rounds, it matters less and less who can make the last demand (because the future discounted payoffs from which the backward induction starts are very small), and expect some convergence. However, we will not study the "limit" of these finite games, but instead consider a new concept, a game with an infinite number of rounds.

11.10   Stationary Strategies

We now analyze the bargaining game with alternating offers with an *infinite* number of rounds, that is, the game goes on forever. Formally, this is a game Γ with perfect information represented by an infinite tree. It can be defined recursively as shown in Figure 11.14. In the first and second round, player I and player II each make a demand x and y, respectively, from the interval $[0, 1]$, which the other player can accept or reject. After rejection, a chance move terminates the game

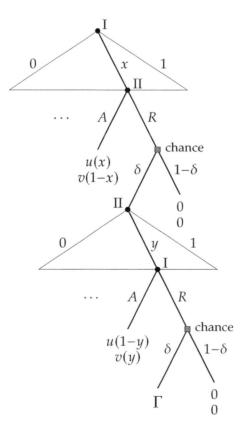

Figure 11.14 The bargaining game Γ with an infinite number of rounds, which repeats itself after any demand x of player I in round one and any demand y of player II in round two after each player has rejected the other's demand. The next round is reached with probability δ each time, and otherwise terminates with payoffs $(0, 0)$.

with probability $1 - \delta$ where both players receive payoff zero. With probability δ, the game continues into the next round where the player who has just rejected the offer can make her counter-demand. With the beginning of the third round, this is again player I, so that this defines the same game Γ that we started with in round one. Correspondingly, Γ is simply appended as the beginning of the subgame that starts after the second chance move that follows player I's rejection of player II's counter-demand y. (Note that infinitely many copies of Γ are substituted at each node that results from any x followed by any y.)

An SPE of this infinite game cannot be defined by backward induction because the game goes on forever, so that there is no final move from which one could start the induction. However, the game is defined recursively and is the same game after two rounds. Hence, it is possible to define strategies of the players that repeat themselves every two rounds. Such strategies are called *stationary* and mean that player I always demands x whenever he can make a demand (in each odd-numbered round, starting with the first round), and player II always demands y (in each even-numbered round, starting with the second round). The SPE condition states that these strategies should define an equilibrium in any subgame of the game. Any such subgame starts either with a demand by one of the players, with the other player's response, or with a chance move.

In addition to the players's demands in each round, the SPE should also specify whether the other player accepts or rejects the offer. After rejection, the resulting expected payoffs automatically shrink by a factor of δ. Consequently, a player's demand should not be so high as to result in rejection, but should instead make it optimal for the other player to accept. The demand itself is maximal subject to this condition, that is, it is, as before, chosen so as to make the other player indifferent between accepting and rejecting, and the other player accepts in the SPE.

We now show that an SPE exists, and how to find it, using a graph similar to the left picture in Figure 11.13. Like there, we draw the three curves of payoff pairs $(u(x), v(1 - x))$ that result from player I's demand x in round one if this demand is accepted by player II, the discounted payoffs $\delta(u(1 - y), \delta v(y))$ if player I accepts player II's demand y in round two, and the third-round twice-discounted payoffs $\delta^2(u(s), \delta^2 v(1 - s))$ if player II accepts player I's demand s in round three. The demands x and y are chosen so as to make the other player indifferent between acceptance and rejection. For the demand s in round three, we want to fulfill the requirement of stationary strategies, that is, $s = x$.

Figure 11.15 illustrates how to fulfill these requirements. Suppose that player I's demand s in round three defines the point $A = (\delta^2 u(s), \delta^2 v(1 - s))$, relatively high up on the inner curve as shown in Figure 11.15(a). In the previous round two, point $B = (\delta u(1 - y), \delta v(y))$ has the same horizontal coordinate as point A because player II tries to make player I indifferent with her demand y, according to the equation

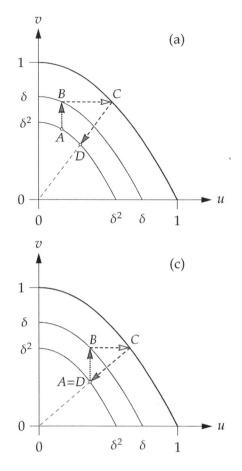

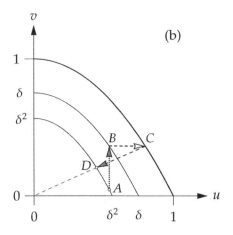

Figure 11.15 Finding stationary strategies in an SPE of the infinite game, by following the arrows from A to B to C to D so that $A = D$.

$$\delta u(1 - y) = \delta^2 u(s). \tag{11.4}$$

In the first round, point C is found similarly, as the point $(u(x), v(1 - x))$ that has the same vertical coordinate as B, because player I demands x so as to make player II indifferent between accepting and rejecting, according to

$$v(1 - x) = \delta v(y). \tag{11.5}$$

The demand x is the same as demand s, as required in stationary strategies, if point C defines the same relative position on the outer curve as A on the inner curve. In Figure 11.15, we have shown point D as $\delta^2 C$, that is, point C shrunk by a factor of δ^2, which is the expected payoff pair two rounds later. In Figure 11.15(a), A and D do not coincide, because $s < x$, that is, the third-round demand s was too small. Figure 11.15(b) shows a point A on the inner curve corresponding to a demand s that is too favourable for player I, where the arrows from A to B to C according to equations (11.4) and (11.5) give a first-round demand x with $x < s$.

In general, the pictures demonstrate that when starting from $s = 0$, these equations give a first-round demand x where $x > s$, and when starting from $s = 1$, a first-round demand x where $x < s$. Moreover, when starting from any s in $[0, 1]$, the resulting demand x is a *continuous* function of s. Because the continuous function $x - s$ is positive for $s = 0$ and negative for $s = 1$, it is zero for some intermediate value s, where $s = x$, which defines a stationary SPE. (This SPE is also unique, see Binmore, 1992, theorem 5.8.4.)

In short, the stationary SPE of the infinite bargaining game is a solution to the equations (11.4), (11.5), and $s = x$. These are equivalent to the two, nicely symmetric equations

$$u(1 - y) = \delta u(x), \qquad v(1 - x) = \delta v(y). \tag{11.6}$$

The first of these equations expresses that player II makes player I indifferent in every even-numbered round by her demand y. The second equation in (11.6) states that player I makes player II indifferent by his demand x in every odd-numbered round. In this stationary SPE, player I's demand x in round one is immediately accepted by player II. As before, the full strategies specify acceptance of any lower and rejection of any higher demand. The actual demand of the other player is always accepted, so the later rounds are never reached.

In the example with utility functions $u(x) = \sqrt{x}$ and $v(y) = y$, the stationary strategies in an SPE are found as follows. They are given by the demands x and y that solve (11.6), that is,

$$\sqrt{1 - y} = \delta\sqrt{x}, \qquad 1 - x = \delta y.$$

The first equation is equivalent to $y = 1 - \delta^2 x$, which substituted into the second equation gives $1 - x = \delta - \delta^3 x$, or $1 - \delta = x(1 - \delta^3)$. Because $1 - \delta^3 = (1 - \delta)(1 + \delta + \delta^2)$, this gives the solutions for x and y as

$$x = \frac{1 - \delta}{1 - \delta^3} = \frac{1}{1 + \delta + \delta^2}, \qquad y = \frac{1 + \delta}{1 + \delta + \delta^2}. \tag{11.7}$$

The pie is split in the first round into $x = \dfrac{1}{1 + \delta + \delta^2}$ for player I and $1 - x = \dfrac{\delta + \delta^2}{1 + \delta + \delta^2}$ for player II. The share $1 - x$ of the pie for player II is lower than her demand y in the second round, as illustrated by Figure 11.15(c), which shows that the relative position of point A on the outer curve is higher than that of point B on the middle curve. This holds because the second-round payoffs are discounted.

Does it matter for our analysis that player I makes the first demand? Of course, if player II made the first demand, one could swap the players and analyze the game with the described method. However, the game where player II makes the first demand, and the corresponding stationary solution, is already analyzed as part of the game Γ where player I moves first. The reason is that the game with player II moving first is the subgame of Γ in Figure 11.14 after the first chance move, which repeats itself as in Γ. All we have to do is to ignore player I's first demand,

and the first chance move and the corresponding discount factor δ. That is, we draw the same picture of curves with discounted payoff pairs as in Figure 11.15, except that we omit the outer curve and scale the picture by multiplication by $\frac{1}{\delta}$.

For the alternating-offers game with player II moving first, the resulting equations are given as before by (11.6). They have the same stationary equilibrium solution. This solution starts with player II making the first demand y, followed by player I's counter-demand x, and so on in subsequent rounds. Because in this SPE the first demand is accepted, player II gets payoff $v(y)$ and player I gets payoff $u(1 - y)$, which is $\delta u(x)$ by (11.6). Hence, all that changes when player II makes the first demand is that player II's payoff is no longer discounted by multiplication by δ, and player I's payoff is discounted instead.

11.11 The Nash Bargaining Solution Via Alternating Offers

In this final section, we show that for very patient players, when the discount factor δ approaches 1, the stationary SPE converges to the Nash bargaining solution. This is true for the example that we have used throughout, with utility functions $u(x) = \sqrt{x}$ and $v(y) = y$, where the split in the Nash bargaining solution is $(x, 1 - x) = (\frac{1}{3}, \frac{2}{3})$. This is also the limit of the stationary solution in (11.7) when $\delta \to 1$ (note the convenient cancellation by $1 - \delta$ in the term $\frac{1 - \delta}{1 - \delta^3}$ that represents x; this term has an undefined limit $\frac{0}{0}$ when $\delta = 1$, which without the cancellation requires using l'Hôpital's rule).

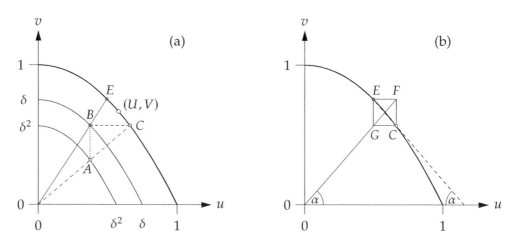

Figure 11.16 Why the player's payoffs in the stationary SPE converge to the Nash bargaining solution (U, V) as $\delta \to 0$. Point $C = \big(u(x), v(1 - x)\big)$ is the split when player I makes the first demand, point $E = \big(u(1 - y), v(y)\big)$ when player II makes it.

The stationary solution to the infinite bargaining game of alternating offers is shown in Figure 11.15(c), with points $B = \big(\delta u(1 - y), \delta v(y)\big)$ and $C = \big(u(x), v(1 - x)\big)$.

These points are shown again in Figure 11.16(a), with a "preview" of the Nash bargaining solution (U, V). Let the point E be given by $E = (u(1 - y), v(y))$, which is the "undiscounted" version of B, that is, it has the same relative position on the outer curve as B on the middle curve. As mentioned at the end of the previous section, these are also the players' payoffs in the stationary SPE when player II moves first with her demand y. As $\delta \to 1$, the points E and C converge to the same limit point. As shown in Figure 11.16(b), the points E and C define a line where the line segment between E and C is a "cord" of the Pareto-frontier. In the limit, the line becomes a *tangent* to the Pareto-frontier. If the Pareto-frontier is differentiable, the tangent is unique, otherwise we obtain in the limit some tangent to the frontier.

We use Figure 11.16(b) to show that as $\delta \to 1$, the points E and C converge to the Nash bargaining solution. (The picture shows that the *midpoint* between E and C is already very close to (U, V) even when δ is not close to 1.) We use the geometric characterization of the bargaining solution in Proposition 11.5. Consider the points E and C as the upper-left and lower-right corner of a rectangle. As shown in Figure 11.16(b), the upper-right corner of that rectangle is the point $F = (u(x), v(y))$, and its lower-left corner is the point $G = (u(1 - y), v(1 - x))$. The equations (11.6) state that $G = \delta F$, that is, the points F and G are on a line through the origin $(0, 0)$, where the line segment between F and G is a diagonal of the rectangle. This line has slope $v(y)/u(x)$, shown as the angle α near the origin. The other diagonal of the rectangle through the points E and C has the same (negative) slope, shown as the angle α when that line intersects the u-axis.

In the limit, when $\delta \to 1$, we have $y \to 1 - x$, and we obtain the same picture as in Figure 11.5 of a "roof" with two sides that have, in absolute value, the same slope. The left side of the roof is a line through the origin, and the right side of the roof is a tangent (of correct slope) to the Pareto-frontier of the bargaining set. This is exactly the condition of Proposition 11.5 that characterizes the apex of the roof as the Nash bargaining solution (U, V).

We summarize the main result of this chapter. The model of alternating offers with a discount factor is an interesting non-cooperative game in its own right. When the number of bargaining rounds is infinite, the game has an SPE in stationary strategies. When the discount factor δ gets close to one, this stationary equilibrium gives a payoff pair that approaches the Nash bargaining solution.

The Nash bargaining solution is originally derived from reasonable "axioms" for a solution to the bargaining problem. The "implementation" of this solution is not spelled out in this axiomatic approach, where it is merely assumed that the players reach a binding agreement in some way.

The alternating-offers game justifies this solution via a completely different route, namely a model of players that interact by making demands and possible

rejections and counter-demands. In the resulting SPE with stationary strategies, each player anticipates the future behavior of the other player. The players immediately reach an agreement, which is the Nash bargaining solution.

11.12   Further Reading

The axiomatic bargaining model is due to John Nash (1950) (and should not be confused with Nash equilibrium). An authoritative textbook on game theory, and cooperative games in particular, is Maschler, Solan, and Zamir (2013), where chapter 15 treats the Nash bargaining solution. It is also described in Osborne and Rubinstein (1994, chapter 15) and Osborne (2004, section 16.3).

In favor of cooperative game theory, the assumption that people keep agreements may be quite justified in practice, without worrying too much about "enforcement mechanisms". The Dutch cultural historian Johan Huizinga (1949) argues that "play", the willingness to follow fairly arbitrary rules for their own sake, is an essential element of culture, where even just pretending to follow the rules of the game is much more tolerated than being a spoilsport and questioning the game itself.

For further details on the Ultimatum game see Osborne (2004, section 6.1.1). The infinite iterated-offers bargaining model with its unique stationary SPE that approaches the Nash bargaining solution is due to Rubinstein (1982). A finite multistage game was introduced earlier by Ståhl (1972).

The original inspiration for this chapter was Binmore (1992, chapter 5) and Binmore (2007, chapters 16 and 17). There, the players are assumed to have different "bargaining power", reflected in different discount factors for the players. Then the stationary strategies in the infinite game with very patient players approach a "generalized" Nash bargaining solution. We have simplified this to equal bargaining powers and equal discount factors, so that the stationary strategies approach the original Nash bargaining solution. For a similar textbook treatment see Osborne (2004, chapter 16).

11.13   Exercises for Chapter 11

Exercise 11.1. Consider a bargaining set S with threat point $(0,0)$. Let $a > 0$, $b > 0$, and let $S' = \{(au, bv) \mid (u,v) \in S\}$. The set S' is the set of utility pairs in S re-scaled by a for player I and by b for player II. Recall that one axiomatic property of the bargaining solution is independence from these scale factors.

Show that the Nash product fulfills this property, that is, the Nash bargaining solution from S obtained by maximising the Nash product re-scales to become the solution for S'.

Note: This is extremely easy once you state in the right way what it means that something maximizes something. You may find it useful to consider first the simpler case $b = 1$.

Exercise 11.2. Consider the following two bimatrix games (i) and (ii):

(i)

I \ II	l	r
T	1 \ -2	3 \ 3
B	0 \ 6	3 \ 1

(ii)

I \ II	l	r
T	3 \ 4	2 \ 1
B	4 \ 3	1 \ 1

For each bimatrix game, do the following:

(a) Find max-min strategies $\hat{x}$ and $\hat{y}$ (which may be mixed strategies) and corresponding max-min values u_0 and v_0 for player I and II, respectively.

(b) Draw the convex hull of the payoff pairs in the 2×2 game, and the *bargaining set S* that results from the payoff pairs (u, v) with $u \geq u_0$ and $v \geq v_0$ with the threat point (u_0, v_0) found in (a). Indicate the Pareto-frontier of S.

(c) Find the Nash bargaining solution (U, V).

(d) Show how the players can implement the Nash bargaining solution for S with a joint lottery over strategy pairs in the 2×2 game.

Exercise 11.3. We repeat the definition of the discretized ultimatum game. Let M be a positive integer. Player I's possible actions are to ask for a number x, called his demand, which is one of the (integer) numbers $0, 1, 2, \ldots, M$. In response to this demand, player II can either accept (A) or reject (R). When player II accepts, then player I will receive a payoff of x and player II a payoff of $M - x$. When player II rejects, both players receive payoff zero.

(a) Draw the game tree for this game, as an extensive game with perfect information, for $M = 3$.

(b) What is the number of pure strategies in this game, for general M, for player I and for player II? What is the number of reduced pure strategies in this game, for player I and for player II?

(c) Determine all subgame-perfect equilibria of the game in pure strategies.

(d) Determine all equilibria (not necessarily subgame-perfect) of this game in pure strategies.

(e) Determine all subgame-perfect equilibria of this game where both players may use behavior strategies, in addition to those in (c).

Exercise 11.4. Consider the following bargaining problem. In the usual way, a "unit pie" is split into non-negative amounts x and y with $x + y \leq 1$. The utility function of player I is $u(x) = x$, the utility function of player II is $v(y) = 1 - (1 - y)^2$. The threat point is $(0, 0)$.

(a) Draw the bargaining set (as set of possible utilities (u, v)) and indicate the Pareto-frontier in your drawing.

(b) Find the Nash bargaining solution. How will the pie be split, and what are the utilities of the players?

Exercise 11.5. Consider the bargaining problem of splitting a unit pie with utility functions $u, v \colon [0, 1] \rightarrow [0, 1]$ as in Exercise 11.4, that is, $u(x) = x$ and $v(y) = 1 - (1 - y)^2$. Assume now that the bargaining behavior is described by the subgame-perfect equilibrium of the standard alternating-offers bargaining game with k rounds. In round 1 player I makes a certain demand. If player II rejects this, she can make a certain counter-demand in round 2, and so on. A rejection in the last round k means that both players get nothing. The last player to make a demand is player I if k is odd, and player II if k is even. When agreement takes place in round i for some i with $1 \leq i \leq k$, then the bargaining set has shrunk by a factor of δ^{i-1}. That is, the utility of what the players can get after each round of rejection is reduced by multiplication by the discount factor δ.

(a) Suppose that the number of rounds is $k = 2$. Draw the picture of shrinking bargaining sets (for some choice of δ) in these two rounds. Find the subgame perfect equilibrium, in terms of player I's demand x, for general δ. What is the demand x when $\delta = \frac{3}{4}$? What are the strategies of the two players in this subgame perfect equilibrium?

(b) Suppose that the number of rounds is $k = 3$. Draw the picture of shrinking bargaining sets in these three rounds. Find the subgame perfect equilibrium, for general δ, and compute the demand of player I in round 1 (you do not need to specify the players' strategies). What is player I's demand x when $\delta = \frac{1}{4}$? (Then x happens to be a rational number.)

(c) Consider now an infinite number of rounds, and look for stationary equilibrium strategies, where player I demands the same amount x each time he makes a demand, and player II demands the same amount y every time she makes a demand. What is this demand x when $\delta = \frac{3}{5}$? What is the demand x when δ tends to 1? Compare this with the result of Exercise 11.4(b).

12

Correlated Equilibrium

In this final chapter, we explain a new equilibrium concept called *correlated equilibrium* that is more general than Nash equilibrium. It allows for randomized actions of the players that depend on an external signal (like a traffic light) that is observed by the players (typically in different ways) so that their actions can be correlated.

The set of Nash equilibria of a game is a set of fixed points and typically disconnected. The set of correlated equilibria is larger, and mathematically simpler because it is a convex polytope. In Section 12.4, we also explain the concept of *coarse correlated equilibrium*, where the players commit to letting the correlation device choose their actions for them, rather than just giving them recommendations. This restricts the players' possibility for deviating and leads to an even larger set (with possibly more beneficial outcomes, see Exercise 12.3).

Because the set of correlated equilibria is a convex polytope, it would be nice to prove that it is not empty without relying on a fixed-point theorem. Indeed, Hart and Schmeidler (1989) gave such an existence proof based on the minimax Theorem 8.4, which we present in Section 12.5 (and which is not normally found in textbooks at this level). The minimax theorem leads to elegant proofs of the existence of stationary distributions for Markov chains (Theorem 12.9) and of correlated equilibria (Theorem 12.11).

The concept of correlated equilibrium is used much less often than Nash equilibrium in the analysis of games, but is important in models of learning in games, mentioned in the final reference Section 12.6.

12.1   Prerequisites and Learning Outcomes

Apart from the basic concepts of strategic form and equilibrium in Chapter 3, the main prerequisite is Chapter 8 with the minimax Theorem 8.4.

After studying this chapter, you should be able to:

- explain correlated equilibrium, its difference to Nash equilibrium, and that correlated equilibria are more than just randomizations over Nash equilibria;

- write down the incentive constraints of a game that define the set of correlated equilibria;

- solve these incentive constraints for small games, like in Exercise 12.1;

- describe the concept of coarse correlated equilibrium and why it requires a commitment by the players;

- outline the mathematical argument for the existence of a correlated equilibrium with the help of the minimax theorem, and why this route is overall simpler than via the existence of a Nash equilibrium;

- describe in more detail the idea behind this existence proof, explained at the end of Section 12.5: the auxiliary zero-sum game with its "correlator" COR and "deviator" DEV, and how COR can "neutralize" any "deviation plan" of DEV.

12.2   Examples of Correlated Equilibria

Towards the end of the last century I visited the Hebrew University of Jerusalem. In the faculty club, I ran into the game theorists Robert Aumann and Michael Maschler. Aumann described how "a conversation with Professor Maschler" led him to the idea of correlated equilibrium. He said they discussed how one would play a mixed strategy in practice: Let the outcome of some random event determine the action that you take. For example, choose A if the sun shines, and choose B if it does not. The probability is roughly known from the weather forecast. But this approach revealed a new possibility: Using the weather for randomization does not produce a mixed strategy as considered by von Neumann or Nash, where players play randomly by flipping a coin on their own. If players let their actions be decided by the weather, then their random choices will not be independent. The concept of correlated equilibrium, due to Aumann (1974), allows for *correlated* randomized actions among the players, and generalizes the concept of Nash equilibrium. (In this chapter, we always use the term "Nash equilibrium" to distinguish it from other equilibrium notions.) Correlated equilibria are the topic of this chapter.

Consider the variant of the game Chicken in Figure 12.1. Each player has two strategies, here named T and B for player I and L and R for player II. Compared to the original Chicken game in Figure 3.3, a player's payoff is increased by 3 if the *other* player plays her second strategy (by Proposition 9.7, or seen directly, this does not affect the best response to any mixed strategy). The game has the two non-symmetric pure Nash equilibria (B, L) with payoffs $1, 5$ to players I and II, and (T, R) with payoffs $5, 1$. In the unique mixed equilibrium, both players use the mixed strategy $(\frac{1}{2}, \frac{1}{2})$, with payoff $\frac{5}{2}$ to each player.

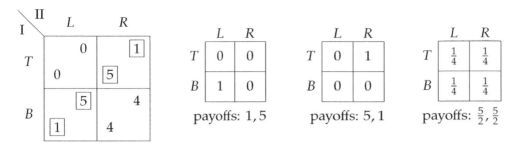

Figure 12.1 A variant of the game Chicken in Figure 3.3. The three panels on the right show the three Nash equilibria of the game as probability distributions on strategy profiles.

The three panels on the right in Figure 12.1 show the three Nash equilibria as probability distributions on the four strategy pairs. A pure Nash equilibrium selects only one strategy pair. The mixed Nash equilibrium selects each strategy pair with equal probability $\frac{1}{4}$.

In a correlated equilibrium, this probability distribution is thought of as a *correlation device* that randomly selects a pure strategy profile, where for each player the strategy in that profile is privately told to the player as a *recommendation* of what to play. Each player is assumed to know how the device works, and can *infer* from his own recommendation what the other players have been recommended to play. The *equilibrium condition* states that it is optimal for each player to choose their recommended action, assuming that the other players follow their recommendations.

For the first pure Nash equilibrium (B, L) in Figure 12.1, the device selects this strategy pair, and recommends player I to play B and player II to play L. Here, each player knows what the other has been told, and his own recommendation is a best response to that strategy, so this is a correlated equilibrium. The same applies to the second pure equilibrium (T, R).

In the third, mixed Nash equilibrium in Figure 12.1, the device chooses each strategy pair with equal probability. This includes strategy pairs that are not themselves Nash equilibria, such as (T, L). If the device selects (T, L), then player I is recommended to play T. However, and this is crucial, player I is *not* told that player II has been recommended to play L. Player I can only infer from the fact that he has been recommended to play T that player II has been recommended to play L and R with equal probability (this is the same when player I has been recommended to play B). Given this randomized choice of player II, player I is indifferent between his two rows, and therefore it is optimal to follow the recommended action. The same holds for any strategy pair that is chosen by the device. Therefore, this is also a correlated equilibrium.

	L	R
T	0	$\frac{1}{2}$
B	$\frac{1}{2}$	0

payoffs: $3,3$

	L	R
T	0	$\frac{1}{3}$
B	$\frac{1}{3}$	$\frac{1}{3}$

payoffs: $\frac{10}{3}, \frac{10}{3}$

	L	R
T	$\frac{1}{3}$	$\frac{1}{3}$
B	$\frac{1}{3}$	0

payoffs: $2,2$

	L	R
T	z_{11}	z_{12}
B	z_{21}	z_{22}

Figure 12.2 Examples of correlated equilibria for the game in Figure 12.1, and their expected payoffs. The general form of a correlated distribution is shown on the right.

In a mixed Nash equilibrium (x, y) of an $m \times n$ bimatrix game, a pure strategy pair (i, j) is chosen with probability $x_i y_j$ because the two players randomize independently. In a correlated equilibrium for an $m \times n$ game, each strategy pair (i, j) has some probability z_{ij}, called the *joint probability* for this pair. Figure 12.2 shows more general probability distributions of this kind. In the left panel, the two strategy pairs (B, L) and (T, L) are chosen with equal probability. No matter which pair is chosen, both players know with certainty the recommendation to the other player. The two pure strategies are best responses to each other, so the equilibrium condition holds. Here, the device chooses the other two strategy pairs (T, L) and (B, R) with probability zero. The joint probability distribution is not independent but *correlated*, that is, it cannot be expressed in the form $z_{ij} = x_i y_j$ because, for example, $z_{11} = 0$ would imply $x_1 = 0$ or $y_1 = 0$, where $x_1 = 0$ implies $z_{12} = 0$ and $y_1 = 0$ implies $z_{21} = 0$, but both z_{12} and z_{21} are positive.

This correlated equilibrium in the left panel in Figure 12.2 gives an expected payoff of 3 to each player, which is not achieved in any Nash equilibrium. As an equal "mixture" between the pure Nash equilibria, it also seems like a fair way to play the game. It could be implemented by a fair coin toss, whose outcome then determines which of the two pure Nash equilibria should be played.

However, the players can achieve even higher expected payoffs with the device shown in the second panel in Figure 12.2. There, the strategy pair (B, R) with payoffs 3 for each player has probability $\frac{1}{3}$, the same as the other two pairs (B, L) and (T, R). The top-left strategy pair (T, L) with the low payoffs $0, 0$ has probability zero. If the device chooses (B, R), then player I (and similarly player II) only knows that the other player has been recommended either pure strategy with probability $\frac{1}{2}$, against which strategy 2 is a best response. If the device chooses (B, L), then player I also infers that player II has been told to play either strategy with probability $\frac{1}{2}$, whereas player II knows that player I has been told to play strategy 2 for sure; vice versa, the same applies when the device chooses (T, L). Hence, this is also a correlated equilibrium, with expected payoffs $(\frac{10}{3}, \frac{10}{3})$. It can be shown that this correlated equilibrium has the largest sum of expected payoffs to the two players.

However, there can also be correlated equilibria that have worse expected payoffs than any Nash equilibrium, as shown in the third panel in Figure 12.2. There, the device chooses the strategy pair (T, L) with its bad payoffs $0, 0$ with probability $\frac{1}{3}$, and the strategy pair (B, R) with its good payoffs $4, 4$ with probability zero. This is a correlated equilibrium for the same reasons as before: Each player infers from his own recommendation either a deterministic or a $(\frac{1}{2}, \frac{1}{2})$ random choice of his opponent, against which the own recommendation is a best response. The expected payoff, however, is $\frac{0+5+1}{3} = 2$ to each player, a sum of payoffs that is worse than in any Nash equilibrium.

12.3 Incentive Constraints

In this section, we formalize the concept of correlated equilibrium, first for two players because the notation is easier. Consider an $m \times n$ bimatrix game (A, B) with payoffs a_{ij} and b_{ij} to players I and II if they play row i and column j, respectively. A *correlation device* is a probability distribution z on pure strategy pairs, with probability z_{ij} for the pair (i, j), where these numbers fulfill

$$z_{ij} \geq 0 \quad (1 \leq i \leq m, \ 1 \leq j \leq n), \qquad \sum_{i=1}^{m} \sum_{j=1}^{n} z_{ij} = 1. \tag{12.1}$$

These probabilities are common knowledge to the players. The device selects a strategy pair according to these probabilities, and privately tells each player only his pure strategy in that pair. Suppose player I is told strategy i. Then the ith row of the device (as in the panels in Figure 12.2) has probabilities $z_{i1}, \ldots, z_{in}$ for the n columns $j = 1, \ldots, n$ of what player II has been told. Given that player I has been told i, the *conditional probability* for column j is therefore

$$\mathsf{prob}(j \mid i) = \frac{z_{ij}}{\sum_{l=1}^{n} z_{il}}. \tag{12.2}$$

This expression is well defined whenever row i is chosen with positive probability by z, because then $z_{il} > 0$ for at least one column l, so that the denominator in (12.2) is positive. This conditional probability distribution is what player I infers (when told to play row i) about the recommended action j to player II. If player I follows the recommended action i, he gets a certain expected payoff. This payoff is optimal for player I if no other row k gives him higher expected payoff, using the same conditional distribution about player II's random choice j (because player I does not get any further information when he deviates from i to k). That is, for all other rows k we require that

$$\sum_{j=1}^{n} a_{ij} \frac{z_{ij}}{\sum_{l=1}^{n} z_{il}} \geq \sum_{j=1}^{n} a_{kj} \frac{z_{ij}}{\sum_{l=1}^{n} z_{il}}. \tag{12.3}$$

Multiplication by $\sum_{l=1}^{n} z_{il}$ on both sides shows that (12.3) is equivalent to

$$\sum_{j=1}^{n}(a_{ij}-a_{kj})z_{ij} \geq 0 \qquad (1 \leq i,k \leq m). \tag{12.4}$$

The m^2 inequalities for all pairs i,k in (12.4) hold also trivially if $i=k$, and also if $\sum_{l=1}^{n} z_{il}=0$ where row i has probability zero under the distribution z, because this implies $z_{ij}=0$ for all j for this i. If $\sum_{l=1}^{n} z_{il}>0$, then (12.4) implies (12.3) and states that player I has no incentive to deviate from the recommended row i to row k. The inequalities (12.4) are therefore called the *incentive constraints* for player I. The corresponding incentive constraints for player II are

$$\sum_{i=1}^{m}(b_{ij}-b_{il})z_{ij} \geq 0 \qquad (1 \leq j,l \leq n) \tag{12.5}$$

and say that player II, when recommended to play column j, has no incentive to deviate to column l.

Definition 12.1 (Correlated equilibrium, two players). For an $m \times n$ bimatrix game (A,B), a *correlated equilibrium* is an $m \times n$ matrix z of joint probabilities for strategy pairs (i,j) that fulfills (12.1) and the *incentive constraints* (12.4) for player I and (12.5) for player II. The matrix z defines a *correlation device* that selects the strategy pair (i,j) with probability z_{ij}, and then privately (without telling the other player) tells i to player I and j to player II. $\qquad \square$

The incentive constraints (12.4) for player I are about each row i of the matrix z. Up to a normalization factor (as it appears in (12.3)), this row of z is player I's "posterior" about the behavior of player II after player I gets the recommendation to play i. This can be seen as an inferred "mixed strategy" of player II, against which row i is a best response; in general, this inferred mixed strategy depends on i. In the same way, the incentive constraints (12.5) for player II are about each column j of z, which describes (up to normalization) an inferred mixed strategy of player I, against which column j is a best response. If z is obtained from a Nash equilibrium via $z_{ij}=x_iy_j$, then these inferred mixed strategies of the opponent are always the same, namely the actual mixed strategies.

Proposition 12.2. *Let (A,B) be an $m \times n$ bimatrix game, let x be a mixed strategy of player I and y be a mixed strategy of player II, and $z \in \mathbb{R}^{m \times n}$ be defined by $z_{ij}=x_iy_j$ for all i and j. Then z is a correlated equilibrium of (A,B) if and only if (x,y) is a Nash equilibrium of (A,B).*

Proof. Clearly, $\sum_{i=1}^{m} x_i=1$ and $\sum_{j=1}^{n} y_j=1$ imply that z is a probability distribution on pairs (i,j) as stated in (12.1). Because $\sum_{l=1}^{n} y_l=1$ and $z_{il}=x_iy_l$, we have $\sum_{l=1}^{n} z_{il}=x_i$ for $1 \leq i \leq m$.

Suppose z is a correlated equilibrium. If $x_i > 0$, then (12.4) implies (12.3), which states that $\sum_{j=1} a_{ij} y_j \geq \sum_{j=1} a_{kj} y_j$, that is, the expected payoff to player I in row i is at least as large as in any other row k. According to the best-response condition (6.11), this means that x is a best response to y. Similarly, (12.5) implies that y is a best response to x. Hence, (x, y) is a Nash equilibrium.

If (x, y) is a Nash equilibrium, these implications hold in reverse: If $x_i = 0$, (12.4) holds trivially, and if $x_i > 0$, the fact that the expected payoff in row i is at least the expected payoff in any other row k implies (12.3) and therefore (12.4). Similarly, y being a best response to x is equivalent to (12.5). Therefore, the incentive constraints hold for both players, and z is a correlated equilibrium. $\square$

The incentive constraints (12.4) and (12.5) involve separately the two payoff matrices A and B but affect together the matrix z. There are $m(m-1)$ nontrivial constraints in (12.4) (for $i = k$ they are trivial) and $n(n-1)$ nontrivial constraints in (12.5). For the game in Figure 12.1, they state

$$
\begin{aligned}
-z_{11} + z_{12} \geq 0, \qquad & z_{21} - z_{22} \geq 0, \\
-z_{11} + z_{21} \geq 0, \qquad & z_{12} - z_{22} \geq 0,
\end{aligned}
\tag{12.6}
$$

that is, both z_{12} and z_{21} have to be at least as large as z_{11} and z_{22}. The set of all joint distributions z on pure strategy pairs as in (12.1) is a simplex of dimension $mn - 1$, here a tetrahedron of dimension 3. The additional incentive constraints (12.4) and (12.5) define the set of all correlated equilibria, which is a subset of that simplex and a polytope.

For our 2×2 game with incentive constraints (12.6), that polytope is shown in Figure 12.3. It has five vertices, including $(z_{11}, z_{12}, z_{21}, z_{22}) = (0, 0, 1, 0)$, $(0, 1, 0, 0)$, and $(\frac{1}{4}, \frac{1}{4}, \frac{1}{4}, \frac{1}{4})$. These are the pure Nash equilibria and the mixed Nash equilibrium in Figure 12.1, marked in Figure 12.3 as BL, TR, and M. The two other vertices of the correlated equilibrium set are $(0, \frac{1}{3}, \frac{1}{3}, \frac{1}{3})$ and $(\frac{1}{3}, \frac{1}{3}, \frac{1}{3}, 0)$, which are the two middle panels in Figure 12.2 (its left panel $(0, \frac{1}{2}, \frac{1}{2}, 0)$ is a convex combination of $(0, 0, 1, 0)$ and $(0, 1, 0, 0)$ and not a vertex). Any such vertex is recognized by having at least three of the inequalities in (12.1) and (12.6) holding as equalities, which then have a unique solution.

Correlated equilibrium has an analogous definition for a game with N players. Recall from Definition 3.1 that a strategy profile $s = (s_1, \ldots, s_N)$ in an N-player game is also written as (s_i, s_{-i}) with strategy s_i of player i and the partial profile s_{-i} of strategies of the remaining players, where the set of these partial profiles is $S_{-i} = \bigtimes_{j=1, j \neq i}^{n} S_j$.

Definition 12.3 (Correlated equilibrium, N players). Let G be an N-player game in strategic form with finite strategy set S_i for player $i = 1, \ldots, N$ and payoff $u_i(s)$ for each strategy profile $s = (s_1, \ldots, s_N)$ in $S = S_1 \times \cdots \times S_N$. A *correlation device* is a

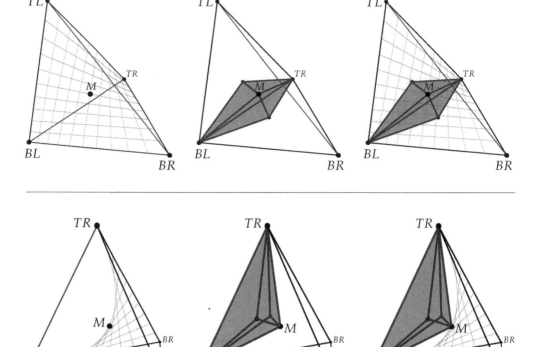

Figure 12.3   Two perspectives of the simplex of joint distributions z for the game in Figure 12.1, with TL, TR, BL, BR (standing for the four pure-strategy pairs) and the mixed Nash equilibrium M. The left pictures show the uncorrelated strategies defined by $z_{ij} = x_i y_j$, the middle pictures the polytope of correlated equilibria defined by the additional incentive constraints (12.6), and the right pictures both sets, whose intersection defines the Nash equilibria BL, TR, M, as stated in Proposition 12.2.

probability distribution z on strategy profiles, that is, a function $z : S \to \mathbb{R}$ so that

$$z(s) \geq 0 \quad (s \in S), \qquad \sum_{s \in S} z(s) = 1 \,. \tag{12.7}$$

The players know z. The device selects the strategy profile s with probability $z(s)$ and then privately tells each player i the respective strategy s_i in s. If for each player $i = 1, \ldots, N$ the *incentive constraints*

$$\sum_{s_{-i} \in S_{-i}} \big(u_i(s_i, s_{-i}) - u_i(t_i, s_{-i}) \big) z(s_i, s_{-i}) \geq 0 \qquad (s_i, t_i \in S_i) \tag{12.8}$$

hold, then z is called a *correlated equilibrium* of G.   □

The incentive constraints (12.8) generalize those in (12.4) and (12.5) for two players. In (12.8), player i is told to play s_i, and can infer from this information the absolute probabilities $z(s_i, s_{-i})$ for all possible partial profiles s_{-i} of the other players. The sum of these probabilities is positive because otherwise player i could not have been told s_i. Division by this sum then gives the conditional probabilities of what the other players have been told, and similar to (12.3) the inequality expresses that player i has nothing to gain be deviating from s_i to any other strategy t_i. This is the equilibrium property.

Proposition 12.2 holds analogously for N players.

Proposition 12.4. *Consider an N-player game G in strategic form as in Definition 12.3 and a mixed strategy σ_i for each player i. Suppose z is a correlation device so that $z(s) = \prod_{i=1}^{N} \sigma_i(s_i)$ for all $s = (s_1, \ldots, s_N) \in S$. Then z is a correlated equilibrium of G if and only if $(\sigma_1, \ldots, \sigma_N)$ is a Nash equilibrium of G.*

Proof. The proof of Proposition 12.2 carries over by applying the best-response condition for N players (see Proposition 6.3). □

A correlated equilibrium z of the game G as in Definition 12.3 is defined by its probabilities $z(s)$ for strategy profiles so that the incentive constraints (12.8) hold. These constraints, and the conditions (12.7) that state that $z \in \Delta(S)$, are linear in the variables $z(s)$ for $s \in S$. Hence, the set of correlated equilibria of G is a polyhedron, and as a subset of the simplex $\Delta(S)$ bounded and therefore a polytope. Mathematically, this is a simpler object than the set of Nash equilibria of G, which is in general not convex. However, so far the only proof that an N-player game G has a correlated equilibrium is via the existence of a Nash equilibrium, which relies on Brouwer's fixed-point theorem. An elegant alternative existence proof of correlated equilibria that uses von Neumann's minimax theorem will be given in Section 12.5.

12.4 Coarse Correlated Equilibrium

Correlated equilibrium needs a correlation device. For example, a traffic light sends correlated signals to the different road users that are recommendations for actions. The equilibrium condition is that it is optimal to follow these recommendations, assuming that the other players do so.

In this section, we consider a concept that goes one step further. The correlation device not only recommends the action, but each player *commits* to following the recommended action (for example, by letting the device choose the action, like in a self-driving car). In order to make this a game, the player's choice comes at an earlier stage, namely whether to participate in the usage of this device.

Suppose the correlation device is a distribution $z \in \Delta(S)$ on strategy profiles for a game G as in Definition 12.3. If all players participate, then some profile s will be chosen by z and used by the players. If player i does not participate, he only knows the marginal probability distribution on the partial profiles s_{-i} of the other players given by z. Then player i can choose his own optimal action $s_i \in S_i$, but does not know what action the device would have chosen. If against the marginal distribution he cannot gain by participating in the scheme, and this applies to every player i, then z is called a *coarse correlated equilibrium*.

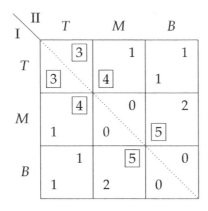

Figure 12.4 Left: 3×3 game which is dominance solvable, with unique Nash and correlated equilibrium (T,T). Right: Distribution on strategy pairs that is a coarse correlated equilibrium with higher payoff than in the Nash equilibrium.

Figure 12.4 shows an example for a symmetric 3×3 game. The game is dominance solvable (see Proposition 3.4): Strategy B is strictly dominated by strategy T (for both players), and after eliminating B, strategy M is strictly dominated in the remaining 2×2 game by T, so that (T,T) is the unique Nash equilibrium of the game. It is also the unique correlated equilibrium, by the following proposition.

Proposition 12.5. *Consider an N-player game where the pure strategy t_i of player i strictly dominates his pure strategy s_i. Then s_i is never recommended to player i in any correlated equilibrium z of the game.*

Proof. The strict domination of s_i by t_i is stated in (3.4), which is equivalent to $u_i(s_i, s_{-i}) - u_i(t_i, s_{-i})$ in (12.8) being negative for all $s_{-i} \in S_{-i}$. Hence, the inequality (12.8) can only hold if $z(s_i, s_{-i}) = 0$ for all s_{-i}. This means that s_i is never part of any strategy profile s chosen by z, and is therefore never told to player i by z. □

Consider now the correlation device shown on the right in Figure 12.4, which selects the strategy profiles (M,B) and (B,M) with probability $\frac{1}{2}$ each. It defines a coarse correlated equilibrium: The expected payoff to each player is $\frac{7}{2}$ if they commit to playing the action chosen by the device. Alternatively, a player may choose to opt out from using the device, and then, assuming the other player uses the device, only knows that the other player will choose M and B with probability $\frac{1}{2}$ each. The respective payoffs to the player who opted out when playing T, M, or B are $\frac{5}{2}$, $\frac{5}{2}$ and 1, respectively, which are lower, so it is optimal for the player to commit to using the device. Moreover, each player's payoff is higher than in the unique Nash and correlated equilibrium.

The following is a formal definition of coarse correlated equilibrium

Definition 12.6. Consider an N-player game, with the notation of Definition 12.3. Consider a correlation device z, known to the players, where each player has in advance the choice to commit to the pure strategy told to him by the device from the profile selected using z, or alternatively to opt out from this and choose any strategy without knowledge of the selection by the device. This is called a *coarse correlated equilibrium* if every player chooses to commit to the selection by the device and has no incentive to unilaterally deviate from this. □

The following states that coarse correlated equilibrium is more general than correlated equilibrium, and also gives in (12.9) the constraints that characterize coarse correlated equilibria.

Proposition 12.7. *Every correlated equilibrium is a coarse correlated equilibrium.*

Proof. Consider a coarse correlated equilibrium z. If each player commits to playing the action that is selected by z, then player i's expected payoff is $\sum_{s\in S} u_i(s)\,z(s)$. If player i opts out, then player i only knows that every partial profile s_{-i} is played with probability $\sum_{s_i\in S_i} z(s_i, s_{-i})$ if the other players commit to playing the action that is selected by z. Hence, the constraints for a coarse correlated equilibrium state that for each player i and each $t_i \in S_i$, player i does not get a higher payoff when playing t_i, that is,

$$
\begin{aligned}
\sum_{s\in S} u_i(s)\,z(s) &= \sum_{s_{-i}\in S_{-i}} \sum_{s_i\in S_i} u_i(s_i, s_{-i})\,z(s_i, s_{-i}) \\
&\geq \sum_{s_{-i}\in S_{-i}} u_i(t_i, s_{-i}) \sum_{s_i\in S_i} z(s_i, s_{-i}).
\end{aligned}
\tag{12.9}
$$

The incentive constraints (12.8) of a correlated equilibrium state that for all s_i and t_i in S_i,

$$
\sum_{s_{-i}\in S_{-i}} u_i(s_i, s_{-i})\,z(s_i, s_{-i}) \geq \sum_{s_{-i}\in S_{-i}} u_i(t_i, s_{-i})\,z(s_i, s_{-i})
\tag{12.10}
$$

which imply (12.9) by summation over all $s_i \in S_i$. □

Coarse correlated equilibria are considered further in Exercises 12.3 and 12.2.

12.5   Existence of a Correlated Equilibrium

In this section, we prove the existence of a correlated equilibrium in an N-player game without assuming the existence of a Nash equilibrium (which relies on Brouwer's fixed-point theorem), using only the minimax Theorem 8.4.

As a warm-up, and because we will use it in the existence proof of a correlated equilibrium, we consider the concepts of a Markov chain Y on a finite set T and a stationary distribution for Y. Recall that $\mathbf{1}$ is the vector with all components equal to 1. If $T = \{1, \ldots, n\}$, then we write $x \in \Delta(T)$ as $x \in \mathbb{R}^n$, $x \geq \mathbf{0}$, $\mathbf{1}^\top x = 1$.

Definition 12.8. Let $n \geq 1$ and $T = \{1, \ldots, n\}$. A *Markov chain* on T is an $n \times n$ matrix Y so that $Y \geq 0$ and $Y\mathbf{1} = \mathbf{1}$. A *stationary distribution* for a Markov chain Y on T is $x \in \Delta(T)$ so that $x^\top = x^\top Y$. $\square$

With T seen as a set of *states*, a Markov chain Y on T specifies probabilistic *state transitions*. From a state chosen as a row of Y, the next state is reached according to the distribution defined by that row. If the rows are chosen according to a stationary distribution x on T, then the condition $x^\top = x^\top Y$ means that every state has the same probability before and after the state transitions defined by Y. An example of a Markov chain Y and its stationary distribution x (see also Exercise 12.5) is

$$Y = \begin{pmatrix} 0.2 & 0.8 \\ 0.5 & 0.5 \end{pmatrix}, \qquad x = \begin{pmatrix} \frac{5}{13} \\ \frac{8}{13} \end{pmatrix}.$$

Theorem 12.9. *Every Markov chain has a stationary distribution.*

Proof. Let Y be a Markov chain on $T = \{1, \ldots, n\}$ and $x \in \Delta(T)$. Because $x^\top Y\mathbf{1} = x^\top \mathbf{1} = 1$, the map $x^\top \mapsto x^\top Y$ is a continuous function $\Delta(T) \to \Delta(T)$, which has a fixed point by Brouwer's fixed-point theorem. This fixed point is a stationary distribution for Y. However, we do not use the fixed-point theorem but instead the minimax Theorem 8.4 for the $n \times n$ zero-sum game with payoff matrix $I - Y$ to the row player, were I is the $n \times n$ identity matrix.

We prove that the game with matrix $I - Y$ has value zero. Let $s \in \Delta(T)$ be a mixed strategy of the minimizing column player in this game. Taking $s = \mathbf{1}\frac{1}{n}$ shows $(I - Y)s = \mathbf{1}\frac{1}{n} - Y\mathbf{1}\frac{1}{n} = \mathbf{0}$, so by playing s the column player pays at most zero in each row, and therefore the value of the game is at most zero.

If the value of the game $I - Y$ is negative, then the column player has a mixed strategy s so that $(I - Y)s < \mathbf{0}$, that is, $s < Ys$. But this cannot be the case: Because $Y\mathbf{1} = \mathbf{1}$, every row Y_a of Y for $a \in T$ is a probability distribution on T, and $Y_a s$ is a weighted average of the components s_b of s, where $s < Ys$ would mean $s_a < Y_a s$ for all $a \in T$. This certainly fails for the largest component s_a of s (as proved in (6.15), for example). Hence, the value of the game is indeed zero.

Therefore, the row player has a max-min strategy $x \in \Delta(T)$ so that $x^\top(I - Y) \geq \mathbf{0}^\top$. If the row vector $x^\top(I - Y)$ has some positive component, then $x^\top(I - Y)\mathbf{1} > 0$, which contradicts $(I - Y)\mathbf{1} = \mathbf{0}$. Hence, $x^\top(I - Y) = \mathbf{0}^\top$, which is equivalent to $x^\top = x^\top Y$, that is, x is a stationary distribution for Y.   □

The next lemma provides a key step in the proof that a finite game has a correlated equilibrium. It states that if each row of a nonnegative square matrix has a sum of at most 1, then increasing the diagonal element in that row so that the sum equals 1 as in (12.13) creates a Markov chain whose stationary distribution fulfills the equation (12.12) that we need subsequently (this works because the diagonal elements in this equation cancel out).

Lemma 12.10. *Let $n \geq 1$ and $T = \{1, \ldots, n\}$. Let $y(a, b) \geq 0$ for $a, b \in T$ so that*

$$\forall a \in T : \quad \sum_{b \in T} y(a, b) \leq 1. \tag{12.11}$$

Then there exists $x \in \Delta(T)$ so that

$$\forall c \in T : \quad x(c) \sum_{b \in T} y(c, b) = \sum_{a \in T} x(a)\, y(a, c). \tag{12.12}$$

Define the $n \times n$ matrix Y with entries Y_{ab} by

$$Y_{ab} = \begin{cases} y(a, b) & \text{if } a \neq b, \\ 1 - \sum_{c \in T,\, c \neq a} y(a, c) & \text{if } a = b. \end{cases} \tag{12.13}$$

Then Y is a Markov chain on T, and $x \in \Delta(T)$ fulfills (12.12) if and only if x is a stationary distribution for Y.

Proof. Clearly, (12.11) implies $\sum_{c \in T,\, c \neq a} y(a, c) \leq 1$ and therefore $Y_{aa} \geq 0$ in (12.13), and thus $Y_{ab} \geq 0$ for all $a, b \in T$. By (12.13),

$$\forall a \in T : \quad \sum_{b \in T} Y_{ab} = 1, \tag{12.14}$$

that is, $Y \geq 0$ and $Y\mathbf{1} = \mathbf{1}$, so Y is a Markov chain on T.

Equation (12.12) holds irrespective of the values $y(c, c)$ for $c \in T$, because the term $x(c)\,y(c, c)$ on both sides of the equation cancels out. We replace $y(c, c)$ by Y_{cc} as defined in (12.13), which does not affect the validity of (12.12). Then the numbers $y(a, b)$ are just the entries Y_{ab} of Y, and $\sum_{b \in T} y(c, b) = 1$ by (12.14), so that (12.12) is equivalent to $x^\top = x^\top Y$, that is, x is a stationary distribution for Y. By Theorem 12.9, such a distribution exists.   □

Consider an N-player game G in strategic form as in Definition 12.3. We show the existence of a correlated equilibrium for G with help of an auxiliary zero-sum game derived from G that we call the *deviation game*. The deviation game has a row player COR (the "correlator") and a column player DEV (the "deviator"). The set of pure strategies of COR is the set S of pure strategy profiles of G. The set D of pure strategies of DEV is defined as

$$D = \bigcup_{i=1}^{N} (\{i\} \times S_i \times S_i) \tag{12.15}$$

where a pure strategy $(i, a, b) \in D$ means that DEV selects a player i and two pure strategies a and b of player i.

The payoffs to the row player COR in the deviation game are given in an $|S| \times |D|$ matrix A with the entries

$$A(s, (i, a, b)) = \begin{cases} u_i(a, s_{-i}) - u_i(b, s_{-i}) & \text{if } s_i = a, \\ 0 & \text{if } s_i \neq a. \end{cases} \tag{12.16}$$

The interpretation is that a mixed strategy of COR is considered as a correlation device, which is altered by the strategy (i, a, b) chosen by DEV as follows: In the strategy profile s chosen by COR, if $s_i = a$ (that is, the strategy s_i in the profile s recommended to player i matches a), then player i deviates from this recommendation a and instead plays b. In the deviation game, the payoff gain for player i when he deviates this way is paid by COR to DEV (if player i is worse off by deviating, then COR receives the difference as his payoff).

The value of the deviation game is at most zero by letting DEV play any pure strategy of the form (i, a, a) (that is, no player in G ever deviates). In that case, according to (12.16), the payoff in row s of A is zero for every $s \in S$.

The value of the deviation game equals zero if and only if the row player COR has a mixed strategy $z \in \Delta(S)$ so that

$$z^\top A \geq \mathbf{0}^\top. \tag{12.17}$$

Then every column (i, a, b) of the game corresponds to exactly one of the incentive constraints (12.8) with $(i, s_i, t_i) = (i, a, b)$. Hence, such an optimal mixed strategy z of COR defines a correlated equilibrium of G. Therefore, G has a correlated equilibrium if and only if the deviation game has value zero. This is the case, as stated in the following theorem. It has a nice intuition that we explain afterwards.

Theorem 12.11. *Consider an N-player game G in strategic form as in Definition 12.3. Then the deviation game between COR and DEV with the matrix A of payoffs in (12.16) to the row player COR has value zero. Any max-min strategy of COR is a correlated equilibrium of G, which shows that G has at last one correlated equilibrium.*

Proof. The value of the deviation game is at most zero. If its value is less than zero, then the column player DEV has a mixed strategy $y \in \Delta(D)$ so that $Ay < \mathbf{0}$, which implies that $z^\top A y < 0$ for every response $z \in \Delta(S)$ of COR. We show that this is not the case, by finding, as a function of y, a mixed strategy x of COR so that $x^\top A y = 0$. The distribution x on S will be a *product distribution* given by a profile of N mixed strategies $x_i \in \Delta(S_i)$ for each player $i = 1, \ldots, N$, that is, $x(s) = \prod_{i=1}^{n} x_i(s_i)$. We write $x(s) = x_i(s_i) x_{-i}(s_{-i})$ with $x_{-i}(s_{-i}) = \prod_{k=1,\ k \neq i}^{n} x_k(s_k)$. It turns out that x can be chosen solely as a function of y, independently of the players' payoffs in the original game G. For such a product distribution x, we have

$$
\begin{aligned}
x^\top A y &= \sum_{(i,a,b) \in D} \sum_{s \in S} x(s)\, y(i,a,b)\, A\big(s, (i,a,b)\big) \\
&= \sum_{(i,a,b) \in D} \sum_{s_{-i} \in S_{-i}} x(a, s_{-i})\, y(i,a,b) \big(u_i(a, s_{-i}) - u_i(b, s_{-i})\big) \\
&= \sum_{i=1}^{N} \sum_{a \in S_i} \sum_{b \in S_i} \sum_{s_{-i} \in S_{-i}} x_i(a)\, x_{-i}(s_{-i})\, y(i,a,b) \big(u_i(a, s_{-i}) - u_i(b, s_{-i})\big) \\
&= \sum_{i=1}^{N} \sum_{s_{-i} \in S_{-i}} x_{-i}(s_{-i}) \sum_{a \in S_i} \sum_{b \in S_i} x_i(a)\, y(i,a,b) \big(u_i(a, s_{-i}) - u_i(b, s_{-i})\big). \quad (12.18)
\end{aligned}
$$

For player i and $s_{-i} \in S_{-i}$, we arrange the last double sum according to the payoffs $u_i(c, s_{-i})$ for $c \in S_i$:

$$
\begin{aligned}
&\sum_{a \in S_i} \sum_{b \in S_i} x_i(a)\, y(i,a,b) \big(u_i(a, s_{-i}) - u_i(b, s_{-i})\big) \\
&= \sum_{a \in S_i} \sum_{b \in S_i} x_i(a)\, y(i,a,b)\, u_i(a, s_{-i}) \;-\; \sum_{a \in S_i} \sum_{b \in S_i} x_i(a)\, y(i,a,b)\, u_i(b, s_{-i}) \\
&= \sum_{c \in S_i} \sum_{b \in S_i} x_i(c)\, y(i,c,b)\, u_i(c, s_{-i}) \;-\; \sum_{a \in S_i} \sum_{c \in S_i} x_i(a)\, y(i,a,c)\, u_i(c, s_{-i}) \\
&= \sum_{c \in S_i} u_i(c, s_{-i}) \left(x_i(c) \sum_{b \in S_i} y(i,c,b) \;-\; \sum_{a \in S_i} x_i(a)\, y(i,a,c) \right). \quad (12.19)
\end{aligned}
$$

We now invoke Lemma 12.10 with $T = S_i$ and $y(a,b) = y(i,a,b)$: The numbers $y(a,b)$ are the mixed strategy probabilities $y(i,a,b)$ of DEV, which implies (12.11). Hence, there are probabilities $x(c) = x_i(c) \in \Delta(S_i)$ so that (12.12) holds, that is, the difference of the sums in (12.19) is zero for all $c \in S_i$. Therefore, for any payoffs $u_i(c, s_{-i})$, the entire sum in (12.19) is zero, and therefore $x^\top A y = 0$ in (12.18).

This shows that the column player DEV does not have a mixed strategy y with $A y < \mathbf{0}$. Therefore, as claimed, the deviation game has value zero. Any max-min strategy $z \in \Delta(S)$ of the row player COR fulfills $z^\top A \geq \mathbf{0}^\top$ and defines a correlated equilibrium of the original game G. □

The proof of Theorem 12.11 has the following intuition. Similarly to the proof of Theorem 12.9, it uses the minimax Theorem 8.4 by showing that the column player cannot simultaneously make all payoff rows negative. Consider any mixed strategy y of the column player DEV in the deviation game, and some player i in the original game G. The pure strategies $a, b \in S_i$ of player i are seen as states of a Markov chain, and the probabilities $y(i, a, b)$ define transition probabilities from state a to state b. For each a, the sum p_a of these probabilities $y(i, a, b)$ over all $b \neq a$ is usually less than 1. Hence, the complementary probability $1 - p_a$ can be considered as the probability that player i does *not* deviate from his recommendation to play a. This interpretation agrees with the entries of the matrix A in (12.16) and with the definition of the deviation game (which in effect says that if DEV does not choose (i, s_i, t_i) for the strategy s_i in s chosen by COR, then player i does not deviate from s_i). With $Y_{aa} = 1 - p_a$ and $Y_{ab} = y(i, a, b)$, this defines the Markov chain Y in Lemma 12.10. Let x be a stationary distribution for Y, which defines a mixed strategy x_i for player i. Crucially, this stationary distribution *neutralizes the Markov chain*. In other words, the "deviation plan" y by DEV is countered, independently for each player i, by a mixed strategy x_i so that there is no observable difference, in terms of probabilities, between all players playing according to the profile $x = (x_1, \ldots, x_N)$ before or after applying the deviations imposed by y. Hence, the gain for deviating is zero, expressed by the equation $x^\top A y = 0$, and hence it is not possible that $A y < \mathbf{0}$.

The "neutralizing behavior" x is chosen as a *function of y*, independently of the payoffs in G. Moreover, x does not represent an optimal behavior of COR in the deviation game (otherwise, it would be a Nash equilibrium of G by Proposition 12.4, with an existence proof by the minimax theorem, which would be too good to be true). In fact, for the optimal strategy y of COR that never asks for any deviation, the Markov chain on each state set S_i is trivial and has every $x_i \in \Delta(S_i)$ as a stationary distribution.

12.6   Further Reading

Correlated equilibrium was introduced by Aumann (1974). We have described its *canonical form* where the correlation device directly generates strategy profiles with the strategies in them as private recommendations to the players, rather than using a device that sends arbitrary messages that are interpreted by the players. The canonical form suffices because only the players' actions matter (Aumann, 1987). As argued by Aumann (1987), correlated equilibrium is the expression of "Bayesian rationality" with a "common prior" about the state of the world, where this prior is in effect the correlation device. In contrast, playing Nash equilibrium requires much stronger assumptions about commonly accepted norms of behavior; see Aumann and Brandenburger (1995) and Aumann and Dreze (2008).

Games similar to the modified Chicken game in Figure 12.1 have been described by Aumann (1974, p. 72) and Myerson (1991, p. 250). Nau, Gomez Canovas, and Hansen (2004) prove that Nash equilibria lie on the relative boundary of the polytope of correlated equilibria, and show a picture similar to Figure 12.3.

The term "coarse correlated equilibrium" is due to Young (2004). The game in Figure 12.4 is taken from Moulin and Vial (1978, p. 205), who introduced the concept and called it the "simple extension" of a correlated equilibrium.

The existence proof of Theorem 12.11 is due to Hart and Schmeidler (1989) and, very similarly and independently, Nau and McCardle (1990). Lemma 12.10 with its construction of the Markov chain Y from y by increasing the diagonal elements is better than scaling the rows of y, which would require a case distinction for all-zero rows, and an, easily overlooked, scaling of the stationary distribution x. A polynomial-time algorithm for computing a correlated equilibrium based on this existence proof, even for a large class of compactly specified games where the strategic form is exponentially large, is described in Papadimitriou and Roughgarden (2008). The computational complexity of Nash and correlated equilibria is discussed by Gilboa and Zemel (1989). In general, correlated equilibria can be found by linear programming and are computationally more tractable than Nash equilibria and fixed-point problems.

Correlated equilibria are important because they appear as limiting behavior of players who *learn* their strategies in repeated plays of the game. The limit distribution on strategy profiles is thereby given by how often they are played in these repeated interactions. For zero-sum games, *fictitious play*, where players use their best response against the current history, converges to a Nash equilibrium (Robinson, 1951). For non-zero-sum games, however, this fails already for 3×3 games, even if the game has a unique equilibrium, such as the game in Exercise 12.2, as shown by Shapley (1963).

In learning based on *regret-matching*, a player checks for any two of his strategies whether or not he should have played one instead of the other, and adapts his behavior accordingly. Hart and Mas-Colell (2000) introduced this learning procedure and showed that it approaches the set of correlated equilibria. It has its origin in the proof of Theorem 12.11, with the idea of applying ficticious play to the auxiliary zero-sum game (Hart and Mas-Colell, 2000, p. 1142). For a survey of learning in games see Young (2004).

12.7 Exercises for Chapter 12

Exercise 12.1. Find the Nash and correlated equilibria of the following zero-sum game (with payoffs to the row player):

I \ II	l	r
T	4	2
B	0	3

Exercise 12.2. Consider the following variant of the Rock-Paper-Scissors game where two equal strategies tie with payoff 0 for both players, but two unequal strategies give payoff 1 to the winner and -2 to the loser.

I \\ II	R	P	S
R	0 \\ 0	1 \\ -2	-2 \\ 1
P	-2 \\ 1	0 \\ 0	1 \\ -2
S	1 \\ -2	-2 \\ 1	0 \\ 0

(a) Find a Nash equilibrium of this game, and its payoff to the two players. Show that when this Nash equilibrium is considered as a correlated equilibrium, all incentive constraints (12.4) and (12.5) are tight (that is, hold as equalities). This fact may suffice to indicate that the Nash and correlated equilibrium is unique; for a different proof see von Stengel and Zamir (2010, Remark 13).

(b) Find a coarse correlated equilibrium of this game that has higher expected payoff to the players than the Nash equilibrium in (a).

Exercise 12.3. For the game in Figure 12.4, the coarse correlated equilibrium z on the right has only two probabilities $a = z_{32}$ for the strategy pair (B, M) and $b = z_{23}$ for the strategy pair (M, B) that are positive; all other components z_{ij} are zero. Find all possible values of a and $b = 1 - a$ so that the correlation devices z that are defined in this way define a coarse correlated equilibrium of the game.

Exercise 12.4. Show that in an N-player game where every player has two strategies, every coarse correlated equilibrium is a correlated equilibrium.

Exercise 12.5. Show that if Y is a Markov chain on two states, then the off-diagonal elements of Y, in their columns, are proportional to the probabilities of a stationary distribution x of Y. That is, if

$$Y = \begin{pmatrix} a & b \\ c & d \end{pmatrix},$$

then $(c \ b) = (c + b)x^{\mathsf{T}}$. Exactly when is x unique? Argue carefully.

References

The page numbers in this book where a reference is cited, for example on page 33, are given after the reference with ↑33.

Adler, I. (2013). The equivalence of linear programs and zero-sum games. *International Journal of Game Theory* 42(1), 165–177. ↑221

Albert, M. H., R. J. Nowakowski, and D. Wolfe (2007). *Lessons in Play: An Introduction to Combinatorial Game Theory*. AK Peters, Wellesley, MA. ↑30, 31, 38

Allais, M. (1953). Le comportement de l'homme rationnel devant le risque: critique des postulats et axiomes de l'école américaine [with English summary]. *Econometrica* 21(4), 503–546. ↑133

Aumann, R. J. (1974). Subjectivity and correlation in randomized strategies. *Journal of Mathematical Economics* 1(1), 67–96. ↑334, 348

Aumann, R. J. (1987). Correlated equilibrium as an expression of Bayesian rationality. *Econometrica* 55(1), 1–18. ↑348

Aumann, R. and A. Brandenburger (1995). Epistemic conditions for Nash equilibrium. *Econometrica* 63(5), 1161–1180. ↑132, 348

Aumann, R. J. and J. H. Dreze (2008). Rational expectations in games. *American Economic Review* 98(1), 72–86. ↑132, 348

Avenhaus, R. and M. J. Canty (1996). *Compliance Quantified: An Introduction to Data Verification*. Cambridge University Press, Cambridge, UK. ↑166

Avis, D., G. D. Rosenberg, R. Savani, and B. von Stengel (2010). Enumeration of Nash equilibria for two-player games. *Economic Theory* 42(1), 9–37. ↑259, 260

Berlekamp, E. R., J. H. Conway, and R. K. Guy (2001–2004). *Winning Ways for Your Mathematical Plays, 2nd ed., Volumes 1–4*. AK Peters, Wellesley, MA. (First edition 1982). ↑x, 31

Bernoulli, D. (1738). Specimen theoriae novae de mensura sortis. *Commentarii Academiae Scientiarum Imperiales Petropolitanae* 5, 175–192. Translated as: Exposition of a new theory on the measurement of risk, *Econometrica* 22(1), 23–36, 1954. ↑133

Bewersdorff, J. (2005). *Luck, Logic, and White Lies: The Mathematics of Games*. AK Peters, Wellesley, MA. ↑31

Biggs, N., E. K. Lloyd, and R. J. Wilson (1976). *Graph Theory 1736–1936*. Clarendon Press, Oxford. ↑203, 353

Binmore, K. (1992). *Fun and Games*. D. C. Heath, Lexington, MA. ↑327, 330

Binmore, K. (2007). *Playing for Real*. Oxford University Press, New York. ↑4, 330

Bouton, C. L. (1901). Nim, a game with a complete mathematical theory. *Annals of Mathematics* 3(1/4), 35–39. ↑5, 31

Braess, D. (1968). Über ein Paradoxon aus der Verkehrsplanung. *Unternehmensforschung* 12(1), 258–268. Translated in Braess, Nagurney, and Wakolbinger (2005). ↑44, 49

Braess, D., A. Nagurney, and T. Wakolbinger (2005). On a paradox of traffic planning. *Transportation Science* 39(4), 446–450. ↑352

Brouwer, L. E. J. (1911). Über Abbildung von Mannigfaltigkeiten. *Mathematische Annalen* 71(1), 97–115. ↑203

Brouwer, L. E. J. (1976). *Collected Works 2: Geometry, Analysis, Topology and Mechanics*. North-Holland, Amsterdam. Edited by H. Freudenthal. ↑203

Bryant, V. (1990). *Yet Another Introduction to Analysis*. Cambridge University Press, Cambridge, UK. ↑204

Chen, X., X. Deng, and S.-H. Teng (2009). Settling the complexity of computing two-player Nash equilibria. *Journal of the ACM* 56(3), Article 14. ↑259

Cheng, S.-F., D. M. Reeves, Y. Vorobeychik, and M. P. Wellman (2004). Notes on equilibria in symmetric games. In: *Sixth Workshop on Game Theoretic and Decision Theoretic Agents at the 3rd Conference on Autonomous Agents and Multi-Agent Systems (AAMAS '04)*. New York. ↑74

Chvátal, V. (1983). *Linear Programming*. W. H. Freeman, New York. ↑221, 260

Cohen, D. I. A. (1967). On the Sperner lemma. *Journal of Combinatorial Theory* 2(4), 585–587. ↑203

Conway, J. H. (1977). All games bright and beautiful. *American Mathematical Monthly* 84(6), 417–434. ↑31

Conway, J. H. (2001). *On Numbers and Games, 2nd ed.* AK Peters, Natick, MA. (First edition 1976). ↑24, 31

Cottle, R. W., J.-S. Pang, and R. E. Stone (1992). *The Linear Complementarity Problem*. Academic Press, San Diego, CA. ↑260

Cournot, A. A. (1838). *Recherches sur les principes mathématiques de la théorie des richesses*. Hachette, Paris. Translated as: *Researches into the Mathematical Principles of Wealth*, A. M. Kelly, New York, 1960. ↑66, 74

Dantzig, G. B. (1963). *Linear Programming and Extensions*. Princeton University Press, Princeton, NJ. ↑220, 260

Dixit, A. K. and B. J. Nalebuff (1991). *Thinking Strategically: The Competitive Edge in Business, Politics, and Everyday Life*. W. W. Norton, New York. ↑74

Euler, L. (1741). Solutio problematis ad geometriam situs pertinentis. *Commentarii Academiae Scientiarum Imperiales Petropolitanae* 8, 128–140 + Plate VIII. Translated as: The solution of a problem relating to the geometry of position (Biggs, Lloyd, and Wilson, 1976, 3–8). ↑203

Fishburn, P. C. (1970). *Utility Theory for Decision Making.* Wiley, New York. ↑133

Fraenkel, A. S. (1996). Scenic trails ascending from sea-level Nim to alpine Chess. In: *Games of No Chance*, edited by R. J. Nowakowski, volume 29 of *MSRI Publications*, 13–42. Cambridge University Press, Cambridge, UK. ↑31

Freudenthal, H. (1942). Simplizialzerlegungen von beschränkter Flachheit. *Annals of Mathematics* 43(3), 580–582. ↑170, 196, 204

Friedman, D. (1998). Monty Hall's three doors: Construction and deconstruction of a choice anomaly. *The American Economic Review* 88(4), 933–946. ↑296

Gale, D. (1960). *The Theory of Linear Economic Models.* McGraw-Hill, New York. ↑220

Gale, D. (1974). A curious Nim-type game. *The American Mathematical Monthly* 81(8), 876–879. ↑31

Gale, D. and A. Neyman (1982). Nim-type games. *International Journal of Game Theory* 11(1), 17–20. ↑31

Gibbons, R. (1992). *Game Theory for Applied Economists.* Princeton University Press, Princeton, NJ. ↑ix, 74, 98, 295

Gilboa, I. and E. Zemel (1989). Nash and correlated equilibria: Some complexity considerations. *Games and Economic Behavior* 1(1), 80–93. ↑349

Grundy, P. M. (1939). Mathematics and games. *Eureka* 2, 6–8. ↑1, 16, 31

Hart, S. and A. Mas-Colell (2000). A simple adaptive procedure leading to correlated equilibrium. *Econometrica* 68(5), 1127–1150. ↑349

Hart, S. and D. Schmeidler (1989). Existence of correlated equilibria. *Mathematics of Operations Research* 14(1), 18–25. ↑x, 333, 349

Herstein, I. N. and J. Milnor (1953). An axiomatic approach to measurable utility. *Econometrica* 21(2), 291–297. ↑133

Hong, S.-K., I.-J. Song, and J. Wu (2007). Fengshui theory in urban landscape planning. *Urban Ecosystems* 10(3), 221–237. ↑44, 49

Huizinga, J. (1949). *Homo Ludens: A Study of the Play-Element in Culture.* Routledge, London. ↑330

Knaster, B., C. Kuratowski, and S. Mazurkiewicz (1929). Ein Beweis des Fixpunktsatzes für *n*-dimensionale Simplexe. *Fundamenta Mathematicae* 14(1), 132–137. ↑169, 184, 187, 203

Kuhn, H. W. (1950). Simplified two-person poker. In: *Contributions to the Theory of Games, Vol. I*, edited by H. W. Kuhn and A. W. Tucker, volume 24 of *Annals of Mathematics Studies*, 97–103. Princeton University Press, Princeton, NJ. ↑297

Kuhn, H. W. (1953). Extensive games and the problem of information. In: *Contributions to the Theory of Games, Vol. II*, edited by H. W. Kuhn and A. W. Tucker, volume 28 of *Annals of Mathematics Studies*, 193–216. Princeton University Press, Princeton, NJ. Reprinted in Kuhn (1997). ↑263, 283, 285, 295

Kuhn, H. W. (1960). Some combinatorial lemmas in topology. *IBM Journal of Research and Development* 4(5), 518–524. ↑204

Kuhn, H. W. (1969). Approximate search for fixed points. In: *Computing Methods in Optimization Problems 2*, edited by L. A. Zadeh, L. W. Neustadt, and A. V. Balakrishnan, 199–211. Academic Press, New York. ↑203

Kuhn, H. W. (ed.) (1997). *Classics in Game Theory*. Princeton University Press, Princeton, NJ. ↑354, 355, 356

Lemke, C. E. (1965). Bimatrix equilibrium points and mathematical programming. *Management Science* 11(7), 681–689. ↑260

Lemke, C. E. and J. T. Howson, Jr (1964). Equilibrium points of bimatrix games. *Journal of the Society for Industrial and Applied Mathematics* 12(2), 413–423. ↑x, 223, 230, 232, 259

Lewis, M. (2016). *The Undoing Project: A Friendship That Changed Our Minds*. W. W. Norton, New York. ↑133

Liu, L. A. (1996). The invariance of best reply correspondences in two-player games. Working paper. City University of Hong Kong, Faculty of Business, Department of Economics and Finance. ↑259

Loomis, L. H. (1946). On a theorem of von Neumann. *Proceedings of the National Academy of Sciences of the United States of America* 32(8), 213–215. ↑220

Luce, R. D. and H. Raiffa (1957). *Games and Decisions: Introduction and Critical Survey*. Wiley, New York. ↑133

Mangasarian, O. L. (1964). Equilibrium points of bimatrix games. *Journal of the Society for Industrial and Applied Mathematics* 12(4), 778–780. ↑260

Maschler, M., E. Solan, and S. Zamir (2013). *Game Theory*. Cambridge University Press, Cambridge, UK. ↑166, 330

Matoušek, J. and B. Gärtner (2007). *Understanding and Using Linear Programming*. Springer, Berlin. ↑221

Monderer, D. and L. S. Shapley (1996). Potential games. *Games and Economic Behavior* 14(1), 124–143. ↑50, 74

Morgenstern, O. (1973). Ingolf Ståhl: Bargaining theory (book review). *The Swedish Journal of Economics* 75(4), 410–413. ↑356

Moscati, I. (2016). Retrospectives: How economists came to accept expected utility theory: The case of Samuelson and Savage. *Journal of Economic Perspectives* 30(2), 219–236. ↑133

Moulin, H. and J.-P. Vial (1978). Strategically zero-sum games: The class of games whose completely mixed equilibria cannot be improved upon. *International Journal of Game Theory* 7(3/4), 201–221. ↑259, 349

Myerson, R. B. (1991). *Game Theory: Analysis of Conflict*. Harvard University Press, Cambridge, MA. ↑133, 348

Nash, J. F., Jr. (1950). The bargaining problem. *Econometrica* 18(2), 155–162. Reprinted in Kuhn (1997). ↑299, 330

Nash, J. (1951). Non-cooperative games. *Annals of Mathematics* 54(2), 286–295. Reprinted in Kuhn (1997). ↑x, 53, 59, 135, 136, 149, 152, 166, 169, 205, 214

Nau, R., S. Gomez Canovas, and P. Hansen (2004). On the geometry of Nash equilibria and correlated equilibria. *International Journal of Game Theory* 32(4), 443–453. ↑348

Nau, R. F. and K. F. McCardle (1990). Coherent behavior in noncooperative games. *Journal of Economic Theory* 50(2), 424–444. ↑349

Neyman, A. (1997). Correlated equilibrium and potential games. *International Journal of Game Theory* 26(2), 223–227. ↑74

Osborne, M. J. (2004). *An Introduction to Game Theory.* Oxford University Press, New York. ↑ix, 74, 76, 166, 330

Osborne, M. J. and A. Rubinstein (1994). *A Course in Game Theory.* MIT Press, Cambridge, MA. ↑133, 295, 330

Owen, G. (1967). Communications to the editor: An elementary proof of the minimax theorem. *Management Science* 13(9), 765–765. ↑220

Papadimitriou, C. H. (1994). On the complexity of the parity argument and other inefficient proofs of existence. *Journal of Computer and System Sciences* 48(3), 498–532. ↑203, 259

Papadimitriou, C. H. and T. Roughgarden (2008). Computing correlated equilibria in multi-player games. *Journal of the ACM* 55(3), Article 14. ↑349

Park, S. (1999). Ninety years of the Brouwer fixed point theorem. *Vietnam Journal of Mathematics* 27(3), 187–222. ↑203

Perea, A. (2012). *Epistemic Game Theory: Reasoning and Choice.* Cambridge University Press. ↑132

Pigou, A. (1920). *The Economics of Welfare.* McMillan, London. ↑49

Pratt, J. W. (1964). Risk aversion in the small and in the large. *Econometrica* 32(1/2), 122–136. ↑133

Robinson, J. (1951). An iterative method of solving a game. *Annals of Mathematics* 54(2), 296–301. Reprinted in Kuhn (1997). ↑349

Rosenthal, R. W. (1973). A class of games possessing pure-strategy Nash equilibria. *International Journal of Game Theory* 2(1), 65–67. ↑49

Roughgarden, T. (2016). *Twenty Lectures on Algorithmic Game Theory.* Cambridge University Press, Cambridge, UK. ↑xii, 49

Roughgarden, T. and É. Tardos (2002). How bad is selfish routing? *Journal of the ACM* 49(2), 236–259. ↑49

Rubinstein, A. (1982). Perfect equilibrium in a bargaining model. *Econometrica* 50(1), 97–109. ↑299, 330

Savani, R. and B. von Stengel (2006). Hard-to-solve bimatrix games. *Econometrica* 74(2), 397–429. ↑260

Savani, R. and B. von Stengel (2016). Unit vector games. *International Journal of Economic Theory* 12, 7–27. ↑260

Schrijver, A. (1986). *Theory of Linear and Integer Programming*. John Wiley & Sons, Chichester, UK. ↑221

Schwalbe, U. and P. Walker (2001). Zermelo and the early history of game theory. *Games and Economic Behavior* 34(1), 123–137. ↑98, 357

Selten, R. (1975). Reexamination of the perfectness concept for equilibrium points in extensive games. *International Journal of Game Theory* 4(1), 25–55. Reprinted in Kuhn (1997). ↑295

Selvin, S. (1975). Letters to the editor: A problem in probability. *The American Statistician* 29(1), 67. ↑296

Shapley, L. S. (1963). *Some topics in two-person games*. Memorandum RM-3672-PR. The Rand Corporation, Santa Monica, CA. ↑349

Shapley, L. S. (1974). A note on the Lemke-Howson algorithm. *Mathematical Programming Study 1: Pivoting and Extensions*, 175–189. ↑223, 259

Siegel, A. N. (2013). *Combinatorial Game Theory*, volume 146 of *Graduate Studies in Mathematics*. American Mathematical Society, Providence, RI. ↑30

Sperner, E. (1928). Neuer Beweis für die Invarianz der Dimensionszahl und des Gebietes. *Abhandlungen aus dem Mathematischen Seminar der Universität Hamburg* 6(1), 265–272. ↑169, 170, 184, 203

Sprague, R. (1935). Über mathematische Kampfspiele. *Tohoku Mathematical Journal (First Series)* 41, 438–444. ↑1, 16, 31

Ståhl, I. (1972). *Bargaining Theory*. The Economic Research Institute at the Stockholm School of Economics, Stockholm. Reviewed in Morgenstern (1973). ↑330

Szpiro, G. G. (2010). *Numbers Rule: The Vexing Mathematics of Democracy, from Plato to the Present*. Princeton University Press, Princeton, NJ. ↑133

Szpiro, G. G. (2020). *Risk, Choice and Uncertainty: Three Centuries of Economic Decision-Making*. Columbia University Press, New York. ↑133

Turocy, T. L. and B. von Stengel (2002). Game theory. In: *Encyclopedia of Information Systems*, volume 2, 403–420. Elsevier Science (USA). ↑295

Tversky, A. (1969). Intransitivity of preferences. *Psychological Review* 76(1), 31–48. ↑133

van Damme, E. (1987). *Stability and Perfection of Nash Equilibria*. Springer, Berlin. ↑74, 166

von Neumann, J. (1928). Zur Theorie der Gesellschaftsspiele. *Mathematische Annalen* 100(1), 295–320. Translated as: On the theory of games of strategy. In: *Contributions to the Theory of Games, Vol. IV*, eds. A. W. Tucker and R. D. Luce, Annals of Mathematics Studies vol. 40, Princeton University Press, 1959, pp. 13–42. ↑205, 206, 213, 215, 216, 220 von Neumann, J. and M. Fréchet (1953). Communication on the Borel notes. *Econometrica* 21(1), 124–127. ↑220

von Neumann, J. and O. Morgenstern (1947). *Theory of Games and Economic Behavior, 2nd ed.* Princeton University Press, Princeton, NJ. ↑103, 104, 133, 220

von Stackelberg, H. (1934). *Marktform und Gleichgewicht.* Springer, Berlin. Translated as: *Market Structure and Equilibrium*, Springer, Heidelberg, 2011. ↑98

von Stengel, B. (1996). Efficient computation of behavior strategies. *Games and Economic Behavior* 14(2), 220–246. ↑295

von Stengel, B. (1999). New maximal numbers of equilibria in bimatrix games. *Discrete and Computational Geometry* 21(4), 557–568. ↑260

von Stengel, B. (2002). Computing equilibria for two-person games. In: *Handbook of Game Theory with Economic Applications*, edited by R. J. Aumann and S. Hart, volume 3, 1723–1759. North-Holland, Amsterdam. ↑259, 260

von Stengel, B. (2010). Follower payoffs in symmetric duopoly games. *Games and Economic Behavior* 69(2), 512–516. ↑98

von Stengel, B. (2021). Finding Nash equilibria of two-player games. Preprint arXiv:2102.04580. ↑166, 259

von Stengel, B. and S. Zamir (2010). Leadership games with convex strategy sets. *Games and Economic Behavior* 69(2), 446–457. ↑350

Wardrop, J. G. (1952). Some theoretical aspects of road traffic research. *Proceedings of the Institution of Civil Engineers* 1(3), 325–362. ↑50

Wythoff, W. A. (1907). A modification of the game of Nim. *Nieuw Archief voor Wiskunde* 7(2), 199–202. ↑31

Young, H. P. (2004). *Strategic Learning and Its Limits.* Oxford University Press, Oxford. ↑349

Zermelo, E. (1913). Über eine Anwendung der Mengenlehre auf die Theorie des Schachspiels. In: *Proc. Fifth Congress Mathematicians, Cambridge 1912*, 501–504. Cambridge University Press, Cambridge, UK. Translated in Schwalbe and Walker (2001). ↑98

Ziegler, G. M. (1995). *Lectures on Polytopes*, volume 152 of *Graduate Text in Mathematics.* Springer, New York. ↑260

Index